Global Practices on Road Traffic Signal Control

About WCTRS

WCTRS (World Conference on Transport Research Society) started in Rotterdam in 1977 as a forum of bridging transport researchers and practitioners. It is registered as an international academic and non-profit society under the Swiss Law. WCTRS provides a forum for the interchange of ideas among transport researchers, managers, policy makers, and educators from all over the world, from a perspective which is multi-modal, multi-disciplinary, and multi-sectoral. The Society has become a primary forum for such international exchanges in transportation. It now includes eminent scholars covering all modes of transport, all approaches from engineering, economics to policymaking as well as from practice to theory, and all regions from developing, emerging to developed countries. This "Diversity" with "Cooperative bridging between differently characterized people" is the most noteworthy "Asset" of WCTRS, which cannot be found anywhere else in the world. For more information, please visit www.wctrs-society.com.

About the WCTRS Book Series

Besides the World Conferences on Transport Research and the WCTRS scientific journals "Transport Policy" and "Case Studies on Transport Policy", the WCTRS Book series is another forum for leading transport professionals from all countries and areas to learn from one another. It focuses on the latest research on emerging issues and opportunities in transportation planning, policy, management, engineering, economics, and sustainability. Books in the series will take multidisciplinary, multimodal and multisectoral approaches to the topics, leading the way in bridging the gap between transportation research and practice.

The individual books are edited by outstanding transport professionals with a close relation to WCTRS. All books are undergoing a thorough scientific review procedure. The overall WCTRS book series is coordinated by an Editorial Board with the following members:

Global Practices on Road Traffic Signal Control

Fixed-Time Control at Isolated Intersections

Edited by

Keshuang Tang

Manfred Boltze

Hideki Nakamura

Zong Tian

ELSEVIER

Elsevier
Radarweg 29, PO Box 211, 1000 AE Amsterdam, Netherlands
The Boulevard, Langford Lane, Kidlington, Oxford OX5 1GB, United Kingdom
50 Hampshire Street, 5th Floor, Cambridge, MA 02139, United States

British Library Cataloguing-in-Publication Data
A catalogue record for this book is available from the British Library

Library of Congress Cataloging-in-Publication Data
A catalog record for this book is available from the Library of Congress

ISBN: 978-0-12-815302-4

For Information on all Elsevier publications
visit our website at https://www.elsevier.com/books-and-journals

Publisher: Joe Hayton
Acquisition Editor: Brian Romer
Development Editor: Lindsay Lawrence
Production Project Manager: Anitha Sivaraj
Cover Designer: Mark Rogers

Typeset by MPS Limited, Chennai, India

Contents

List of contributors

Ghassan Abu-Lebdeh Department of Civil Engineering, College of Engineering, American University of Sharjah, Sharjah, United Arab Emirates

Wael Alhajyaseen Qatar Transportation and Traffic Safety Center, College of Engineering, Qatar University, Doha, State of Qatar

Konstantinos Ampountolas School of Engineering, University of Glasgow, Glasgow, United Kingdom

Shriniwas Arkatkar Department of Civil Engineering, Sardar Vallabhbhai National Institute of Technology, Surat, India

Ashish Bhaskar School of Civil Engineering and Built Environment, Queensland University of Technology, Brisbane, QLD, Australia

Manfred Boltze Institute for Transport Planning and Traffic Engineering, Technische Universität Darmstadt, Darmstadt, Germany

Nicolas Chiabaut Université de Lyon, IFSTTAR/ENTPE, LICIT, Rue Maurice Audin, Vaulx-en-Velin, France

Andy H.F. Chow Systems Engineering and Engineering Management, City University of Hong Kong, Kowloon Tong, Hong Kong; Centre for Transport Studies, University College London, London, United Kingdom

Edward Chung Department of Electrical Engineering, Hong Kong Polytechnic University, Hong Kong, P.R. China

Martin Fellendorf Institute of Transport Planning and Traffic Engineering, Graz University of Technology, Graz, Austria

Bernhard Friedrich Institute of Transportation and Urban Engineering, Technische Universität Braunschweig, Braunschweig, Germany

Fatih Gundogan ISBAK Inc., Istanbul, Turkey

Miho Iryo-Asano Graduate School of Environmental Studies, Nagoya University, Nagoya, Japan

Jin-Tae Kim Korea National University of Transportation, Uiwang, Republic of Korea

Youngho Kim Korea Transport Institute, Sejong, Republic of Korea

Keping Li College of Transportation Engineering, Tongji University, Shanghai, P.R. China

Meng Li Department of Civil Engineering, Tsinghua University, Beijing, P.R. China

Dongbo Liu Traffic Management Research Institute, Ministry of Public Security, Wuxi, P.R. China

Wei Liu Chongqing Jiaotong University, Chongqing, P.R. China

Tom Mathew Department of Civil Engineering, Indian Institue of Technology Bombay, Mumbai, India

Monica Menendez Division of Engineering, New York University Abu Dhabi (NYUAD), Abu Dhabi, United Arab Emirates

Sudeshna Mitra Civil Engineering Department, Indian Institute of Technology, Kharagpur, India

Yetis Sazi Murat Faculty of Engineering, Civil Engineering Department, Pamukkale University, Denizli, Turkey

Hideki Nakamura Department of Civil and Environmental Engineering, Nagoya University, Nagoya, Japan; Graduate School of Environmental Studies, Nagoya University, Nagoya, Japan

Takashi Oguchi Institute of Industrial Science, The University of Tokyo, Tokyo, Japan

Thomas Riedel Verkehrs-Systeme AG, Winterthur, Switzerland

Chaitrali Shirke School of Civil Engineering and Built Environment, Queensland University of Technology, Brisbane, QLD, Australia

Alex Stevanovic Department of Civil, Environmental Engineering, Florida Atlantic University, Boca Raton, FL, United States

Keshuang Tang College of Transportation Engineering, Tongji University, Shanghai, P.R. China

Zong Tian Department of Civil and Environmental Engineering, University of Nevada, Reno, Reno, NV, United States

Axel Wolfermann Department of Civil Engineering, Hochschule Darmstadt University of Applied Sciences, Darmstadt, Germany

Xiaoyan Xie LVMT, UMR-T 9403, École des Ponts, IFSTTAR, UPEM, UPE, Champs-sur-Marne, France

Biao Yin LVMT, UMR-T 9403, École des Ponts, IFSTTAR, UPEM, UPE, Champs-sur-Marne, France

Introduction

1

Keshuang Tang[1], Zong Tian[2], Manfred Boltze[3] and Hideki Nakamura[4]
[1]College of Transportation Engineering, Tongji University, Shanghai, P.R. China, [2]Department of Civil and Environmental Engineering, University of Nevada, Reno, Reno, NV, United States, [3]Institute for Transport Planning and Traffic Engineering, Technische Universität Darmstadt, Darmstadt, Germany, [4]Graduate School of Environmental Studies, Nagoya University, Nagoya, Japan

1.1 Motivation

Road traffic signals are a most important instrument of traffic management and operations. The design of signalized intersections and signal programs has a serious impact on capacity, traffic safety, economic efficiency, and environmental compatibility. Statistics indicate that approximately 40% of total traffic accidents, approximately 80% of vehicle delays on urban roads, and approximately 20% of vehicle emissions occur at signalized intersections (WHO, 2016; FHWA, 2016; Kan et al., 2018).

In all countries, road traffic signals are used to control traffic. Many countries have developed comprehensive experience and knowledge in how to design signalized intersections and signal programs. Based on that, several national guidelines (TRB, 2010; FHWA, 2008; UK DOT, 2006; FGSV, 2015; JSTE, 2006; AUSTROADS, 2003), reports, and papers have been published, dealing with signal timing procedures, methods, and optimization models for a specific country, region, or situation. However, there is no international standard on road traffic signal control. While significant differences between countries exist, international harmonization and learning from each other is difficult, so far, because of the different languages and limited accessibility of national regulations for road traffic signal control.

To contribute to the international exchange of knowledge, this book summarizes the practice in road traffic signal control in several countries and regions around the globe. Individual reports are authored by outstanding national experts from 16 countries, providing a comprehensive overview and comparison of international practice on road traffic signal control. Based on that, the book allows an in-depth study and comparison of global practices. Strengths and weaknesses of specific regulations can be identified, and results may be used for the improvement and international harmonization of road traffic signal control.

Global Practices on Road Traffic Signal Control. DOI: https://doi.org/10.1016/B978-0-12-815302-4.00001-7

1.2 Scope

This book particularly addresses the background, history, state-of-the-art practice, and working examples for fixed-time control at isolated intersections in 16 selected countries. The selected countries include the United States of America, Canada, Germany, Austria, the United Kingdom, France, Switzerland, Turkey, Australia, New Zealand, Japan, China, India, South Korea, Qatar, and the United Arab Emirates (UAE).

In addition, common principles of road traffic signal control, international comparisons, and in-depth discussions on the differences across the selected countries, as well as helpful suggestions from a global perspective, are also provided in the book. Relevant knowledge on software, standards, legislation, equipment suppliers, and maintenance requirements for road traffic signal control are covered to some extent, without too much emphasis.

1.3 Purpose

This book can act as a fundamental reference book on international practice of road traffic signal control for international researchers and practitioners, with a focus on fixed-time control at isolated intersections. It is neither a summary of findings previously published in literature, nor a handbook or guideline. Potential applications of the book are explained below.

- To provide a comprehensive overview of the state-of-the-art practices of road traffic signal control in various countries and regions.
- To gain deep understandings of the reasons, underlying background and special considerations for existing differences among the selected countries and regions.
- To provide a global summary of useful practical experiences for fixed-time control at isolated intersections.
- To help the proper selection of signal timing procedures, methods, and parameters for fixed-time signal control.
- To offer helpful suggestions on common principles and improvement countermeasures of road traffic signal control from a global perspective.

Potential audiences of the book include practicing engineers, academic staff and students, hardware and software engineers in signal control manufactures, and maintenance service providers of signal control.

1.4 Book structure

This book consists of 15 chapters in total. This chapter highlights the motivation, scope, purpose, and structure of the book. Chapter 2, Principles of road traffic signal control, addresses major principles of road traffic signal control, which are independent from specific geographical areas, and it provides definitions for basic terms in traffic signal control. Chapters 3–14 are country-specific or region-specific chapters, in which the state-of-the-art practice on fixed-time control at isolated intersections in 16 countries are introduced and discussed. Items include background, history, control strategies, signal timing procedures and methods, signal phasing, critical parameters, special considerations, examples, and state-of-the-art implementations. The sequence of the country-specific chapters is arranged based on geographical areas, that is, America, Europe, Oceania, and Asia, respectively. Chapter 15, Summary, summarizes worldwide successful experiences and common knowledge, and it provides suggestions for the improvement of road traffic signal control. The titles and authors of each chapter are listed in Table 1.1.

Table 1.1 Chapter titles and Author Information

Chapter #	Author	Country	Job Title	Affiliation
Chapter 1: Introduction	Keshuang Tang*	China	Professor	Tongji University
	Zong Tian	United States	Professor	University of Nevada, Reno
	Manfred Boltze	Germany	Professor	Technische Universität Darmstadt
	Hideki Nakamura	Japan	Professor	Nagoya University
Chapter 2: Principles of Road Traffic Signal Control	Manfred Boltze*	Germany	Professor	Technische Universität Darmstadt
	Hideki Nakamura	Japan	Professor	Nagoya University
	Zong Tian	United States	Professor	University of Nevada, Reno
Chapter 3: The United States of America and Canada	Zong Tian*	United States	Professor	University of Nevada, Reno
	Aleksandar Stevanovic	United States	Associate Professor	Florida Atlantic University
	Wei Liu	China	Professor	Chongqing Jiaotong University
Chapter 4: Germany and Austria	Axel Wolfermann*	Germany	Professor	Darmstadt University of Applied Sciences
	Bernhard Friedrich	Germany	Professor	Braunschweig University of Technology
	Martin Fellendorf	Austria	Professor	Technical University of Graz
Chapter 5: The United Kingdom	Andy H.F. Chow*	Hong Kong (China)	Assistant Professor	City University of Hong Kong
	Konstantinos Ampountolas	United Kingdom	Senior Lecturer	University of Glasgow
Chapter 6: France	Biao Yin	France	Research Associate	École des Ponts ParisTech
	Nicolas Chiabaut	France	Professor	Université de Lyon
	Xiaoyan Xie*	France	Research Associate	École des Ponts ParisTech
Chapter 7: Switzerland	Thomas Riedel*	Switzerland	Manager	Verkehrs-Systeme AG
	Monica Menendez*	United Arab Emirates	Associate Professor	New York University Abu Dhabi

(*Continued*)

Table 1.1 (Continued)

Chapter #	Author	Country	Job Title	Affiliation
Chapter 8: Turkey	Yetis Sazi Murat*	Turkey	Professor	Pamukkale University
	Fatih Gundogan	Turkey	Vice President	Istanbul IT and Smart City Technologies Inc.
Chapter 9: Australia and New Zealand	Chaitrali Dipak Shirke	Australia	Research Associate	Queensland University of Technology
	Ashish Bhaskar*	Australia	Senior Lecturer	Queensland University of Technology
	Edward Chung	Hong Kong (China)	Professor	Hong Kong Polytechnic University
Chapter 10: Japan	Hideki Nakamura*	Japan	Professor	Nagoya University
	Miho Iryo-Asano	Japan	Associate Professor	Nagoya University
	Takashi Oguchi	Japan	Professor	The University of Tokyo
Chapter 11: China	Keshuang Tang*	China	Professor	Tongji University
	Keping Li	China	Professor	Tongji University
	Meng Li	China	Associate Professor	Tsinghua University
	Dongbo Liu	China	Vice Director	Traffic Management Research Institute, the Ministry of Public Security
Chapter 12: India	Shriniwas Arkatkar*	India	Associate Professor	Sardar Vallabhbhai National Institute of Technology
	Tom Mathew	India	Professor	Indian Institute of Technology Bombay
	Sudeshna Mitra	India	Associate Professor	Indian Institute of Technology Kharagpur
Chapter 13: South Korea	Jin-Tae Kim*	South Korea	Associate Professor	Korea National University of Transportation
	Youngho Kim	South Korea	Research Associate	Korea Transport Institute
Chapter 14: Qatar and the United Arab Emirates	Wael Alhajyaseen*	Qatar	Assistant Professor	Qatar University
	Ghassan Abu-Lebdeh	United Arab Emirates	Professor	American University of Sharjah
Chapter 15: Summary	Keshuang Tang*	China	Professor	Tongji University
	Manfred Boltze	Germany	Professor	Technische Universität Darmstadt
	Zong Tian	United States	Professor	University of Nevada, Reno
	Hideki Nakamura	Japan	Professor	Nagoya University

Note that * indicates corresponding authors of the chapters.

1.5 Notation and terminology

As mentioned above, there is no international standard on road traffic signal control, so far. For different countries and regions, individual standards and guidelines have been developed in different languages, each of them using their own terms, definitions, and symbols. In some cases, terms used in one language may not even exist in another language, because the method of signal programing is different.

With Chapter 2, Principles of road traffic signal control, this book aims to contribute to the harmonization of terms used in traffic signal control. The following individual chapters have been written taking this as a basis. However, to ensure the correct representation of practices in road traffic signal control in each case, different notations and differences in terminology could not be fully eliminated.

Bibliography

Australasian Road Transport and Traffic Agencies (AUSTROADS), 2003. Guide to Traffic Engineering Practice Series: Traffic Signals, Australia. AUSTROADS.

Department of Transport, the United Kingdom (UK DOT), 2006. Traffic Advisory Leaflet: General Principles of Traffic Control by Light Signals, the United Kingdoms. UK DOT.

Federal Highway Administration, US Department of Transportation (FHWA), 2008. Traffic Signal Timing Manual. FHWA.

Federal Highway Administration, US Department of Transportation (FHWA), 2016. 2015 Status of the Nation's Highways, Bridges, and Transit: Conditions & Performance. FHWA.

FGSV, 2015. Richtlinien für Lichtsignalanlagen—RiLSA (German Standard for Traffic Signals). FGSV-Verlag, Cologne, Edition 2015 (in German).

Japan Society of Traffic Engineers (JSTE), 2006. Manual on Traffic Signal Control. JSTE, Revised Edition (in Japanese).

Kan, Z., Tang, L., Kwan, M., et al., 2018. Estimating vehicle fuel consumption and emissions using GPS big data. Int. J. Environ. Res. Public Health 15 (4), 566.

Transportation Research Board (TRB), 2010. Highway Capacity Manual 2010. TRB.

World Health Organization (WHO), 2016. Global Status Report on Road Safety 2015. WHO.

Principles of road traffic signal control

2

Manfred Boltze[1], *Hideki Nakamura*[2] *and Zong Tian*[3]
[1]Institute for Transport Planning and Traffic Engineering, Technische Universität Darmstadt, Darmstadt, Germany, [2]Graduate School of Environmental Studies, Nagoya University, Nagoya, Japan, [3]Department of Civil and Environmental Engineering, University of Nevada, Reno, Reno, NV, United States

2.1 Basic functionality of traffic signals

2.1.1 Areas of application

Road traffic signals are used to control traffic and to avoid conflicts between road users which are using the same road space (conflict area), for example at intersections or crossings. For that purpose, traffic signals are indicating specific traffic streams of road users the permission to pass a marked (or virtual) stop line or to stop and wait. "Instructions conveyed by traffic signals shall take precedence over those conveyed by road signs regulating priority." [United Nations, 1968, § 5 (1)].

Traffic signals are used to control traffic at road intersections or at pedestrian crossings. There are several specific implementation areas for traffic signals, such as traffic control at bottlenecks (e.g., at road construction sites), safeguarding of railway crossings, ramp metering at highway entrances, or at the exit of emergency service stations. Traffic signals are also used for lane signalization (which is not covered in this book).

Traffic signals can also be used to prioritize specific road-user groups, to meter the traffic entering an area or road section, and to harmonize traffic flow by coordination of traffic signals in an arterial.

Traffic signals are a very important instrument to manage traffic. They are installed to improve traffic safety, traffic flow, environmental conditions, or economic efficiency of traffic operations.

2.1.2 Signal heads and signal indications

A *GREEN light* is a positive signal authorizing vehicles to proceed.

A *RED light* is a negative signal indicating that road users are not allowed to proceed; they must stop and wait at the stop line.

A *YELLOW light* (also called an amber light) gives road users notice that the RED light will be signaled soon. It indicates that road users have either to stop at

Global Practices on Road Traffic Signal Control. DOI: https://doi.org/10.1016/B978-0-12-815302-4.00002-9

the stop line before the RED light appears, or they have to pass and clear the conflict area as soon as possible. However they decide, drivers are not allowed to pass the stop line under a RED light.

Simultaneous *RED and YELLOW lights* may be used to inform road users that the GREEN light will appear soon. The purpose of this signal is to avoid start-up loss time at the beginning of GREEN.

Signal heads for motorized road traffic usually have three fields (RED, YELLOW, GREEN) which may be arranged vertically (RED on top) or horizontally (RED on the right or on the left, depending on the driving side). In some special cases, also signal heads with two fields or only one field can be applied. As the signals may address only specific traffic streams, the signal lights may show symbols for the respective road-user group (e.g., pedestrians or cyclists) or for the respective driving direction (e.g., left-turn arrow).

Depending on national regulations, the *position of signal heads* may be on the near side or the far side of the intersection approach. In any case, the visibility of signal heads must be ensured, and all road users must be clearly aware which signals are relevant for them.

The *size of signal heads* must allow for clear visibility and recognition by road users. Commonly, the diameter of each signal head is about 200 mm (e.g., in Germany) or 300 mm (e.g., in the United States and China). In approaches with higher speed, on interurban roads or in other cases with a higher need of perceptibility, also a larger diameter is used (e.g., 450 mm on expressways in Japan). The technical design of the signal heads must avoid reduced perceptibility by backlight or phantom light which may be caused by sunlight. Hoods and contrast shields may help to increase perceptibility.

Even if there is only one lane, the *number of signal heads* installed on every intersection approach should be at least two. More signal heads may be useful if there are additional lanes, and they are necessary if there is separated control of traffic streams on that approach.

Signal groups may be formed if the same signal is always shown to all streams in that signal group (group-based control). For example, there may be one signal group for all streams of one approach (approach-based control). In North America's ring-barrier-based phasing structure, a signal group is equivalent to a *signal phase*.

In many countries, additional *special signals* are used, such as flashing GREEN, flashing YELLOW, flashing RED, special signals for pedestrians or for public transport vehicles, and countdown signals. These are mentioned in the following chapters on national traffic signal control regulations.

2.1.3 Signal timing parameters

Signal programs (usually called *signal plans* in North America) determine the signalization of all traffic streams at an intersection or crossing over time. A signal program for a schematic four-leg intersection is shown in Fig. 2.1.

The *GREEN time* should allow all vehicles to depart which arrive in the respective traffic stream within one cycle. Therefore, the GREEN time depends on the

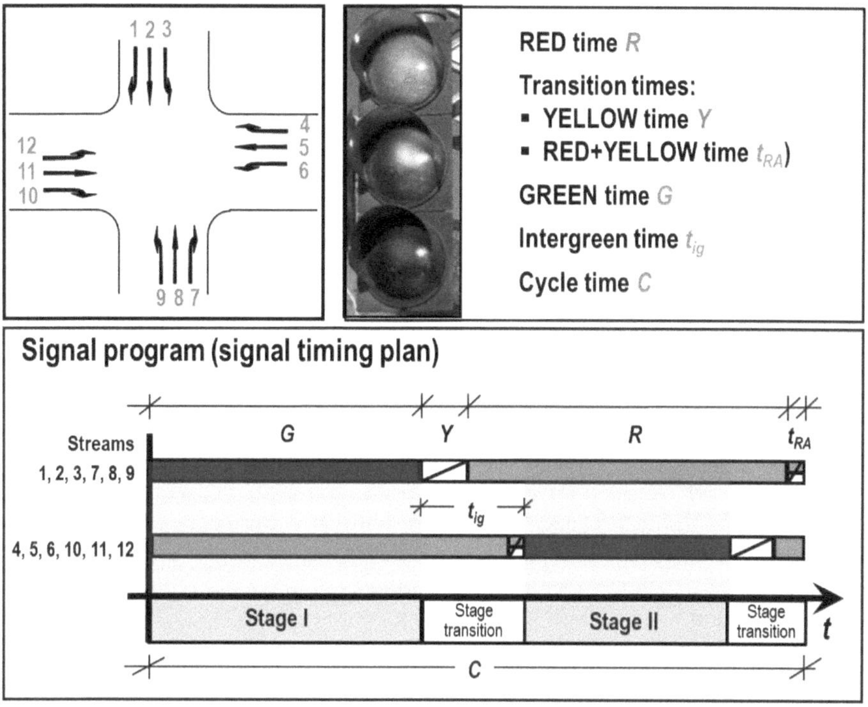

Figure 2.1 Traffic signal control terms (explained for a schematic four-leg intersection).

cycle time, on any lost times (e.g., intergreen, start-up lost time) and on the saturation flow rate, which describes the number of vehicles which can theoretically pass the stop line within 1 hour of continuous GREEN time. A minimum GREEN time needs to be defined (e.g., 5 s), and a maximum GREEN time is usually indirectly defined by the maximum cycle time and minimum GREEN time for other stages.

The *YELLOW time* must correspond with driving dynamics (vehicle speed and braking performance), since, after beginning of YELLOW, drivers need to decide either to stop safely in front of the stop line or to proceed and pass the stop line before the beginning of RED. Usually, the YELLOW time is set to a value between 3 s and 5 s, depending on the speed limit on the approach to the traffic signal. Traffic signals shall not be operated without a defined speed limit which is compatible with the YELLOW time.

The *RED time* is not explicitly calculated, but it results from the other signal durations. The minimum RED time and maximum RED time can be defined.

The *RED and YELLOW time* is usually set to 1 or 2 seconds. A too short or missing RED and YELLOW time may lead to start-up loss time at the beginning of GREEN. A too long RED and YELLOW time leads to vehicles starting too early which may cause accidents. In several countries this signal is not used at all.

Usually, the same signalization is repeated in a cycle with a duration defined as *cycle time*. During one cycle all traffic streams are given GREEN at least once.

Too short cycle times may lead to oversaturation and congestion because more stage transitions per hour cause more loss time per hour. In principle, longer cycle times increase capacity. However, too long cycle times unnecessarily increase average delay and may lead to more traffic accidents because of RED light violations (specifically by pedestrians). The long GREEN times which result from a long cycle time may also lead to unused green time and unnecessarily long delays.

Within one cycle, two or more *stages* (also named phases) can be defined as periods in which GREEN is given to one or more specific traffic streams, while all others have to wait. All streams given GREEN in one stage must be *compatible streams* (i.e., they do not have a joint conflict area) or they must be at least *partly compatible streams*. For example, for right-hand traffic, the partly compatible streams are left-turn streams and the opposing straight-through streams, left-turn streams and the opposing right-turn streams which exit in the same direction, left-turn streams and pedestrian streams crossing at the intersection exit, right-turn streams and pedestrian streams crossing at the intersection exit.

The *number of stages* is at least two. More than two stages can be useful to separate specific traffic streams, for example, left-turn movements. In general, more stages allow a safer traffic operation. However, more stages also increase the number of stage transitions (see below) and the cycle time and reduce the capacity of the intersection.

The *stage transition* (from one stage to another) must ensure that all road users of the ending stage have cleared the conflict area before any road user of the beginning stage reaches the same conflict area. In this context, the concept of intergreen times or the concept of a ring-barrier structure is applied.

The *intergreen time* is defined as the time between the end of GREEN for one traffic stream and the beginning of GREEN for another stream which is not compatible with the first stream. The intergreen time must allow the last vehicle of the ending stream to clear the conflict area before the first vehicle of the beginning stream can reach the same conflict area. For signal groups, the intergreen time must fulfill this condition for all possible combinations of individual streams. A sufficient intergreen time is most important for traffic safety. On the other hand, too long intergreen times reduce capacity. The optimal duration of intergreen does not depend on cycle time nor traffic volumes.

The intergreen time consists of a *crossing time* (time between the end of GREEN and the moment when the last vehicle of the ending stream crosses the stop line) and a *clearing time* (time the last vehicle needs from the stop line until it has completely cleared the conflict area), and it may be reduced by an *entering time* (time the first vehicle of the beginning stream needs from the stop line to reach the conflict area). The clearing distance and the entering distance are explained in Fig. 2.2. The necessary intergreen time can be determined for each combination of traffic streams if they have a joint conflict area. In a simplified approach, the intergreen time can be determined only for the critical pair of traffic streams within the two consecutive stages.

Ring-barrier structure is a standard practice in North America. Unlike the European practice in determining intergreen times, the yellow and red clearance

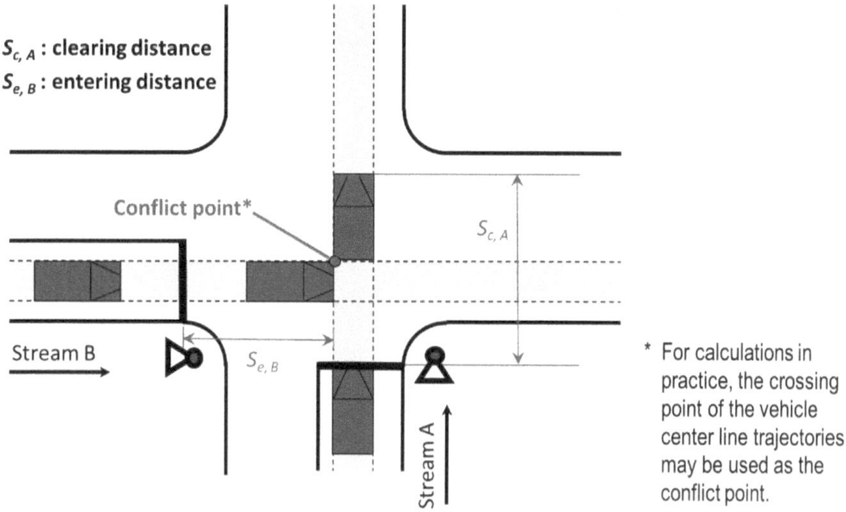

Figure 2.2 Clearing distance and entering distance.

intervals are used and calculated for each signal phase (or called signal group under European's stage-based control scheme), thus they are not variable when the phase sequence is changed. The red clearance interval is required to completely clear the traffic stream(s) served by the phase out of the intersection (e.g., the far side of the crosswalk) before the next conflicting phase starts. Therefore, the sum of yellow and red clearance intervals is generally longer than the intergreen time.

2.1.4 Traffic signal control methods and operating modes

On the *macroscopic control level*, certain signal programs are selected and activated for a specific time of the day or for a specific situation (e.g., traffic or environment). Widely applied macroscopic control methods are the *time-dependent (or time-of-day) signal program selection* and the *traffic-responsive signal program selection*. A more sophisticated alternative is *traffic-responsive frame signal program generation*.

On the *microscopic control level*, there are various *control modes*, which differ in the degree of variability of different control parameters (cycle time, stage number, stage sequence, green times, offset time). In *fixed-time control* none of these parameters in the signal program is flexible. In *actuated control* at least one of the parameters in the signal program is actuated according to the current situation (e.g., green time adaptation based on traffic volume detection). The *signal program generation* allows forming a signal program under full flexibility of all parameters. Usually, this can only be used at isolated (single) intersections.

Besides traffic signal control methods for isolated intersections, there are also methods for groups of traffic signals on an arterial or in an area. Such *coordinated traffic signal control* aims at optimizing traffic flow by allowing vehicles to pass through a series of intersections with minimal delay and stops. For that purpose, the *offset* of the signal programs at consecutive intersections must be designed according to vehicle speed and intersection spacing. However, coordinated traffic signal control is not covered in this book.

2.2 Consideration of various impacts on different road-user groups

2.2.1 General remarks

When designing traffic signals, usually, all kinds of traffic modes should be considered: walking, cycling, public transport, and motorized private transport. Furthermore, the freight transport, especially heavy vehicle traffic, requires special examination. Depending on local conditions, further modes may need specific considerations, for example, motorcycles as a predominant mode in the case of Vietnam.

The design of the traffic signal control shall consider the various impacts on the different road-user groups as comprehensively as possible. This includes the principles of barrier-free mobility and sustainability.

In some cases, the advantages regarding a specific impact or road-user group can only be achieved together with disadvantages for other road-user groups. In such cases, a comprehensible consideration shall be done before decision-making. Comprehensibility and transparency are important conditions for the acceptance of decisions by citizens. Therefore, decisions to design traffic signal control need rules for a fair balance of different impacts, for the treatment of goal conflicts, and therefore for the prioritization of different road-user groups.

Due to their general advantages regarding environmental compatibility, walking, cycling, and public transport shall be promoted. However, any absolute priority for one single traffic mode is not appropriate. Instead, in each case, the overall impacts of measures at traffic signals must be considered to avoid inadequate negative impacts on other traffic modes and other relevant parameters.

Planning instruments for traffic signal control shall be developed and applied which allow to determine all major impacts and to reflect them in the course of consideration.

2.2.2 Main goals and relevant criteria

There are *four main goals* within the framework of traffic management, which are generally relevant for all traffic modes: satisfaction of mobility needs, increase of traffic safety, reduction of environmental pollution, and improvement of economic efficiency (Boltze and Dinter, 1994). Among these goals, due to the number of premature deaths caused by traffic-borne air pollution and by traffic accidents, the prevention of emissions with adverse health effects and the improvement of traffic safety are gaining increased attention.

Satisfaction of mobility needs: Mobility is a basic need of human beings. The accessibility of a city determines its attractiveness not only for residents but also the economy. Therefore, it is necessary to design signalized intersections and to develop signal timing programs which improve the quality of traffic flow (delay, number of stops, queue length) and to increase the level of comfort while using different traffic modes. Improved mobility through traffic signal control directly affects other goals such as a reduction of environmental pollution and improvement of economic efficiency. Criteria related to this goal may include delay for vehicles (motorized and nonmotorized), delay for individuals, number of stops, queue lengths, and comfort-related criteria.

Increase of traffic safety: While in some developed countries, safety measures have been well taken into consideration when designing signalized intersections and developing signal timing programs, many developing countries have not yet realized the consequences and safety impacts of inadequate signal designs and control. Further efforts are still necessary to reduce the number of traffic accidents, injuries, and deaths. Criteria related to this goal may include the number of traffic accidents, the number of slight and severe injuries, and the number of fatalities.

Reduction of environmental pollution: To protect human health, traffic signal control programs should aim at reducing the negative impacts of traffic through noise emission and air pollution. In special cases, traffic signal control may also influence land consumption. Criteria related to this goal may include the emissions of air pollutants, measurements of the ambient air quality, noise level, energy consumption, and CO_2 emissions.

Improvement of economic efficiency: The main focus of this goal is on the efficient use of infrastructure and vehicles. Overall traffic quality, such as indicated by the average travel speed in a road network, also has some relation to economic efficiency, but this is usually covered by the goal "satisfaction of mobility needs." Criteria related to this goal may include average travel speed, the saturation degree of green times, and the cost for implementation and operation of the signalized intersection.

Mode choice has a significant impact on the sustainability of the overall transport system. Therefore, besides the direct impacts of traffic signal control measures, indirect impacts influencing the medium- and long-term mode choice behavior must also be considered.

To deteriorate traffic flow for motorized private vehicles in urban areas is not an appropriate means to control mode choice. Inappropriate restraints for motorized traffic will not only lead to congestion and delays for drivers, but also harm human health significantly by avoidable emissions of noise and air pollutants, specifically for local residents, pedestrians, and cyclists (Boltze and Jiang, 2017).

2.2.3 Multicriteria and multimodal impact assessment

In many cases, in the design of traffic signal control, contrary requirements and goal conflicts cannot be avoided. As far as significant negative impacts are recognized, they should be verified technically, elaborated transparently, and considered carefully by comparing with the advantages of the measure. For measures which would promote one traffic mode, besides the positive impact on this specific mode, also the negative impacts on other traffic modes should be displayed and quantified, where possible.

For measures to promote pedestrian and cycle traffic, specifically the following parameters must be considered:

- delay, number of stops, and queue lengths in motorized individual traffic;
- delay for public transport;
- noise level and emission of air pollutants by motorized private transport and heavy vehicles (where possible also ambient air quality);
- fuel consumption and CO_2 emissions by motorized private transport and heavy vehicles.

For measures to promote public transport, specifically the following parameters must be considered:

- delay, number of stops, and queue lengths in motorized individual traffic;
- average and maximum delay for pedestrians and cyclists;
- noise level and emission of air pollutants by motorized private transport and heavy vehicles (where possible also ambient air quality);
- fuel consumption and CO_2 emissions by motorized private transport and heavy vehicles.

For measures to promote motorized private transport and heavy vehicles, specifically the following parameters must be considered:

- average and maximum delay for pedestrians and cyclists;
- delay for public transport;
- impacts on mode choice.

To achieve a multimodal impact assessment, the same parameters regarding different traffic modes should be aggregated. Under consideration of the physical fundamentals, most of them can be aggregated without further adjustments (e.g., emissions). For the evaluation of average delay, a person-related aggregation would be appropriate. Based on the average vehicle occupancy, the overall average delay for all road users at the intersection could be calculated (Hunter et al., 2011). Different weights could be assigned to the delay at different traffic modes in order to consider political preferences, comfort aspects for each individual traffic mode (e.g., differences in comfort or weather protection), or other aspects which are not reflected by the chosen criteria (e.g., operational aspects of public transport). (Boltze and Jiang, 2017)

Generally, it is difficult to aggregate the impacts on different goals and parameters. However, finally it is necessary to find a fair balance of, for example, delay for different road users, number of traffic accidents, and health impacts of air pollution and noise. Necessary quantitative weights may be concluded from relevant literature. However, the rates used for monetarization are often controversial and they always reflect political preferences. Therefore, to allow a comprehensive assessment of traffic signal control measures a concept to combine and weight all the objectives and criteria should be developed and politically approved.

2.3 Situation-responsive traffic signal control principles

Traffic signal control should be flexible and adaptive. It is not just fixed rules that should be applied, but instead—according to available technical options—the specific current situation should be considered (e.g., traffic volumes, environmental situation, day of the week, time of day).

Variations in traffic volumes are already considered, if an adaptive traffic signal program is implemented which changes according to time and traffic volumes, and if the above-mentioned suggestions for performance measurement and fair balance are applied.

Prioritization of transit vehicles aims to consider the occupancy of the transit vehicle and its position compared with the schedule. Priority can be given to transit vehicles which are behind schedule, but not to early vehicles.

In situations with critical air pollution levels (impending or already-existing exceedance of threshold values), signal programs should be applied which particularly reduce emissions of air pollutants. However, in uncritical situations, signal programs could be used which give lower weight to these environmental criteria.

At places and times with special demand for noise protection (e.g., at hospitals or at night in residential areas), signal programs should be applied which particularly prevent noise emissions. In such situations, green waves and other measures to avoid stops and start-ups of vehicles have to be combined with further measures to influence route and speed choice.

A concept for an active utilization of traffic signal control to reduce noise and air pollution shall be developed and implemented. Examples for traffic signal control measures which should be considered in critical environmental situations are access metering for critical parts of the network, adaptation of signal coordination (green waves) to further reduce the number of stops, reducing the degree of prioritization for transit vehicles, dynamic green time adaption to avoid heavy vehicle stops, and others. Some of these measures should be combined with other traffic control measures, such as regulations for driving directions, route recommendations, or speed limits. As mentioned above, the impacts of these measures on other road users must be considered comprehensively (Boltze and Jiang, 2017).

2.4 Quality management in planning, implementation, and operation

Quality management for traffic signals is the systematic quality inspection and quality improvement related to the design of road infrastructure, the traffic control, and the technical components of the traffic signal under consideration of the planning and strategic frame conditions. Quality management affects all stages of the life cycle of traffic signals, for example, planning, implementation, and operation. It aims at achieving a continuously high quality regarding traffic safety, environmental impacts, and traffic flow. And it ensures that the requirements and strategic goals of a super-ordinated transport planning and traffic management are fulfilled or adapted to changes. Quality management also supports the operational availability and the flawless operation of all traffic signal components. It is a major precondition of an efficient quality management that the objectives of traffic signal control, the desired levels of service for all road-user groups, and the ways of treating goal conflicts are clearly defined. The technical standards and the responsibilities of all involved stakeholders must also be clarified (FGSV, 2015).

Quality management in the *stage of planning* must include a careful and comprehensive definition of the function of the traffic signal, of local conditions, and objectives of the traffic signal control. It must be ensured that all elements of the signal programs and all details of the control method are determined in a correct way. All details of the traffic signal and its operation must be documented, including procedures to switch the traffic signal on and off, procedures to switch from one signal program to another, and testing procedures.

Quality management in the *stage of implementation* includes the primary implementation as well as the follow-up implementations of all data and programs for traffic signal control on all system levels. Quality management must cover the complete testing of all implemented data and programs, including procedures to switch the traffic signal on and off, to switch among programs, and the system behavior in case of incidents of any kind. It is most important that the traffic signal documentation is well understandable and that it describes exactly what is implemented in the operated system.

Quality management in the *stage of operation* requires frequent measuring of the actual impacts of traffic signal control. The aim is to identify deficiencies at an early stage, to verify the effectiveness of implemented measures and finally to allocate the available resources to where they allow the most benefits. The impact assessment (or performance measurement) should not be a one-time activity, but rather a continuous task (Boltze and Jiang, 2017). For impact assessment during operation the following data may be used: traffic accident data, process data from control devices, operation and disruption data, as well as information from previous experience and from observations at the traffic signal (FGSV, 2015). Information from other systems may also be useful, such as speed profiles from transit operation systems or data obtained from suppliers of navigation systems. Based on existing standards for the quality management of traffic signals—for example, in Germany RiLSA (FGSV, 2015)—a concept for the continuous impact assessment during operation should be developed, optimizing the efforts for data collection. The results of the impact assessment shall be reported to decision makers in defined cycle or event-driven to allow corrections regarding priorities and rules of consideration.

References

Boltze, M., Dinter, M., 1994. The project FRUIT – a goal-oriented approach to traffic management in Frankfurt am Main and the Rhein-Main Region. Traffic Eng. Control 35 (7/8), 437–444.

Boltze, M., Jiang, W., 2017. Berücksichtigung verschiedener Verkehrsteilnehmergruppen und Kriterien bei der Optimierung von Lichtsignalanlagen – Ein Vorschlag zum verkehrspolitischen Rahmen für die Gestaltung der Lichtsignalsteuerung in Städten (Consideration of Different Road User Groups and Criteria in Traffic Signal Optimisation – A Proposed Policy Framework for the Design of Urban Traffic Signal Control). Straßenverkehrstechnik (61), Bonn (Germany), September 2017, pp. 615–621 (in German).

Hunter, B., Wolfermann, A., Boltze, M., 2011. Multimodal assessment of signalized intersections considering the number of travellers. In: TRB 90th Annual Meeting Compendium of Papers DVD, Washington, DC, USA, 23–27 January 2011.

FGSV, 2015. Richtlinien für Lichtsignalanlagen – RiLSA (German Standard for Traffic Signals). FGSV-Verlag, Cologne, Edition 2015 (in German).

United Nations, 1968. Convention on Road Traffic. United Nations, Vienna, Austria.

United States and Canada

3

Zong Tian[1], *Alex Stevanovic*[2] *and Wei Liu*[3]
[1]Department of Civil and Environmental Engineering, University of Nevada, Reno, Reno, NV, United States, [2]Department of Civil, Environmental Engineering, Florida Atlantic University, Boca Raton, FL, United States, [3]Chongqing Jiaotong University, Chongqing, P.R. China

3.1 Background

This chapter documents the general practices on traffic signal control in the United States and Canada. Most of the discussions are based on applications in the United States, as the practice in Canada is very similar, except for the use of measurement systems: Imperial (United States) versus metric (Canada). While the focus of this book series is on isolated and fixed-time controlled intersections, which are not common in the United States or Canada, the chapter also includes relevant information such as general timing procedures for both isolated and coordinated signals that may be of interest to readers from other countries.

3.1.1 *History of traffic signal control*

Traffic signal control in the United States has been in place for over 100 years, with the first electronic signal installation in Cleveland, Ohio, in 1914. In the first two decades of the 20th century, semaphore traffic signals were widely used. The early modern traffic signal with three-color display was created and implemented in Detroit, Michigan, in 1920. The signal was controlled manually as a tower with 12 floodlights in the center of an intersection (United States Access Board, 2003).

For the following 30 years, manually controlled traffic signals were gradually phased out, and traffic signal control made a revolution with the rise of digital computers. Using the early electromechanical controllers and pressure-sensitive detectors, the prototype modern automatic and actuated traffic signal control emerged around the late 1950s and quickly evolved over the past 50 years, due to substantial advances in electrical and computer technologies. Currently, traffic signals and relevant devices in the United States and Canada have been standardized from design, implementation, and operation through national or industrial manuals and guidelines (Hatcher et al., 2014). The emerging signal control technologies, such as adaptive or intelligent signal control, have been attracting more and more attention recently

Global Practices on Road Traffic Signal Control. DOI: https://doi.org/10.1016/B978-0-12-815302-4.00004-2

(Stevanovic, 2010). A major shift of traffic signal control is to focus on multimodal services, including not only motorized vehicles, but also other road users such as pedestrians, bicycles, and public transit.

3.1.2 The state-of-the-art traffic signal control

After more than a century of development, traffic signal control in North America has become a relatively mature system in terms of design and operations. Traffic signals are well recognized as one of the vital control elements of traffic systems. While most state Departments of Transportation (DOTs) do manage and operate some of the signals along state highways, the majority of the signals are maintained and operated by local jurisdictions. Metropolitan Planning Organizations generally provide funding to design and operate the signals, but the cities or counties are responsible for the day-to-day maintenance and operations.

Currently, there is an estimated total of 330,000 signalized intersections in the United States, approximately one per 1000 inhabitants. The number of signals in Canada is estimated at about 25,000. Most signalized intersections are equipped with microprocessor-based controllers and traffic detection devices that can be used for fixed-time, fully actuated, semiactuated, or actuated-coordinated controls (NTOC, 2012). Unlike most developing countries, fixed-time controlled intersections represent only a small portion of signals, which are mostly seen in a city's downtown area. The majority of the signals are either fully actuated or actuated-coordinated under time-of-day (TOD) signal plans.

3.1.2.1 Current nation-wide guidelines

In the United States, the *Manual on Uniform Traffic Control Devices* (*MUTCD*) (Federal Highway Administration, 2012) is the most widely accepted national standard when installing and managing traffic control devices on all public roads. Canada has its own version of the manual (Transportation Association of Canada TAC, 2014). The MUTCD provides details on traffic signal installation, particularly specifications on signal display (e.g., different types of left-turn controls), basic signal timing parameters (e.g., yellow, red clearance, and pedestrian timing), and signal warrants. The signal warrants determine when a traffic signal is justified based on given traffic conditions at an intersection. The MUTCD has gone through several revisions, with the last major update completed in 2009 followed by two revisions in 2012.

The other important document in the United States is the *Signal Timing Manual* (STM), with the first edition published in 2008 and the second edition published in 2015. The STM is considered the only comprehensive guidebook for signal timing practices in the United States. Both editions were produced through the National Cooperative Highway Research Program (NCHRP) (Urbanik et al., 2015). It should be noted that the two versions of the STM do not fully overlap, and both represent an excellent source of information.

Other manuals or guidelines worthy of mention include *Guidelines for Timing Yellow and All-Red Intervals at Signalized Intersections* (NCHRP-Report 731)

(McGee et al., 2012) and *MnDOT Traffic Signal Timing and Coordination Manual* (Minnesota Department of Transportation, 2013).

3.1.2.2 Major signal control manufacturers and control systems

Most traffic signals (more than 95%) in the United States and Canada currently operate under the conventional TOD control mode, although there has been a growing trend toward the use of adaptive or other advanced signal control technologies over the past two decades. Conventional TOD systems require periodic manual timing development and update to cope with traffic demand changes. The discussions in this chapter are pertinent to such conventional TOD systems. There are several widely adopted traffic signal controller standards. One is from the National Electrical Manufacturers Association (NEMA), who developed a controller standard known as the NEMA standard. Two generations of the NEMA standard have been applied—namely, the TS-1 (1976) and the TS-2 (1992). The TS-2 is further divided into Type 1 and Type 2 controller standards. The other standard was developed by the California Department of Transportation (Caltrans) and the New York Department of Transportation (NYDOT) in 1979. The original standard was developed under the Type 170 platform in 1979. This standard has been updated through many recent platforms, including Type 179, Type 170E, and Type 2070. To provide an open architecture hardware and software platform that can support a wide variety of Intelligent Transportation Systems (ITS) applications, the ATC Transportation Controller (ATC) Standard v5.2b was published in 2006. The ATC Standards are being developed and maintained under the direction of the ATC Joint Committee (JC), which is made up of representatives from the American Association of State Highway and Transportation Officials (AASHTO), the Institute of Transportation Engineers (ITE), and the NEMA.

The major traffic signal controller manufacturers in the United States and some of their most representative products are summarized in Table 3.1. Most of the traffic signal controllers in use in the United States and Canada are manufactured by one of these companies.

Table 3.1 Major traffic signal controller manufacturers and their products

Company name	Controller products
Econolite	Cobalt/2070C, 2070E
McCain	170 ATC/170E/2070 ATC/2070/ATC NEMA/FLeX
Naztec/Trafficware	Series 900/ATC Traffic/Series 2070
Peek	ATC CBD/ATC 2000/ATC-1000/3000E M3000E
Siemens USA (EAGLE Traffic Control Systems)	2070 Series/M50, M60 Series
Intelight	2070 ATC/X3 NEMA ATC

3.1.2.3 Traffic signal control software

Traffic signal control software refers to control applications at different levels and for different purposes. It can be categorized as local controller software (also called firmware), central management software, and timing development/optimization software.

The local controller software provides control logic based on external inputs and programmable parameters. Such software is usually provided by controller manufacturers. The central management software brings traffic network data into a single repository for a real-time, integrated view of traffic signal operations. Through National Transportation Communications for ITS Protocol (NTCIP) (Herrick, 1999), the central management software can monitor signals and control them remotely. Most central management software have three major functions—graphical display, database management, and information reporting. The majority of such central management software is developed by controller manufacturers and specifically used on their products, for example, ATMS.now by Trafficware, Centracs ATMS by Econolite, TACTICS by Seimens, QuicNet by McCain, and MaxView by Intelight. There are also third-party central software packages such as TranSuite by TransCore.

3.1.2.4 Signal heads and signal display

The *MUTCD* contains guidance on traffic signal display types and positions for the most common applications. The allowable signal indications include (circular or arrow): green, yellow, red, flashing red, flashing green, and flashing yellow. Each of the signal indications represents a specific meaning or control type. The typical position and arrangements of signal faces are shown in Fig. 3.1. The details can be found in Chapter 4D of the *MUTCD*.

Protected/permissive phasing is a treatment widely used in the United States and Canada to improve both safety and efficiency. In this mode, the protected left-turn arrow is immediately followed by a permissive green ball, still allowing vehicles making left-turns when acceptable gaps become available within the opposing through traffic stream. The permissive/protected mode follows a similar principle, with an advantage that the left-turn phase may be skipped if all the left-turn vehicles can depart during the permissive period, resulting in less lost time.

For pedestrian signals, the two most common types of displays have a countdown timer for the "Flashing don't walk" (FDW) interval and the other one without any countdown timer. Nevertheless, signal timing parameters are the same for both pedestrian signal displays. Fig. 3.2 shows the typical pedestrian signal indications.

3.2 Selection of control strategies

3.2.1 Warrants for the installation of traffic signals

The MUTCD requires an engineering study of traffic conditions, pedestrian characteristics, and physical characteristics of an intersection before determining whether

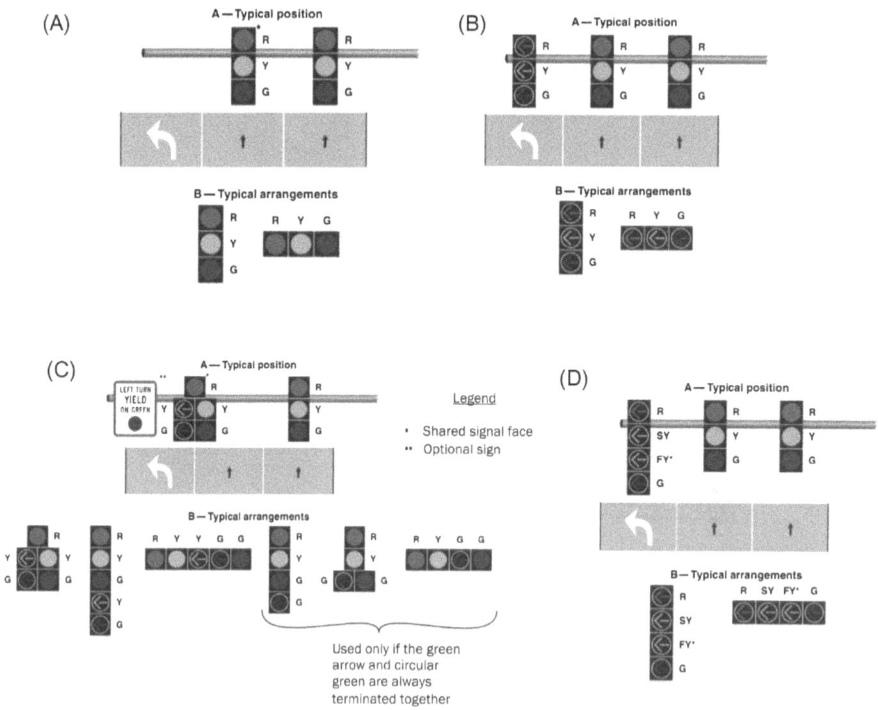

Figure 3.1 Typical position and arrangements of signal heads ((A) for permissive left-turn control; (B) for protected left-turn control; (C) for protected/permissive left-turn control; and (D) for protected/permissive left-turn control with flashing yellow arrow).
Note: *FY** denotes Flashing yellow arrow.
Source: Adapted from Chapter 4D of the MUTCD.

installation of a traffic signal is justified. The various factors are considered in the following nine signal warrants:

- Warrant 1—Eight-Hour Vehicular Volume
- Warrant 2—Four-Hour Vehicular Volume
- Warrant 3—Peak Hour
- Warrant 4—Pedestrian Volume
- Warrant 5—School Crossing
- Warrant 6—Coordinated Signal System
- Warrant 7—Crash Experience
- Warrant 8—Roadway Network
- Warrant 9—Intersection near a Grade Crossing

A signal installation is not justified if none of the signal warrants is met. On the other hand, meeting one or more of these signal warrants can only justify a signal installation, but a traffic signal may not be mandatory. The details of the signal warrant analysis can be found in Chapter 4C of the MUTCD.

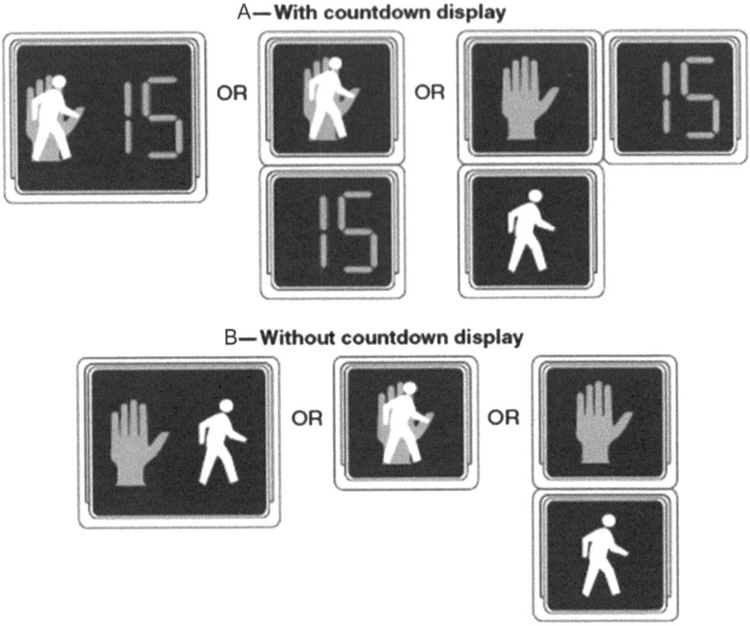

Figure 3.2 Typical pedestrian signal indications.
Source: The MUTCD.

3.2.2 Selection of actuated or fixed-time control

The selection of actuated or fixed-time control is primarily dependent on the available equipment and its traffic characteristics. Actuated signal control requires traffic demand data from detectors located on intersection approaches. Fixed-time control does not require such data or detectors and is the least costly to deploy. In the United States and Canada, fixed-time control is not common and is usually seen in central business districts (CBDs) where the signals are often closely spaced and/or in grid networks.

3.2.3 Selection of isolated or coordinated control

Regardless of actuated or fixed-time control, signalized intersections can either operate as isolated or coordinated. The decision to use coordination is influenced by a variety of factors, including intersection spacing, traffic demand level, and user preference. Coordination should be used to meet specific objectives, and it is not appropriate in every situation. While coordination can reduce travel time, stops, delay, and queues for the coordinated movements, there may be negative impacts on the uncoordinated movements.

The MUTCD includes a guidance that traffic signals within 1/2 mile (0.8 km) of one another along a major route or in a network of intersecting major routes should be coordinated, preferably with interconnected controller units. Information presented in the FHWA report *Signal Timing on a Shoestring* (Henry, 2005) reveals that both simple and complex procedures can be used to identify which intersections to coordinate. In general, when intersections are close together, it is advantageous to coordinate them, particularly if volumes between the intersections are high. Signals are normally not in coordination (running free) during low-volume time periods such as night time.

3.2.4 Selection of left-turn control type

A decision-making process of whether a separate left-turn phase is needed, as well as its type of protection (protected/permissive or protected only), must be determined before signal timing considerations. There is a structured evaluation procedure shown in Fig. 3.3 (modified from Exhibit 4-16 of the STM) that promotes consistent application of left-turn signal phasing. The additional information needed for the flow chart includes two major traffic safety measures: critical left-turn-related crash counts and speed-related minimum-sight-distances. This procedure was derived from a variety of sources to identify the least-restrictive left-turn operational mode that can meet both desired operational and safety objectives. A separate evaluation is needed for each left-turn movement at an intersection.

The major drawback of such flow-chart-type guidelines is the lack of a balanced consideration of the influencing factors. For example, a single factor, such as two or more left-turn lanes, will lead to a *protected* left-turn control type, while other factors (e.g., volume and speed) may well justify the use of other control types. Due to this drawback, agencies have also adopted guidelines that use a multicriteria decision-based approach (Ozmen and Tian, 2009). Such guidelines take all the quantifiable parameters and variables into consideration to derive a likelihood score for each left-turn control type, for example, 30% on protected, 50% on protected/permitted, and 20% on permitted.

3.3 Signal timing procedures

3.3.1 Necessary input data for signal timing

The data needed for developing signal timing plans can be categorized into the following groups: traffic characteristics (vehicular turning movement demands, composition, and pedestrian/bicycle demands), intersection geometry, existing phasing and timing records, and crash history. Site visits are highly recommended to capture the characteristics that are not obvious on the data sheets. Other traffic characteristics (e.g., speed, travel time, and queues) may be beneficial for the signal timing plan development process. Not only will these data be useful when developing the signal timing plans, but these traffic characteristics can also be used as performance measures to evaluate the effectiveness of signal timing changes.

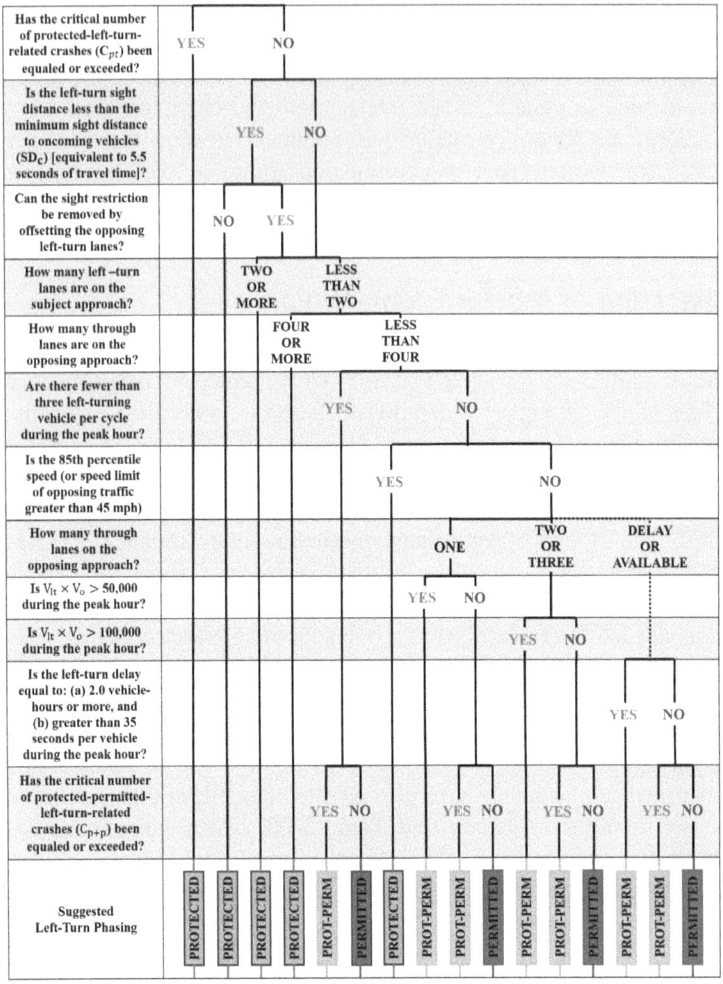

V_{lt}=left-turn volume on subject approach (vehicles per hour)
V_o=Through Plus right-turn volume on approach opposing subject left-turn movement (veh per hour)
Prot-Perm=Protected-Permitted (Desirable) or Protected

Source: Adapted from *the Manual of Traffic Signal Design*, 2nd Edition (11), *The Traffic Signal Book* (12), and the *Traffic Engineering Manual* (13).

Figure 3.3 Left-turn control type selection.
Note: (1) Critical crash count depends on the number of left-turn lanes and the number of years during which crashes are counted. The STM (Exhibit 4-17) has the details. (2) The sight distance = speed × 5.5 s (5.5 is the safe headway). V_{lt}, left-turn volume on subject approach (vehicles per hour). V_o, through Plus right-turn volume on approach opposing subject left-turn movement (vehicles per hour). PROT-PERM, protected-permitted (desirable) or protected.
Source: Reprinted with permission from NCHRP Report 812 – Signal Timing Manual, 2015, by the National Academy of Sciences, Courtesy of the National Academies Press, Washington, D.C.

Intersection geometry can have a major influence on timing development, thus signal timing practitioners should take an active role in new intersection design or intersection improvements, as the best signal timing plan cannot overcome poor geometrics. The existing signal timings can help the practitioners better understand current field conditions and provide suggestions for improvement. Signal timing is one of the many factors that may contribute to crashes. Therefore practitioners should be aware of the potential safety impacts of signal timing changes.

3.3.2 Outcome-based signal timing process

Outcome-based signal timing emphasizes a timing development process to target desired outcomes. It encourages practitioners to consider all road users and clearly establish priorities to serve these users—personal vehicles, pedestrians, bicycles, and public transit. Various performance measures can be established to evaluate the effectiveness of the timing plans. Fig. 3.4 depicts the general timing development steps as well as the elements to be considered in each step. Note that not all the steps and elements are pertinent to timing fixed-time isolated signals.

3.4 Signal timing concepts and parameters

3.4.1 Typical phasing plans

To better understand the signal phasing structure, it is necessary to relate traffic movements to signal phases using the sample intersection shown in Fig. 3.5.

Movements describe user actions at an intersection. At a conventional four-leg signalized intersection, it is possible to have 12 one-way vehicular movements and four two-way pedestrian movements. Note that a single movement can be accommodated by multiple lanes (e.g., through movement in two lanes), or multiple movements can be accommodated by a single lane (e.g., a lane that serves both through and right-turn vehicles).

A *traffic signal phase* is a timing process within the signal controller that serves one or more traffic movements at the same time (where *traffic* may represent all travel modes). Unlike the stage-based timing structure, a signal phase is exclusively associated with a specific movement. Each signal phase includes a green interval, a yellow interval, and an optional red clearance interval. A practitioner must first assign phase numbers to the movements at a signalized intersection before specifying signal timing values for those phases.

The NEMA dual-ring-based phasing structure shown in Fig. 3.6 represents the standard practice of structuring major phases for standardized four-leg intersections in North America. This standardized design evolved from the interval-based phasing structure, which is still commonly seen in Europe and other countries. Compared to the interval-based phasing structure, the ring-barrier structure provides conveniences when dealing with standard intersections. However, the ring-barrier design may have its limitations when operating signalized intersections with

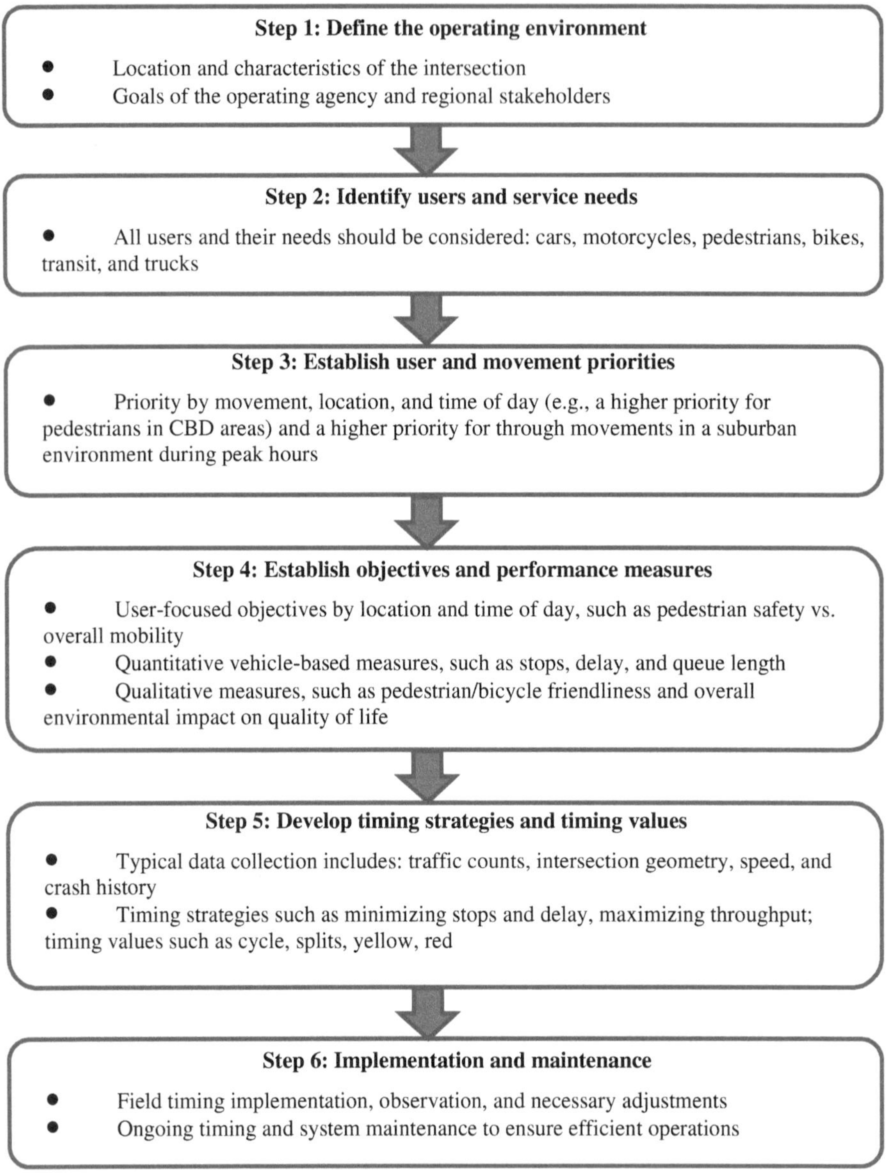

Figure 3.4 Outcome-based signal timing process.
Source: Adapted from the STM.

complex traffic movements, multimodal operations, and/or unconventional geometric characteristics.

A typical four-leg intersection usually needs a maximum of eight phases. The phases are arranged in two rings and two barriers. The phases within the same ring

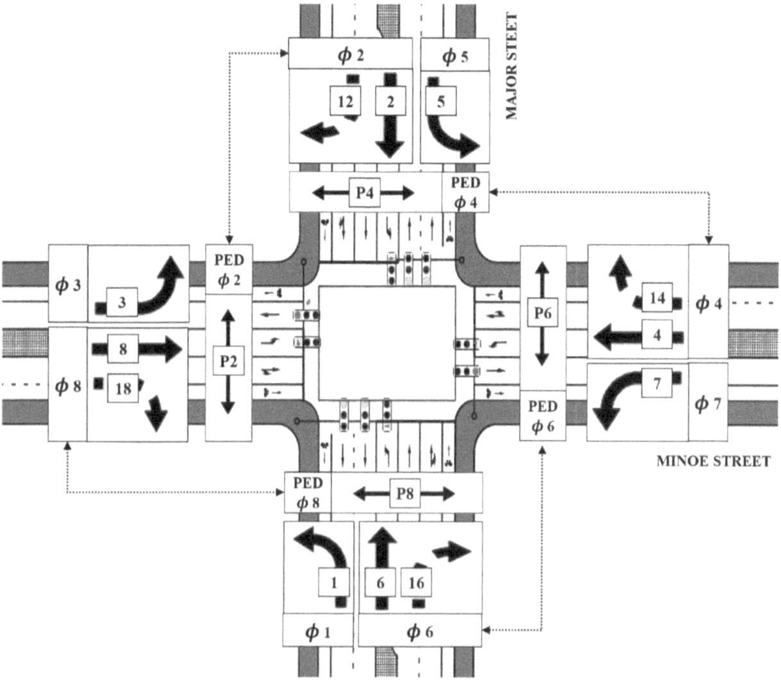

Figure 3.5 Typical movement and phase numbering (with protected left-turns).
Note: The pedestrian phase is generally associated with (and will time concurrently with) the adjacent vehicle phase (as illustrated by the dashed lines).
ϕx, phase number and *Px*, pedestrian phase.
Source: Reprinted with permission from NCHRP Report 812 – Signal Timing Manual, 2015, by the National Academy of Sciences, Courtesy of the National Academies Press, Washington, D.C.

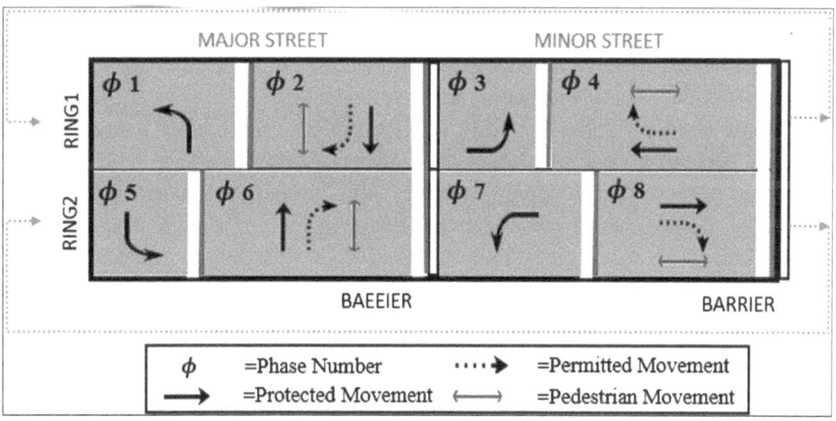

Figure 3.6 NEMA ring-barrier-based phasing diagram.

on either side of the barrier are conflicting phases, while the phases in different rings in the same barrier are compatible phases. The barriers force the phases in the two rings to terminate at the same time, indicating a right-of-way shift between the two crossing streets. Although variations are allowed, a practitioner should attempt to follow these conventions:

- Even phases are typically associated with through movements.
- Odd phases are typically associated with left-turn movements.
- Pedestrian phases are typically set up to run concurrently with the even-numbered vehicular phases. They are generally assigned to the same phase number as the parallel vehicular phases.

To avoid confusion, it is a common practice to maintain a consistent phase numbering scheme within a specific jurisdiction. For example, phase 2 could always be defined as the major street through movement in the northbound (or eastbound) direction or alternatively, as the coordinated phase regardless of direction.

3.4.2 Phase sequence

The example shown in Fig. 3.6 has the left-turn phases leading the through phases. This is the most common phase sequence in North America; however, the sequence of the left-through phase pairs can be freely switched without causing any conflicts. The selection of a phase sequence must consider efficiency, safety, and driver expectations. For isolated, fixed-time control, the phase sequence will only affect operations when left-turn or right-turn lanes and pockets are too short to cause queue spillbacks. For signal coordination, phasing sequence has a profound impact on progression bandwidth.

3.4.3 Treatment of pedestrian and bicycle traffic

A pedestrian phase consists of three intervals: Walk, FDW, and steady Don't Walk. The WALK interval typically begins at the start of the concurrent vehicular green interval and is used so that pedestrians can react to the start of the phase and move into the crosswalk. The FDW interval follows the WALK interval and informs pedestrians that the phase is about to end. Typically, the FDW ends at the beginning of the yellow interval of a concurrent vehicle phase. This feature is necessary for countdown pedestrian signals. MUTCD guidance states that the WALK interval should be at least 7 seconds long. In areas with high pedestrian traffic (such as school zones, CBDs, and sports and event venues), longer WALK intervals should be considered.

Calculation of the FDW interval is based on the following equation, assuming a standard walking speed of 3.5 ft/s:

$$\mathrm{FDW} = \frac{\mathrm{crosswalk}}{3.5 - (R + Y)}$$

where Y = length of the yellow interval (s); R = length of the red interval (s); crosswalk = distance of the crosswalk (ft).

When the pedestrian push button is not at the start of the crosswalk, the MUTCD requires that pedestrians walking at a lower speed (3 ft/s) can get to the opposite curb during WALK and pedestrian clearance (i.e., FDW + Y + R). Thus the WALK interval must satisfy the following:

$$\text{WALK} \geq D/3 - (Y + R) - \text{FDW}$$

where D = distance between the push button and the opposite curb (ft).

Bicycle movements are often treated similarly to pedestrians, as the use of bicycle phases is not common in North America. At intersections with significant bicycle traffic and stop-bar detectors, the minimum green time (combined with the yellow and red clearance intervals) should accommodate a bicycle crossing the intersection. A practitioner does not generally need to accommodate a bicycle solely within the minimum green time. Instead, bicycles should be accommodated within a minimum phase length that includes a minimum green time, yellow change interval, and red clearance interval.

3.5 Determination of signal timing parameters

As isolated fixed-time control only represents a very small portion of all of the signals in the United States and Canada, a formal analytical procedure for the timing development has not been well established. In most cases, the signal timing calculation is mainly affected by the requirements of pedestrian timings, thus vehicular traffic demand may not play a major role when developing signal timings for isolated fixed-time signals. Establishing clear objectives and outcomes is always important during the timing development process. For example, if the objective is to minimize overall intersection vehicle delay, more green time should be allocated to major vehicular traffic movements. On the other hand, if the objective is to balance the level of service for all of the vehicular movements, green times should be allocated in such a way to achieve similar degrees of saturation on all movements. In any case, the timing plans should also accommodate the need for pedestrian movements. Wherever possible, oversaturation should be avoided for individual movements.

Besides the well-known Webster's method for signal timing calculation, the critical movement analysis (CMA) method is commonly adopted by practitioners. CMA is a simplified technique that allows a practitioner to quickly calculate the signal timing parameters. In many cases (such as signals at new streets), CMA is the appropriate level of analysis given the lack of accuracy in future volume forecasts. It can also provide a first-cut, reasonable check of software results. Given the fact that the Webster's method has been well documented elsewhere, this chapter only describes the CMA method process.

3.5.1 Required input parameters

The following input parameters are necessary for designing a signal timing plan using the CMA method:

Saturation flow rate

Saturation flow rate is one of the key parameters for calculating intersection capacity. While many factors affect the saturation flow rate, the CMA method only requires an estimate on its value on a per-lane basis. Additionally, each movement/phase is assumed to have the same saturation flow rate.

Basic timing parameters

Basic timing parameters refer to yellow and red clearance, lost time, and phase and green minimums. These parameters are determined based on intersection geometry, speed, and driver expectations, which are generally independent of traffic volumes.

Pedestrian times

Pedestrian times refer to the phase times needed to accommodate pedestrian crossings at an intersection. The concurrent vehicle phase that handles the pedestrian crosswalk must be at least the sum of WALK, FDW, yellow, and red clearance intervals.

3.5.2 Critical movement analysis method

The CMA method involves the following steps:

Step 1: Gather intersection geometry and traffic volumes

Information regarding intersection lane configuration, lane width, and turning traffic movement volumes must be available. An estimate on the saturation flow rate, averaged for all movements, must also be available.

Step 2: Determine critical phase/movement pairs and critical volume

Critical phases/movements refer to those that demand the most time for service. For a standard eight-phase signal, each side of the barrier (or each street) can only have two critical phases, and they must be in the same ring. The sum of those critical phase volumes (per lane) is the *critical volume* of an intersection.

Step 3: Estimate cycle length

The cycle length should satisfy both the needs of pedestrian crossings and traffic demand levels. The typical cycle length range is between 60 and 140 seconds. A cycle length longer than 180 seconds is rare. In general, the resulting volume-to-capacity (v/c) ratio should be less than 1.0. Table 3.2 illustrates the estimated intersection capacity based on various cycle lengths.

Step 4: Determine phase splits

The total effective green time, calculated by subtracting the total lost time from the cycle length, is first allocated among the critical phases based on their volume ratios. Then, the phase splits can be determined based on the effective green times. The same principle is applied to the noncritical phases; however, each side of the barrier must satisfy the condition that the total critical phase splits equal the total noncritical phase splits. Using phases 1, 2, 5, and 6 as examples, phase 5 + phase 6 = phase 1 + phase 2.

Table 3.2 Estimated capacity and *v/c* ratio

Cycle (s)	Capacity (pcu/h/L)	Volume at 0.80 *v/c* ratio
60	1320	1056
70	1389	1111
80	1440	1152
90	1480	1184
100	1512	1210
110	1538	1231
120	1560	1248

Note: Saturation flow rate is assumed to be 1800 pcu/h/L; total lost time for an eight-phase signal is 16 s (lost time per phase is 4 s); Capacity = (cycle − total lost time)/cycle × 1800.

Step 5: *Timing adjustments*

The CMA method yields a timing plan that provides a balanced service to all critical phases/movements, that is, similar to the Webster's method with equal *v/c* ratios. In practice, the timing is often adjusted to meet certain expected outcomes. For example, when *v/c* ratios are low, more green times could be allocated to movements that have higher volumes. Such an adjustment generally results in lower overall intersection delays.

3.6 Example

The same intersection used in other chapters (shown in Fig. 3.7) is used to illustrate the signal timing development process using the CMA method mentioned above. In addition to the information provided in the figure, the following information and assumptions were used:

- Saturation flow rate is assumed to be 1800 pcu/h/L for all movements (note: this value is less precise than what is shown in the example intersection).
- Minimum green times are 5 and 10 seconds for left-turn and through phases, respectively.
- Yellow (3.5 seconds) and red clearance (1.5 seconds) intervals have been obtained elsewhere following appropriate guidelines.
- Pedestrian WALK (7 seconds) and FDW (variable based on crosswalk width) intervals have been obtained elsewhere following appropriate guidelines: $FDW_{N-S} = 21$ seconds; $FDW_{E-W} = 18$ seconds.
- Peak hour factor (PHF) = 1.0.
- Protected left-turn control applies to all four left-turn movements.
- Lost time per phase is 4.0 seconds.

Step 1: *Gather intersection geometry and traffic volumes*

Information is already shown.

Step 2: *Determine critical phase/movement pairs and critical volume*

The intersection needs a total of eight phases as defined in Table 3.3: 1—NBL, 2—SBT, 3—EBL, 4—WBT, 5—SBL, 6—NBT, 7—WBL, 8—EBT. Right-turn movements do not need additional phases.

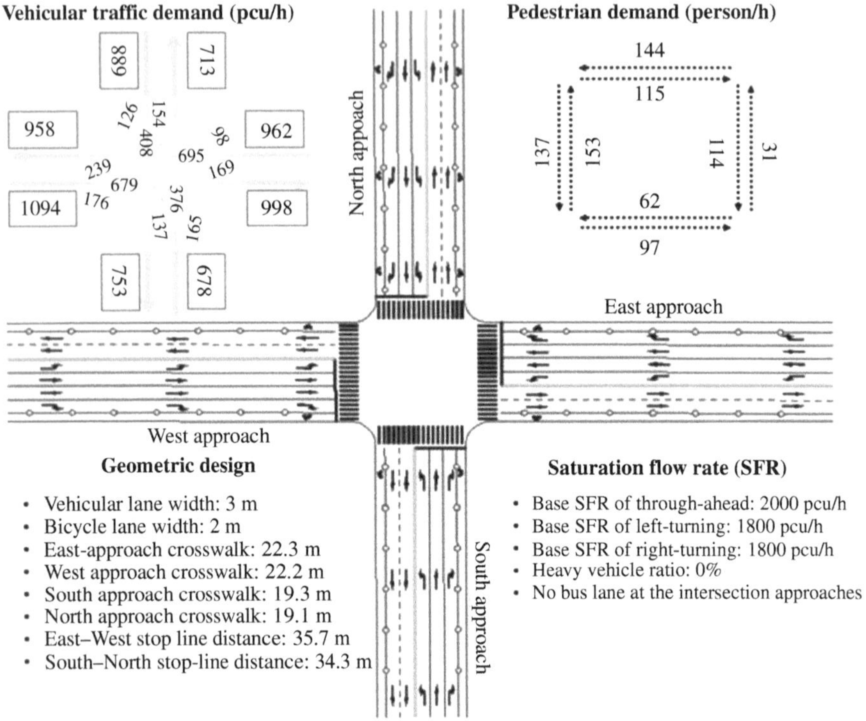

Figure 3.7 Sample intersection.

Table 3.3 Timing parameter calculation using the critical movement analysis method

Direction/ parameters	Phase	Movement	Volume (pcu/h/L)	Volume (%)	Effective green[a] (s)	Phase split[b] (s)
North–South	1*	NBL	137	9	9.6	13.6
	2*	SBT	408	28	28.7	32.7
	5	SBL	154		11.1	15.1
	6	NBT	376		27.2	31.2
East–West	3*	EBL	239	16	16.8	20.8
	4*	WBT	695	47	48.9	52.9
	7	WBL	169		13.1	17.1
	8	EBT	679		52.6	56.6

Note: * denote critical phases.
[a]Effective green is not the actual green interval.
[b]Phase split is the sum of green interval, yellow interval, and red clearance interval.

All volumes need to be converted into volume per lane. In the north–south direction, the total volume for phases 1 and 2 is 137 + 408 = 545; the total volume for phases 5 and 6 is 154 + 376 = 530; thus phases 1 and 2 are critical.

In the east–west direction, the total volume for phases 3 and 4 is $239 + 695 = 934$; the total volume for phases 7 and 8 is $169 + 679 = 848$; thus phases 3 and 4 are critical.

The intersection critical volume is the sum of the critical phase pairs, which is $545 + 934 = 1479$ pcu/h/L.

Step 3: *Estimate cycle length*

Cycle length needs to satisfy both pedestrian crossing times and vehicle needs. For pedestrian timing consideration, the minimum cycle length is calculated to be 83 seconds:

N–S: 5 (Green) + 3.5 (Yellow) + 1.5 (Red) + 7 (WALK) + 21 (FDW) + 3.5 (Yellow) + 1.5 (Red) = 43 seconds

E–W: 5 (Green) + 3.5 (Yellow) + 1.5 (Red) + 7 (WALK) + 18 (FDW) + 3.5 (Yellow) + 1.5 (Red) = 40 seconds

Total = 43 + 40 = 83 seconds.

For consideration of vehicular volumes, the intersection must have sufficient capacity to serve the critical volume. From Table 3.2, it is determined that a 120-second cycle length is an appropriate solution for an expected v/c ratio of 0.95.

Step 4: *Determine phase splits*

The volume ratios for the critical movements/phases are as following:

Phase 1 = 137/1479 = 9%; phase 2 = 408/1479 = 28%

Phase 3 = 239/1479 = 16%; phase 4 = 695/1479 = 47%

The effective green times for the critical movements/phases are calculated as follows (total effective green time is 120 − 16 = 104 seconds):

Phase 1 = 9% × 104 = 9.6 seconds; phase 2 = 28% × 104 = 28.7 seconds

Phase 3 = 16% × 104 = 16.8 seconds; phase 4 = 47% × 104 = 48.9 seconds

For the remaining noncritical phases, the same principle is applied. It once again uses phases 1, 2, 5, and 6 as an example: phase 5 + phase 6 = phase 1 + phase 2. The volume ratio for phase 5 = 154/(154 + 376) = 29%; thus the effective green time for phase 5 = 29% × (9.6 + 28.7) = 11.1 seconds; the effective green time for phase 6 = (9.6 + 28.7) − 11.1 = 27.2 seconds.

Step 5: *Timing adjustments*

The timing plan presented above satisfies all phase minimums. It provides a balanced service to all the critical phases/movements with an equal v/c ratio of 0.95 for all the critical phases. No further adjustment is necessary before potential field implementation. Table 3.3 shows the results.

Bibliography

Federal Highway Administration, 2012. Manual on Uniform Traffic Control Devices for Streets and Highways. US Department of Transportation.

Hatcher, G., Burnier, C., Greer, E., Hardesty, D., Hicks, D., Jacobi, A., et al., 2014. Intelligent transportation systems benefits, costs, and lessons learned: 2014 update (FHWA-JPO-14-159). <https://www.itsknowledgeresources.its.dot.gov/its/bcllupdate/pdf/BCLL_2014_Combined_JPO-FINAL.pdf>.

Herrick, G.C., 1999. The NTCIP Guide: National Transportation Communications for ITS Protocol (version 2) (No. NTCIP 9001v02.05).

Henry, R.D., 2005. Signal Timing on a Shoestring. Federal Highway Administration, No. FHWA-HOP-07-006.

McGee, H., Moriarty, K., Eccles, K., Liu, M., Gates, T., Retting, R., 2012. *Guidelines for Timing Yellow and All-Red Intervals at Signalized Intersections* (No. 731). Transportation Research Board, Washington, DC.

Minnesota Department of Transportation, 2013. MnDOT traffic signal timing and coordination manual. Retrieved from <http://www.dot.state.mn.us/trafficeng/publ/signaloperations/2013_Signal_Opt_and_Timing_Manual.pdf>.

NTOC, 2012. 2012 National traffic signal report card: technical report. Retrieved from <http://library.ite.org/pub/e265477a-2354-d714-5147-870dfac0e294>.

Ozmen, O., Tian, Z., 2009. Multi-criteria decision based left-turn signal control guideline. Transp. Res. Rec. 2128, 96–104.

Stevanovic, A., 2010. *Adaptive Traffic Control Systems: Domestic and Foreign State of Practice*. Transportation Research Board, Washington, DC.

Transportation Association of Canada (TAC), 2014. Manual of Uniform Traffic Control Devices for Canada, *fifth ed.* National Committee on Uniform Traffic Control.

Urbanik, T., Tanaka, A., Lozner, B., Lindstrom, E., Lee, K., Quayle, S., et al., 2015. Signal Timing Manual, second ed. Transportation Research Board.

United States Access Board, 2003. Traffic signal controllers and APS technologies. Retrieved from <https://www.access-board.gov/guidelines-and-standards/streets-sidewalks/public-rights-of-way/guidance-and-research/interfacing-aps-with-traffic-signal-control-equipment/traffic-signal-controllers-and-aps-technologies>.

Germany and Austria

4

Axel Wolfermann[1], Bernhard Friedrich[2] and Martin Fellendorf[3]
[1]Department of Civil Engineering, Hochschule Darmstadt University of Applied Sciences, Darmstadt, Germany, [2]Institute of Transportation and Urban Engineering, Technische Universität Braunschweig, Braunschweig, Germany, [3]Institute of Transport Planning and Traffic Engineering, Graz University of Technology, Graz, Austria

4.1 Background and fundamentals

4.1.1 History of traffic signal control in Germany

The history of traffic signal control in Germany and Austria dates back to the 1920s. The first signals were installed in Hamburg, Berlin, and Vienna. As traffic volumes grew in cities, more policemen were required to control traffic flow at large intersections, and methods to replace policemen were needed. Soon, traffic lights were connected to central controllers via cable. After the Second World War the first traffic control centers were installed.

For several decades two different types of traffic signals were used: traffic signals with colored lights and a special kind of rotary signal ("Heuer-Ampeln"), which proved to be sufficient for small standard intersections and remained in use until 1972 (Stephan, 2009).

The first "green wave" (corridor coordination) was implemented in Hamburg in the 1950s. The cycle time of coordinated arterials was usually adapted by time-of-day, but was adjusted according to traffic demand in some cities. In the 1980s the limits of car-focused mobility and transport planning became obvious and public transport priority at intersections was introduced. Not much later, research on model-based (adaptive) network control started in Germany (Keller et al., 2002). The application remained experimental for many years and is still the exception in Germany. Network optimization similar to SCOOT or SCATS is not widely used, as strategically defined coordination is commonly preferred to network optimization.

4.1.2 Signals and traffic rules

In Germany and Austria, traffic operates on the right-hand side at a standard speed limit of 50 km/h within built-up areas. Some signalized arterials are operated at 60

Global Practices on Road Traffic Signal Control. DOI: https://doi.org/10.1016/B978-0-12-815302-4.00005-4

or 70 km/h, which is the maximum legal speed at signalized intersections. If a general speed limit of 30 km/h is assigned to part of the street network ("Tempo 30 Zone," typically in residential zones), traffic signals are prohibited and intersections are regulated by the yield-to-right rule. This regulation states that the vehicle coming from the right side has priority.

Vehicular traffic control signals are located at the near side of the intersection and consist of three vertically aligned discs of 200 mm diameter (300 mm at 70 km/h approach speed or in the case of particular safety needs) with the red light on top, yellow (amber) in the middle and green on the bottom (Fig. 4.1). Usually signal heads are next to or above the relevant lane and repeated above (or on the left side in the case of left-turning traffic). Each vehicular signal head should be repeated by an identical signal head in case of lamp failure and to improve visibility. The standard vehicular signal sequence is red, red and yellow, green, yellow, red (Fig. 4.2A). In Austria, a flashing green of 4 seconds follows

Figure 4.1 An intersection in Germany.

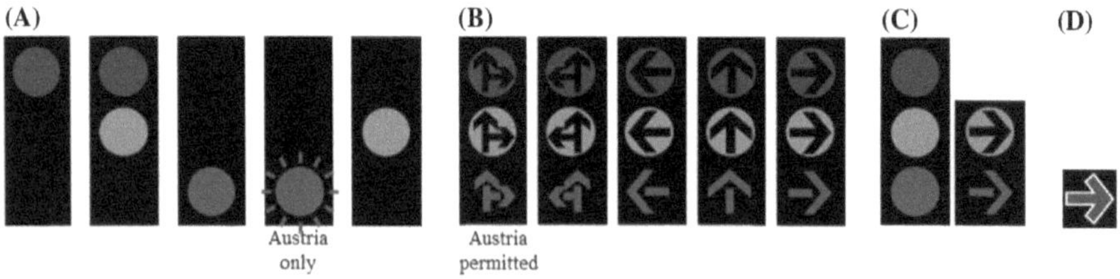

Figure 4.2 Signal heads. (A) Vehicular signal sequence (in Austria 4 s flashing *green* before *yellow*). (B) Signal heads with arrow symbols (protected turning; in Austria through/right signal, right movement is permitted). (C) Permitted-protected right turn signals. (D) Static *green* arrow for right-turn on red.

each green signal. In particular cases (namely semiprotected turns) only partial signal sequences are required.

Standard signals are indicated by a full circle. Arrow symbols (left, right, straight, or a combination of two of them, Fig. 4.2B) indicate a protected stage in Germany (also pedestrian movements have to be protected). Arrow symbols require separate lanes for the indicated streams. While in Austria a single arrow also indicates protected movements, a combination of a right and straight arrow may indicate a permitted movement with parallel pedestrians. This indication is only justified by historical reasons and difficult to understand, particularly for foreign road users (Hebenstreit et al., 2013).

If right-turning vehicles which are signalized together with the through movement obtain an additional green within the cycle, a yellow and green arrow is used in addition to the normal signal. These arrows remain dark during the remaining part of the cycle (Fig. 4.2C). If right turns are to be stopped during part of the cycle (e.g., due to through trams passing on the right), a red and a yellow arrow are used, which remain dark during the remaining cycle.

Lag times for left-turning vehicles are indicated by a green arrow which is located behind the conflict area on the opposite left-hand side of the intersection (diagonally opposite, commonly installed at the signal pole of the opposing signal). Protected-permitted lead times require in addition a yellow flashing arrow above the green arrow to warn drivers of oncoming traffic and pedestrians during the permitted part of the cycle.

Right turn on red is prohibited. It can be permitted by a static green arrow next to the red signal on the right-hand side (Fig. 4.2D). This sign was widespread in East Germany since 1978 and introduced in 1994 in the former Western states after the Reunification. Due to low compliance of drivers (i.e., not stopping at the stop line and yielding to conflicting traffic including bicycles and pedestrians), however, safety concerns regarding pedestrians and cyclists remain prevalent (Maier et al., 2015) and restrict the use to carefully selected locations (low traffic volumes, good visibility, etc.). The Austrian ministry is thinking of introducing right-turn-on-red neglecting the negative experiences made in Germany. Whether right-turn-on-red should be permitted to cyclists is a matter of debate at the moment.

Traffic signals for pedestrians are located on the far side of the crossing and consist of a standing red person on top and a walking green person at the bottom (Fig. 4.3A). Due to its great popularity, the former East German pedestrian signal ("Ampelmännchen") remains in use in many cities. Some cities designed other unconventional persons for selected traffic lights, even though they are not contained in guidelines or regulations. Next to equestrian facilities, horse symbols instead of or in addition to pedestrian symbols can be used.

Cyclists generally have to obey the vehicle signals, because they are regarded as vehicles and, thus, have to use the carriageway (if no compulsory cycle lane or cycle track is signposted). If cyclists are required to cross with the pedestrians, the pedestrian symbols are replaced by joint pedestrian/cyclist symbols (in the

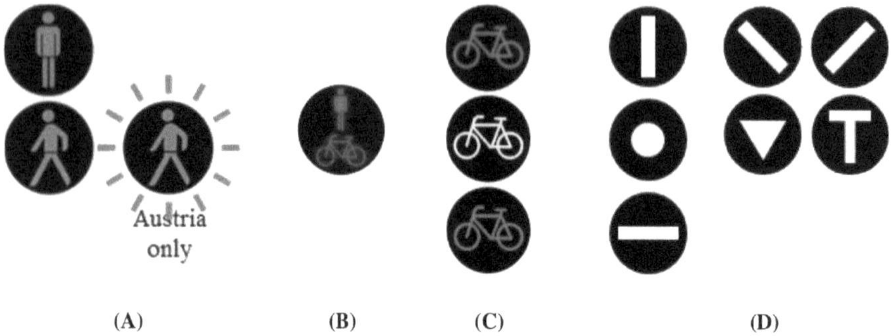

Figure 4.3 Pedestrian, cyclist, and public transport signal heads. (A) Pedestrian signals; in Austria *green* is also followed by 4 s *green* flashing. (B) Combined pedestrian/cyclist signal. (C) Bicycle signals. (D) Signals for public transport on separate lanes.

respective travel direction; Fig. 4.3B). Sometimes it is desired to control cyclists separately from cars and pedestrians (due to separate cycle lanes, otherwise unclear signalization for cyclists, separate bike stages, two-lane turning traffic with parallel cycle lanes, or to reduce intergreen times, cf. Section 4.4.2). In this case cyclists have separate signals on the near side of the intersection with red, yellow, and green signal heads with a bicycle symbol (Fig. 4.3C). Signal heads can be small (i.e., 100 mm diameter) with the bicycle symbol above the signals. Separate signals for cyclists require at least a protective strip or separate bike lanes/tracks.

Trials with countdown signals for pedestrians are operative in some cities. Due to the widespread use of traffic actuation and limited proof for safety improvements, they are not included in the Guidelines for Traffic Signals. Some cities use two red pedestrian heads on top of each other for better perception or a "wait for signal" sign during red time.

Acoustic and tactile signals for pedestrians are introduced at all new or refurbished intersections with the aim of covering all intersections in the future. Acoustic signals are commonly traffic-actuated to reduce noise and due to their impact on intergreen times (cf. Section 4.4.2).

Trams and buses on separate lanes are signaled by white signs (Fig. 4.3D). A vertical or diagonal bar indicates green for through or turning traffic, respectively, a horizontal bar indicates red, a dot indicates yellow. A triangle means "permitted green," that is, yield to conflicting traffic. To enable buses and trams to use the waiting time at stops in front of traffic lights as passenger service time, a "door closing" (T) signal can be used to indicated the impending signal change to green. If public transport is prioritized, feedback symbols are used to signal the driver that the public transport stage will be activated.

The centralization of signal control in traffic control centers emerging since the 1950s resulted in dependence on compatible systems. Cities relied on one supplier for all connected signals to ensure interoperability, which limited flexibility and also competition between suppliers. Cities, in consequence, called for standardization of interfaces between signal controllers and central control centers, which resulted in an industry initiative called "Open Communication Interface for Road Traffic Control Systems" (OCIT). In order to align the interests of cities and push the standardization process into a direction beneficial to the customers of signal controllers (i.e., public administrations), several cities, including big cities like Berlin, Munich, Cologne, and Frankfurt/Main, founded in 1999 the "Open Traffic Systems City Association" (OCA e.V.). As of 2018 it has 38 member cities and federal states in Germany, Austria, and Switzerland. OCIT currently consists of two main components, OCIT-Outstations (the interface between local signal controllers and central control systems) and OCIT-Instations (the interfaces between control computers and external signal control software and management systems) and is operative in many European cities.

As connectivity between vehicles and the infrastructure progresses and signal control needs to be interoperable with further intelligent transport systems (ITS), OCIT will have to be embedded into more encompassing traffic management standards. An important question relates to the role of public administrations in the standardization process of ITS. On the municipal side, the OCA developed a framework for traffic management systems called "Open Traffic Systems" (OTS), into which OCIT is integrated. Research is currently ongoing on how to align this process originating from signal control into standardization processes driven by vehicle manufacturers and related to connected vehicles and ITS and developed in international projects (Kroen, 2013).

4.1.3 Guidelines and legal regulations

4.1.3.1 Germany

Legally, traffic signals are mentioned in §37 of the Road Traffic Regulations (Straßenverkehrs-Ordnung StVO) and by the Administrative Road Traffic Regulations accompanying the StVO (Verwaltungsvorschrift zur StVO). The regulation defines warrants for the installation of traffic signals (cf. Section 2.1) and transition times (i.e., yellow time and red/yellow time). The state-of-practice of signal control is defined in the "Guidelines for Traffic Signals" (RiLSA) published by the German Road and Transport Research Association (Forschungsgesellschaft für Straßen- und Verkehrswesen, FGSV).

FGSV is a German nonprofit organization with a technical/scientific focus. It brings together professionals from academia and the public and private sectors.[1] FGSV publishes recommendations related to all aspects of road traffic. These recommendations are assigned to one of four categories: guidelines (level 1/level 2, R1/R2) and knowledge documents (level 1/level 2, W1/W2). Guidelines are aligned with all relevant boards in FGSV. After agreement by federal committees they are introduced as an obligatory standard for federal roads. State authorities usually

[1] www.fgsv.de/en/research-association.html

adopt these standards for the state roads as well. Knowledge documents are less obligatory and record the state-of-the-art on a specific topic.

RiLSA is a level 1 guideline which documents its importance and high degree of obligation. The most recent release was published in 2015. It is currently translated to English. The "Directives for the Design of Urban Roads" (RASt 2006, R1), and the "Directives for the Design of Rural Roads" (RAL 2012, R1) are two other guidelines which address intersection layout also for signalized intersections. Further details for specific transport modes (also mentioning signal control) are described in "Recommendations for cycle facilities" (ERA 2010, R2), "Recommendations for pedestrian facilities" (EFA 2002, R2), and "Recommendations for public transport facilities" (EAÖ 2013, R2).

A number of knowledge documents related to special signal control topics have been published (e.g., "Advice on signal control centres as part of urban traffic management," 2018, W1, "Advice on public transport prioritisation in urban traffic management," 2018, W1, "Advice on traffic signal control in street networks," H LiS 2014, W1, "Advice on quality management of traffic signals," H QML 2014, W1, "Advice on the signalisation of bicycles," HSRa 2005, W1).

Procedures to determine the traffic flow quality at signalized intersections are described in the German Highway Capacity Manual (HBS 2015, R1). Both for RiLSA and HBS separate publications provide practical examples for their application. Important advice on the determination of road categories in networks is provided by the "Guidelines on Integrated Network Design" (RIN 2008, R1).

The relevant laws and technical standards are illustrated in Fig. 4.4.

Signal control in Germany is coordinated by two authorities: the road traffic authority is responsible for any road regulation, including traffic signals. The planning and operation of signals usually lie in the hands of the road administration, which can be an urban or state administration, depending on the type of road and size of city. Large municipalities control all traffic lights in their jurisdiction, including on state roads. With rare exceptions, the police nowadays is only

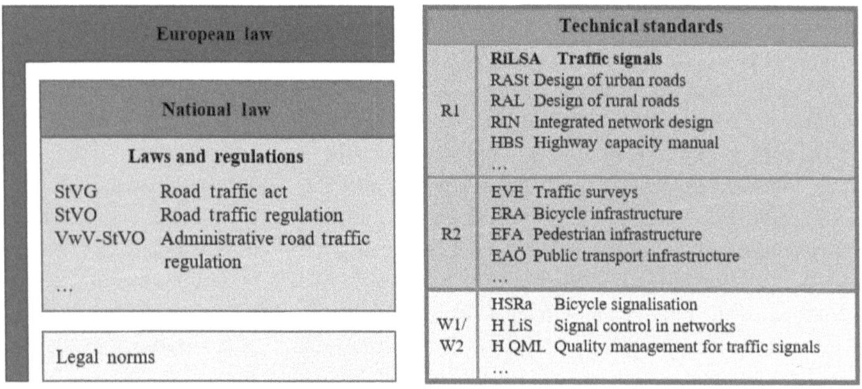

Figure 4.4 Legal framework for traffic signal control in Germany.

involved if traffic signals have to be switched off and traffic has to be manually controlled, or accident hotspots have to be assessed and improved. For the design of intersections and the decision on the type of control of intersections often the planning departments of public administrations are involved.

Signal programs, especially vehicle-actuated control, are commonly developed by specialized consultants based on requirements imposed by politics and the public administration. The control logic is usually transferred to signal manufacturers who implement or compile the software on their hardware. After implementation the signal controllers must be certified by the local administration.

4.1.3.2 Austria

Basically the legal framework for traffic signals and the implementation of signal control is very similar in Germany and Austria. The Austrian Road Regulation Act (StVO Austria) is similar to the German version. Traffic signals are addressed in §38 and §39 of the Austrian StVO with differences on signal sequence as mentioned. The Austrian equivalent to FGSV, the FSV[2] (Austrian Research Association for Roads, Railways and Transport) is organized similarly to FGSV. The publications are called RVS (Planning, Construction and Maintenance of Roads). None of the signal-related guidelines are currently available in English. Since the regulations for traffic signals have not changed recently most of the signal-related guidelines are 20 years old (RVS 05.04.31–34, cf. FSV, 1998a–d). There is one more recent document on the evaluation of traffic signals, which will be addressed in the next section.

4.1.4 Quality management

4.1.4.1 Germany

Safety assessments of accident hotspots, including signalized intersections, are a standard procedure. Furthermore, safety audits are conducted to avoid safety deficiencies of traffic infrastructure proactively. The related recommendations have recently been changed to guidelines (FGSV, 2018). Also, the training of safety auditors is formalized and recommendations for auditing of existing infrastructure, including intersections, are developed.

Quality management intends to constantly improve signalized intersections. RiLSA recommends defining quality by means of performance indicators not only related to safety, but also to traffic flow quality (i.e., delay, number of stops), environmental aspects (e.g., emissions, noise, if not derived from other performance indicators), and controller performance. The targeted quality has to be checked against the realized quality on a regular basis, that is, at least every 1 or 2 years, or when deficiencies become evident for each signalized intersection and also for the network. Performance indicators have to be based on the state-of-the-art as it is

[2] www.fsv.at

laid down in regulations (e.g., for environmental impact of traffic) and guidelines (e.g., the German Highway Capacity Manual, HBS). Thus this quality management combines to a large extent the existing methods and indicators, but streamlines them into a consistent and systematic approach for the continuous improvement of signalized intersections.

RiLSA currently defines only a basic procedure for quality management, because authorities responsible for this quality management considered a comprehensive procedure infeasible with the resources available in the near future. More detailed advice, though, is provided in an additional document (FGSV, 2014) which is based upon research projects (Boltze and Reusswig, 2005).

4.1.4.2 Austria

Based on previous investigations by the OCA in Austria a guideline on quality management of traffic signals was introduced (RVS 05.04.35, FSV, 2013). According to this guideline the road authority is responsible for road maintenance including a regular evaluation of traffic signals. A regular quality assessment is necessary to guarantee effectiveness, functional quality, and keeping the desired control targets. Traffic safety, operational safety, and quality of traffic flow are being evaluated. The quality control process contains (1) analysis of predefined quality performance indicators, (2) definition of quality measures, (3) implementation of measures, and (4) impact analysis. All steps are documented in a database accessible to all involved staff members.

There is a list of quality performance indicators classified into safety and traffic flow. The safety indicators look at accident numbers, severity, accident cost, type, and level of cause by traffic signal control. Visibility of lane markings along with signal heads and dazzling effect by direct sunlight are checked as well. Crossings at yellow (amber) and other conflict situations are accounted. Usually, the safety and design of signal-controlled junctions correlate, which requires a close view on the signal layout map in case of accident costs above average.

The traffic flow is described by performance indicators described in Chapter 6, France, such as average delay per vehicle or person, queue length, number of stops, level of coordination at arterials, dwell times at nearby public transport stops, and prioritization of public transport vehicles. These values are analyzed by vehicle type. Traffic flow-related performance indicators correspond fairly well with environmental values as stops increase air pollution and noise. Other performance indicators are operator-related, such as operational availability (number of interruptions, level of standard operation) and operating conditions (technical state of the controller, poles, signal heads, etc.).

For a quality assessment of a single intersection all performance indicators are collected and analyzed by aggregating them into eight groups. Each group receives a standardized grade between 0 and 3. These eight grades are presented in a radial diagram (Fig. 4.5) and compared between signalized intersections in order to identify a priority list of intersections which require an upgrade.

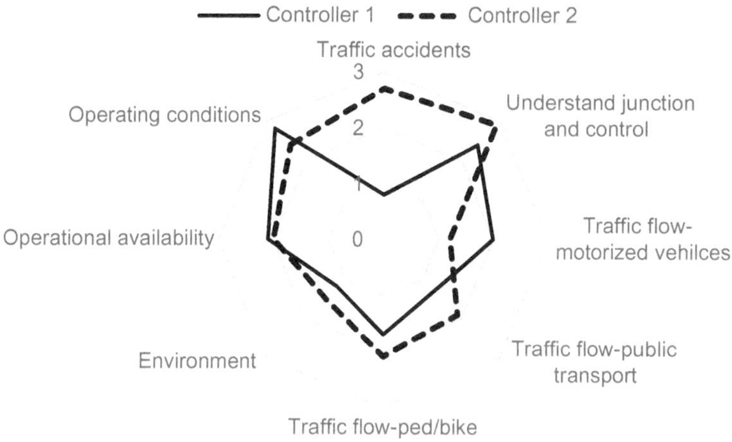

Figure 4.5 Standardized radial diagram to assess the quality of traffic control in Austria (0 bad; 3 excellent).

4.2 Control strategy

4.2.1 Signal control and other intersection types

German intersections, particularly in urban areas, have to serve cars, public transport (mostly buses, trams), cyclists, and pedestrians—the demands of which often conflict with each other. Cities decide on priorities and sometimes also on standard procedures for signal control in their jurisdiction.

Urban arterials in major cities are usually signal-controlled due to the ability to coordinate adjacent intersections, assign priorities to particular road users (often public transport vehicles, but also pedestrians at dedicated pedestrian crossings, and recently sometimes cyclists along cycle routes), and control traffic flow.

Since the self-responsibility of every traffic participant is a primary principle in the Austrian as well as the German Road Traffic Act, both Acts are especially concerned with dangerous situations. As a governing principle traffic signals should be installed only if other control measures are not suitable. Firstly, traffic signals have to improve traffic safety; secondly, traffic signals should make traffic flow more efficient. Not every traffic signal will improve traffic safety. Therefore, intersection design, lane marking, and signal control must match. Safety research in both countries highlights several requirements for traffic signals to be safe:

- Left-turning traffic should be protected wherever possible. For large intersections and high travel demand this is even the best solution in terms of capacity. Only at low speed or low volumes, left-turning traffic can be operated together as semicompatible movement with oncoming traffic and parallel pedestrians (Scholz and Ortlepp, 2010).
- Traffic signals should be operated 24/7. Research in large cities has proved that accident rates increase significantly as traffic signals are switched off during night time (Ortlepp, 2010).

- As with all traffic facilities, signalized intersections should be designed in conformance with the road traffic design standards. Special attention has to be given to vulnerable road users (i.e., pedestrians and cyclists) and their visibility, as many accidents with fatalities involve poor visibility. In urban situations, low turning speeds should be encouraged by small radii and appropriate lane widths.

Roundabouts are installed widely in Germany and Austria, especially in rural areas, villages, and small cities. As a standard assumption, single-lane roundabouts (four legs, 25–40 m diameter) accommodate an average annual average daily traffic (AADT) of up to 25,000 vehicles/day. At higher demand turbo roundabouts or bypasses for right-turning movements are usually considered. Although roundabouts are more economical in terms of operation and land usage (Wirth et al., 2009), signal control is still considered to be the most favorable form of control in German and Austrian cities due to the ability to control movements (coordination, metering) according to general planning objectives.

Traffic signals are frequently used along school routes with traffic actuation for pedestrians and fast response times. Pedestrian signals are also recommended for roads with high vehicle and pedestrian volumes. These signals follow the same principles as signals at intersections regarding signal sequence, intergreen times, and transition times.

The contribution of traffic signals to emission reduction has been discussed over the past two decades. The first step was the installation of corridor coordination, in order to improve traffic flow and minimize the number of stops. Later, network level adaptive control was tested (cf. Section 4.1.1). If traffic adaption is based on real-time models, the control is called "model-based" (as opposed to rule-based). Research on the merits and challenges of model-based signal control highlights benefits for traffic flow improvements and emission reduction as compared to rule-based or fixed-time control, but also underlines the high demands on installation and operation (Boltze et al., 2011). Particularly, the rebound effects of traffic flow improvements for motorized vehicles on traffic generation and modal split have not been researched in Germany yet.

4.2.2 Fixed-time control, actuation and coordination

The most common traffic controls in Germany and Austria are:

- Isolated fixed-time signal control (pretimed) with different signal timings by time-of-day. Usually about 5–10 signal settings are computed according to traffic demand per weekday, weekend, and holidays.
- Fixed-time signal control with coordination, an identical cycle time per intersection and fixed offsets. Coordinated signals are predominately used on arterials with low demand of public transport vehicles.
- Rule-based actuated signal control with differences in variable or fixed cycle length and type of actuation, such as early truncation or late extension only, stage skipping, or additional stages to control public transport and pedestrians only on request are predominately applied in Germany and Austria.

The road traffic regulation encourages the use of corridor coordination ("green wave"), which is common in Germany. However, due to irregular network layouts (regular grids are rare) the application is often limited to parts of the arterial network or one direction only. So far, coordination is predominantly used for motorized vehicles (or pedestrian crossings) and sometimes aligned with public transport (e.g., trams on separate lanes). As part of the effort to foster cycling, trials for the coordination of cyclists are realized, including information systems that are intended to reduce platoon diffusion (Mensik and Beyer, 2013).

The control logic of rule-based actuated signals is mostly based on optimized fixed-time signal settings. Usually stages and sometimes individual signal groups (American phase) are activated, extended, or truncated on request. Although signal manufacturers and consultancies offer code libraries for various real-time controllers, implementing rule-based actuated control requires still individual engineering work. Only in rural areas standard intersections are quite common with four legs, vehicular movements, and parallel pedestrians. In Central European urban areas with road layouts dating back to medieval times standard layouts are rare. The combination of different local layouts, the need sometimes to prioritize conflicting public transport movements and additional signals for bicyclists represent a wide variety of options which cannot be standardized easily.

In German and Austrian cities about one signalized intersection is installed per 1000 inhabitants. In large cities, which are predominately densely populated, and rural areas the number of traffic controllers is less (Berlin 3.4 million residents, ~2100 signalized intersections; Vienna 1.8 million residents, ~1300 signalized intersections; Karlsruhe 310,000 inhabitants, ~310 signalized intersections; Graz 290,000 inhabitants, 295 signalized intersections; Darmstadt 160,000 inhabitants, ~160 signalized intersections). There is neither a statistic in Germany nor in Austria which accounts for the number of actuated and fixed-time controllers nationwide. However, individual surveys indicate that medium-sized cities (250,000–500,000 inhabitants) have the highest share of actuated control due to public transport signal priority or pre-emption. There is less public transport in smaller cities which could be prioritized. In very large cities mass transportation is organized predominately either underground or aboveground, which reduces the need for prioritization. Still, the majority of signals are operated in actuated mode.

Traffic actuation for emergency services is usually limited to the vicinity of fire stations or main emergency routes. Traffic actuation is very flexible and specifically dedicated towards traffic composition and layout of each intersection. Some signal programs are fully adaptive, that is, signal groups are individually signaled without clearly separated stages. Therefore, the main design element of signal programs is signal groups. Stages are used for simplification purposes, but are not strictly necessary (cf. Section 4.4 for details).

Pedestrian volumes are rarely sufficiently high to warrant separate pedestrian stages or even scramble crossings. The increased delay limits the advantages also due to reduced acceptance of the red signal by pedestrians. Thus pedestrians receive green in most cases together with parallel through (and commonly also permitted right turning) traffic.

All signals have special transition programs which enable switching them off in a safe manner. These programs usually lead to a state with dark signals on the priority road and yellow flashing in the subordinate approaches. In case a signal switches off completely, the right-of-way is controlled by right-of-way signs which have to be erected at all signalized intersections.

4.2.3 Special constellations

The basics described in the following sections can be applied to traffic signals in all situations, as well as for the signalization of bottlenecks (e.g., at road works) and signalized roundabouts. Intersections can also partly be signal-controlled, for example, to reduce delay for entering vehicles from a minor street with very low demand. Large intersections with a wide median are split into separate, but coordinated, intersections. Exceptional regulations apply to ramp-metering, namely a shorter minimum green time. RiLSA provides for all these special constellation recommendations.

4.3 Intersection layout

4.3.1 General

Proper intersection layout is crucial for safe and efficient (signalized) intersections. It also interacts with the signal program in several ways:

- The number of lanes determines the capacity;
- The assignment of lanes (permitted directions) determines possible signal groups;
- The intersection geometry determines intergreen times and may influence saturation flow rates;
- Bicycle facilities and bike signalization have to be aligned;
- Center islands for pedestrians influence their minimum green times, intergreen times and possible signal groups.

In order to make intersections easy to understand and clearly arranged, standard layouts as described in the design guidelines (i.e., RASt, RAL) are recommended. Urban and nonurban intersections (outside of built-up areas) differ in several points. Urban intersections usually:

- Are designed for speeds up to 50 km/h, thus not requiring consideration for driving dynamics (beyond braking behavior);
- Have small turning radii to foster low speeds;
- Include tram lines and separate bus lanes which are indicated using public transport signal heads (cf. Fig. 4.3D);
- Contain up to 64 signal groups because of the variety of modes and movements with separate lanes.

Nonurban intersections outside of build-up areas usually:

- Are designed for speeds up to 70 km/h and require deceleration lanes and larger radii for turning traffic;

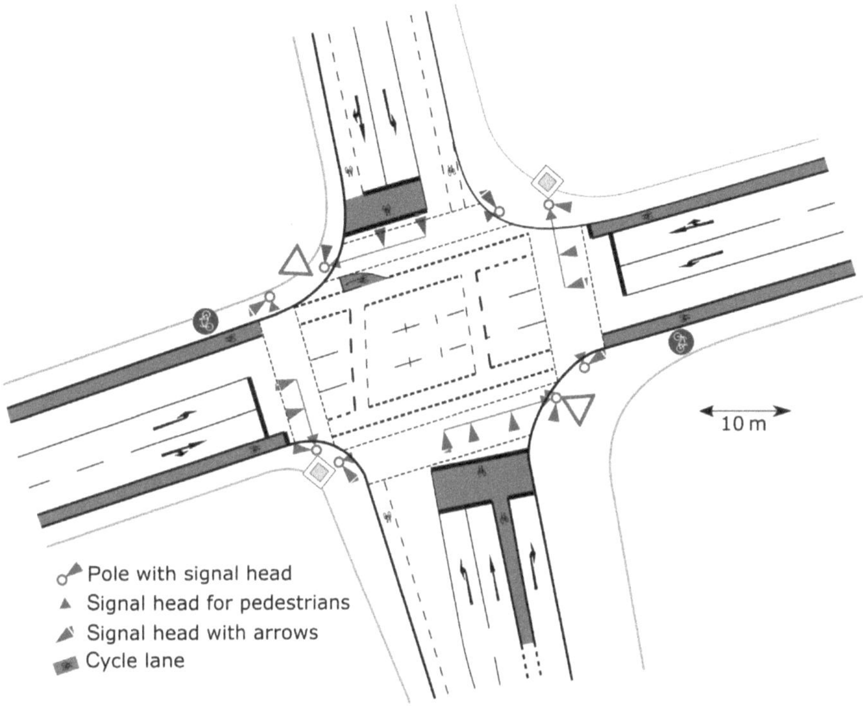

Figure 4.6 Illustration of a four-leg signalized intersection.

- Are less frequented by pedestrians, thus, pedestrian crossings might not be required at all approaches; pedestrian signal groups are commonly actuated by push-buttons;
- Have separate bicycle and pedestrian facilities.

Generally, intersections should be as compact as possible to increase visibility and reduce intergreen times. Stop lines should only be set back if turning curves require it. The number and assignment of lanes is determined by safety considerations (e.g., protected turning traffic, cycle facilities), capacity, and public transport (e.g., separate lanes). The design of a standard medium-sized four-leg urban intersection is shown in Fig. 4.6. Details are described in the following sections.

More complex intersections have to accommodate also public transport. An example is provided in Fig. 4.7.

4.3.2 Vehicle lanes

The standard lane width in Germany and Austria is between 3.0 m and 3.5 m (exceptional up to 3.75 m). Movements receiving green at different times (different signal groups) naturally require separate lanes. Since left-turning traffic should be protected, it should also have a separate lane except geometrically small intersections, approaches with low volumes or intersections with a split cycle.

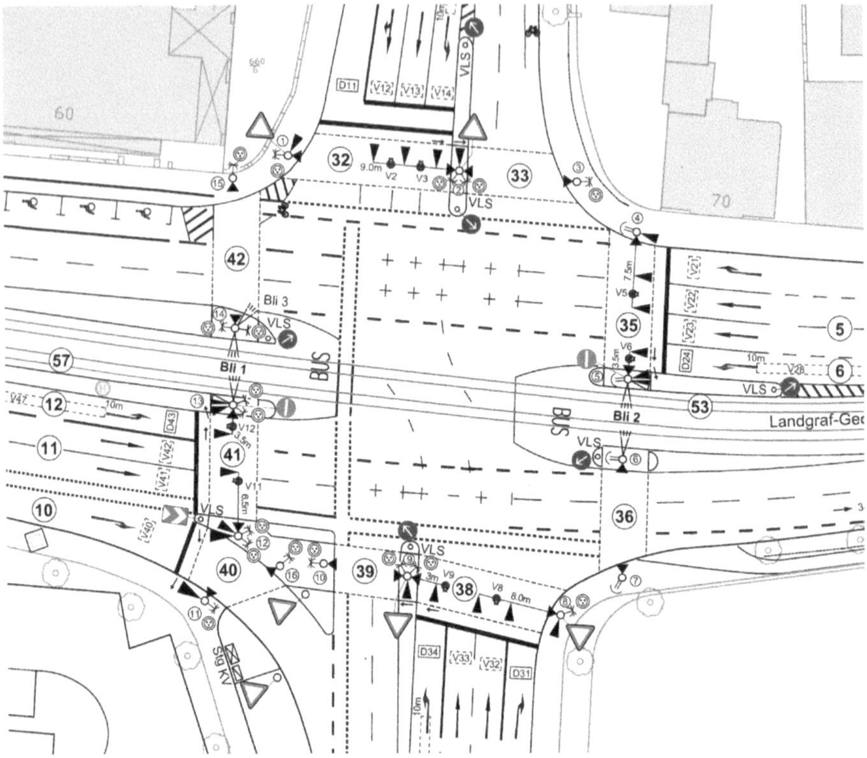

Figure 4.7 Complex urban intersection with detectors and public transport lanes.
Source: Wissenschaftsstadt Darmstadt.

Right-turning traffic will also need a separate lane if parallel pedestrian volumes are high or safety considerations require protected right turns (e.g., parallel tram, high right-turn volumes).

Channelization for right-turning traffic increases turning speeds and leads to an additional conflict area with pedestrians. Thus additional islands tend to decrease safety for pedestrians and should be avoided where pedestrians are commonly present, that is, particularly in urban environments.

4.3.3 Bicycle traffic at urban intersections

Legally, cyclists are considered as vehicles and have to use the road, if no compulsory cycling facilities are provided. Changes of the road traffic regulation since 1997 have shifted the focus increasingly toward rules for better cycling infrastructure, that is, being more liberal concerning the addition of bike lanes and protection strips and more restrictive concerning minimum requirements for cycling facilities (e.g., minimum widths and consideration of pedestrians next to bike paths). While current road and intersection designs often lag behind these changes, here the state-of-the-art is described.

Cyclists generally have to obey vehicle signals. If separate signals are provided for cyclists, separate cycle lanes or paths have to be provided, too. Only if they are clearly separated from the carriageway on separate bike paths or mixed pedestrian/bike paths and the pedestrian signal contains a bike symbol (Fig. 4.3), do cyclists have to obey the pedestrian/bike signal. If cyclists have to obey the normal vehicle signals, they have to be considered in the intergreen time calculation (cf. Section 4.4.2).

Many accidents involving fatalities in urban contexts in Germany and Austria happen between (permitted) right-turning vehicles (often lorries) and through cyclists. Visibility between these two movements is, thus, crucial for safe intersections. Cycle facilities should be close to the road and not obscured by vegetation, road signs or other road furniture. If the intersection layout cannot improve safety satisfactorily, turning movements should be protected by separate signal groups.

Safety for cyclists is also improved by providing sufficient space for waiting cyclists in front of vehicles. Thus cyclists enter the intersection first and are well visible to motor vehicles. This is achieved by advance stop lines for cyclists and extended waiting areas in front of vehicles at least in the minor road (cf. Fig. 4.6). The space should be sufficiently large to accommodate cyclists which are expected to arrive during red in the peak program and also cater for cargo bikes and bikes with trailers, that is, 3–5 m long. In order for cyclists to pass waiting vehicles and reach the advanced stop line/waiting area, a separate cycle lane (or protection strip, if space is limited, cf. dashed lines in Fig. 4.6), should be provided. It should be noted, though, that advance stop lines for bicycles (i.e., set back stop lines for motor vehicles) might increase intergreen times.

Left-turning cyclists generally have two options: direct turning like the vehicles or indirect turning by using parallel bike crossings or the pedestrian crossings (which means they have to dismount). Direct turning requires the passing of (at least) one vehicle lane by cyclists. At high-volume intersections including high approach speeds most cyclists will prefer indirect turning. Because different groups of cyclists prefer different levels of speed and perceive mixed traffic differently, the best quality for cyclists will be achieved by providing both options.

Direct-turning cyclists arriving during red can use an advanced waiting area to stop in front of left-turning vehicles. If bicycle volumes are high, a separate bike lane for turning bikes is appropriate, if not more than two vehicle lanes have to be crossed to reach it. For indirect-turning cyclists, a waiting area at the far side of the intersection next to the bike through lane provides direction and a safe space for those not comfortable with direct left-turning (cf. Fig. 4.6 on the east approach turning south).

For particular high volumes of left-turning cyclists and larger intersections, another option is a "bike lock" which exists also for buses (cf. Fig. 4.8): cars and bikes have staggered stop lines. Cyclists can enter the area between the stop lines and cross from right to left between these two stop lines and align behind the first stop line while cars receive a red signal at the advanced stop line. The signal groups

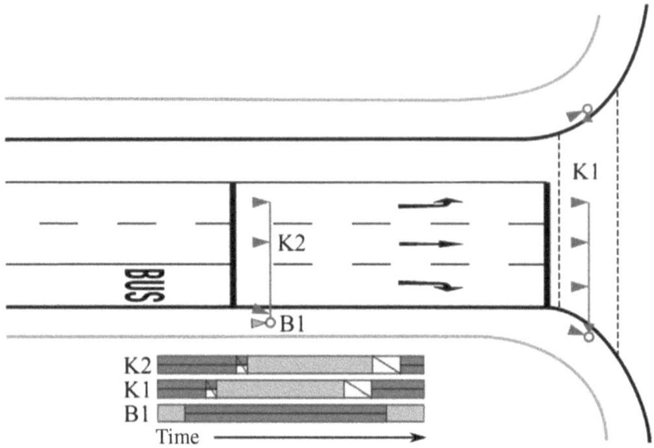

Figure 4.8 Bus lock with fixed timing according to RiLSA (FGSV, 2015).

of the advanced and normal stop line are coordinated in a way that bikes can start first at the normal stop line and cars can pass without stopping again at the normal stop line.

4.3.4 Public transport

Public transport can either use general vehicle signals or have its own signals (cf. Section 1.2), if a separate lane exists. Dynamic public transport stops can be created for trams or buses on a central lane by stopping vehicles during passenger service time in front of the bus stop with two signals (yellow-red-dark). These signals can be coordinated with a subsequent intersection, thus reducing the delay for both public transport and vehicles.

As described for cyclists, "bus locks" can be deployed for buses traveling on the right on a separate lane, but having to turn left at the intersection (Fig. 4.8).

4.3.5 Pedestrians and accessibility

Pedestrian crossings are generally about 4 m wide. They should be as close to the intersection as possible to avoid detours for pedestrians (which might not be accepted) and put them into a position where turning vehicles can clearly see them and observe their right-of-way. In the case of permitted right-turning traffic on mixed lanes with through traffic, a distance of about one vehicle length between the through lane and pedestrian crossing is a good compromise between good visibility (safety) and capacity of through traffic (which might otherwise be blocked by turning vehicles that have to wait for pedestrians).

Median islands for pedestrians should be sufficiently large, if pedestrians might have to stop there. This includes space for strollers and wheelchairs, that is, at least

2 m, preferably 2.5 m length and 4 m width. If cyclists use the pedestrian crossing, a length of 3 m is preferable.

EU regulations and the German Constitution demand that no one is to be discriminated against due to disabilities. The Law for Equal Treatment of Persons with Disabilities clarifies that this includes public infrastructure. Large investments were made to provide barrier-free traffic facilities. More details are provided in the Building Codes of the 16 German states. Thus signalized intersections have to accommodate the needs of hearing, seeing, and mobility-impaired people. Several standards exist which define the following elements: down ramps/low curbs, tactile paving, tactile and acoustic signals, and push buttons with integrated tactile information on intersection layout. So far, many intersections still need refurbishment to comply with accessible design.

4.4 Signal timing procedure

4.4.1 Overview and input data

The signal timing in Germany and Austria is based on signal groups. All signal heads belonging to one signal group show the same symbols at all times. A signal group might give right-of-way to different movements (e.g., right and through movement). Pedestrians have separate signal groups. Bicycles can optionally have separate signal groups. All calculations on cycle length, intergreen time, and green time are based on signal groups. Signal timing is described in the Guidelines for Traffic Signals (RiLSA).

Stages combine compatible or semicompatible movements. Signal groups can receive green during different stages. It should be noted that signal groups belonging to the same stage do not necessarily have to start or end exactly at the same time. From a capacity point of view signal groups should start as early and end as late as intergreen times permit; however, due to safety issues vehicular signal groups may have a late start to allow pedestrians a head-start.

Stages provide better transparency and reduce complexity particularly for traffic actuated programs. They result, on the other hand, in less flexibility and inefficiencies. Fixed-time control, however, is not flexible and, thus, is based on stages ("stage-based control" as opposed to "signal group-based control").

The signal timing procedure for fixed-time signal control in Germany and Austria consists of the following six steps:

1. Collection of traffic demand data for all relevant timing periods (design volumes).
2. Decision on intersection layout (number of lanes and designation, geometry; cf. Section 4.3) and signal groups.
3. Calculation of intergreen times (Section 4.4.2).
4. Decision on stage sequence (Section 4.4.3).
5. Calculation of cycle time and decision on green split (Section 4.4.4).
6. Quality check, including length of turning lanes and permitted movements (Section 4.4.5).

Required documents and data include the following:

- Traffic counts for all movements and separated (at minimum) into bikes, passenger cars, and heavy vehicles for a representative day in at least 15-minutes intervals, which are commonly illustrated by intersection flow diagrams for each period either in passenger car unit (PCU)/h or vehicles/h (details on survey procedures and methods are provided in "Recommendations for traffic surveys," FGSV, 2012);
- A scaled drawing of the intersection layout with available space and relevant infrastructure, street furniture, buildings, entries/exits, lane markings, islands, etc.;
- Information on public transport services at the intersection;
- Information on pedestrian requirements (school routes, approximate volumes).

Saturation flow rates can be approximated by 2000 vehicles/h multiplied by a safety factor (0.8–0.9) or have to be determined according to the procedure described in the German Highway Capacity Manual (HBS, see equations later). A basic headway of 1.8 seconds is assumed (i.e., 2000 vehicles/h) and adjusted by factors considering for vehicle mix (hence, a conversion into passenger car units is not required), turning radius (if radius of lane center line is less than 20 m), lane width (if less than 3 m), and grade (if more than $\pm 2\%$). If more than one factor (lane width, radius, grade) is applicable, only the maximum factor is applied. HBS provides linear diagrams to determine these factors.

$$\begin{aligned} s &= \frac{3600}{h} \\ h &= f_{HV} f_1 f_2 \, 1.8 \text{ s} \\ f_{HV} &= \frac{V_{\text{car}} + 1.75 V_{\text{LorryBus}} + 2.5 V_{\text{TracTrail}}}{V_{\text{car}} + V_{\text{LorryBus}} + V_{\text{TracTrail}}} \\ f_1 &= \max(f_{lw}, f_r, f_g) \\ f_2 &= \min(1, f_g) \end{aligned} \tag{4.1}$$

where s = saturation flow rate (vehicles/h); h = time headway (s/vehicles); f = modification factor (HV: heavy vehicles, lw: lane width, r: turning radius, g: approach grade); V = vehicle volume (vehicle/h) (car: light vehicles, Lorry/Bus: heavy vehicles without tractor-trailer; TracTrail: tractor-trailers and similar).

4.4.2 Intergreen times

Intergreen times are defined as the time between the end of green of one signal group and the beginning of green of a conflicting signal group. The intergreen time depends on the time, the last clearing vehicle needs after the end of green to pass the stop line and to reach the conflict point in the intersection minus the time needed by the first vehicle of the starting signal group to reach the same conflict point. The basic equations for intergreen calculation are provided in Eqs. (4.2)–(4.4), which are contained in RiLSA.

$$t_{ig} = t_{cr} + t_{cl} - t_e \tag{4.2}$$

$$t_{cl} = \frac{l_{veh} + l_{cl}}{v_{cl}} \tag{4.3}$$

$$t_e = \frac{l_e}{v_e} \tag{4.4}$$

where t_{ig} = intergreen time (s); l_{veh} = vehicle length; t_{cr} = crossing time (s); l_{cl}/l_e = clearance/entering distance; t_{cl} = clearance time (s); v_{cl}/v_e = clearance/entering speed; t_e = entering time (s), t_{ig}.

Lengths are determined from the intersection layout as shown in Fig. 4.9. The conflict point is defined by:

- The intersection of center lines of lanes (vehicles) or the point where the lane edges meet, if two movements use the same exit;
- The intersection of lane center line and near edge of the conflict area (pedestrians entering);
- The edge of the conflict area of pedestrian crossings (pedestrians clearing).

Crossing time, speeds, and assumed vehicle length depend on the type of clearing movement. They are defined in RiLSA as six cases provided in Table 4.1. It is assumed that the first entering vehicle reaches the stop line at the onset of green at a speed of 40 km/h (entering speed). Intergreen times are rounded up to the next full second at the end.

Intergreen times have to be computed for all possible conflicts between two signal groups. This can be illustrated by a conflict tree (Fig. 4.10). The base (on the left) is an ending signal group, which assigns green time to one or several lanes (here two lanes), on which one or several movements (here through or right turn) are possible with either cars or bicycles. These movements may have a conflict with one or several of the movements of the beginning signal group determined in the same way (cars are assumed to enter faster than bikes, thus the distinction of vehicle types is usually not necessary). Some of these movement sequences (clearing movement followed by entering movement) either have no

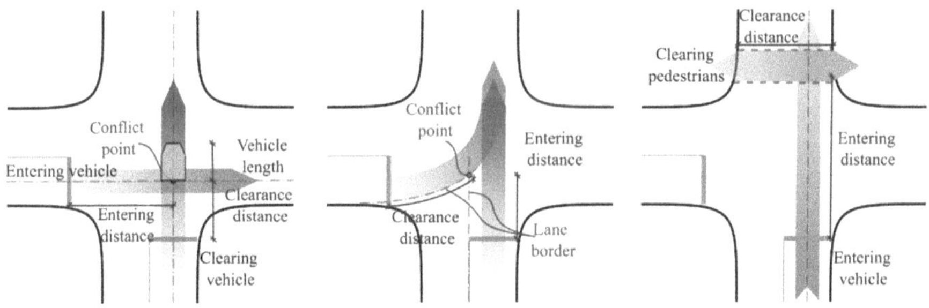

Figure 4.9 Illustration of conflict points and clearance/entering distances.

Table 4.1 Parameters for intergreen calculations according to RiLSA (FGSV, 2015)

Clearing movement	Crossing time	Vehicle length	Clearance speed
Through car	3 s	6 m	10 m/s
Turning car	2 s	6 m	5 or 7 m/s (depending on radius)
Public transport (without stopping)	3 or 5 or 7 s (depending on v_{max})	15 m (tram)	v_{max} (in m/s)
Public transport (with stop in front of intersection)	0 s	15 m (tram)	Accelerated movement
Bike	1 s	0 m	4 m/s
Pedestrian	0 s	0 m	1–1.5 m/s

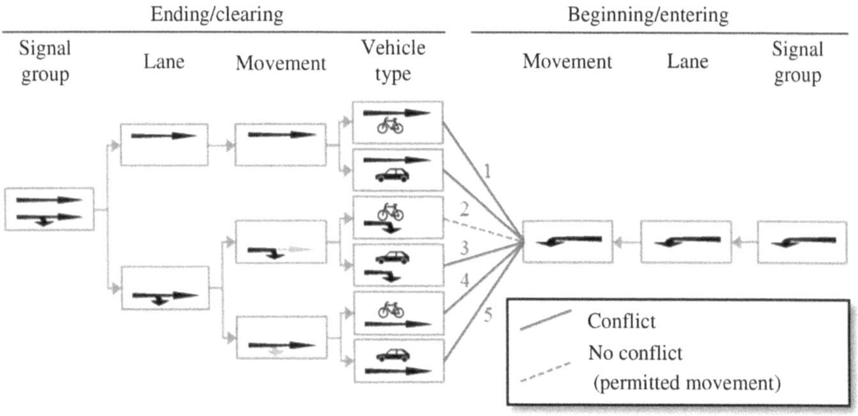

Figure 4.10 Conflict tree to illustrate intergreen calculation (based on Wolfermann, 2009).

conflict or are assumed to be compatible (here right-turning cyclists are deemed compatible with left-turning opposing vehicles). For each conflict (in the figure numbered from 1 to 5) an intergreen time has to be calculated. The maximum intergreen time of all conflicts is determining for the signal group sequence. Bear in mind that bicycles have to be considered in all cases, where they are not legally forced to use separate lanes or tracks. For large intersections (i.e., clearing distances above about 18 m) intergreen times for bikes become longer than those for cars.

The determining intergreen time for each signal group sequence is noted in the intergreen matrix, which is a matrix with all ending signal groups as rows and all beginning signal groups as columns (an example is provided in Section 4.4.5). This intergreen matrix is stored in the controller and valid for all possible signal programs at the intersection.

A simplified intergreen matrix with the times in seconds replaced by marks represents the conflict matrix of all incompatible signal groups which must never receive green at the same time.

4.4.3 Signal phasing

Signal groups have to be defined with consideration of lane designations. Movements belonging to different signal groups must have a separate lane. The combination of signal groups to stages is based on minimum critical traffic demand (sum of flow ratios, as in ring-barrier-based phasing). Stage sequence is influenced by the sum of intergreen times. Stages and stage sequence should in general be designed in a way to reduce the cycle time. Sometimes, however, other requirements can dictate the stage sequence, for example, coordination of adjacent pedestrian signals or coordination with neighboring intersections.

A common signal phasing plan for medium- to large-sized intersections is a four-stage sequence with protected lag–lag left-turn phasing. Permitted-protected left-turn is possible with lagging left-turning being seen as less problematic (from a safety perspective). Leading left-turn before permitted left-turn has to be ended by a yellow left-arrow signal and has to be secured by the intergreen time between left-turn and opposing through traffic. Lagging permitted-protected left-turn requires a green arrow only ("diagonal green" due to the location of the signal head on the far left side of the intersection). Because signal heads are generally mounted on the near side of the intersection, no "yellow trap" exists as left-turn vehicles in the intersection do not see the through signals. Split phasing is possible.

Right-turning vehicles commonly share a lane with through traffic and are permitted with pedestrians and through cyclists. If traffic volumes are high or if many conflicts with parallel pedestrians or cyclists occur and cannot be mitigated by other means, or if trams have separate through tracks on the right, right-turning vehicles should be protected.

Cyclists usually have to use the (motor) vehicle signals. If required for safety or capacity reasons, they can be assigned to a separate signal group to either protect them, give them a head-start, or end their green time early to reduce lost times due to long intergreen times.

If pedestrians receive green in the same stage as permitted turning vehicles, they have to receive a head start in a way that the first pedestrian enters the street before the first entering vehicle arrives at the pedestrian crossing. Permitted movements having the right-of-way must never receive green when a subordinate movement is already green (e.g., through traffic after conflicting permitted left turn). Thus a leading protected-permitted left turn has to be stopped first before through traffic and permitted left-turn receive green.

4.4.4 Cycle length and green times

Cycle time is calculated by motor vehicle demand. It is limited to 30–90 seconds (in exceptions: 120 seconds). This reduces maximum waiting times. Delay for pedestrians and cyclists should not exceed 85 seconds, also due to decreasing acceptance (cf. Section 4.4.5). The cycle time is determined according to Eq. (4.5).

$$C = \frac{\sum_i t_{ig,i} + \sum_j t_{g,\min,j}}{1 - \sum_k \frac{q_k}{s_k}} = \frac{\sum_i t_{ig,i} + \sum_j t_{g,\min,j}}{1 - \sum_k b_k} \tag{4.5}$$

where C = cycle time (s); q_k = relevant vehicle demand (vehicles/h) of stage k; t_{ig} = intergreen time (s); s_k = relevant saturation flow rate (vehicles/h) of stage k; $t_{g,\min}$ = green times independent of cycle time (s); b_k = maximum flow rate of stage k.

For each stage, the maximum flow rate of all signal groups belonging to a stage is computed. $\sum_k q_k/s_k$ depicts the sum of the maximum flow rates of all stages for which the green time is dependent on the cycle time. If signal groups are assigned to more than one stage the critical path is used.

The cycle time can also be computed following Webster's suggestion to minimize delay for motor vehicles, that is, replacing the numerator by $1.5\left(\sum_i t_{ig,i} + \sum_j t_{g,\min,j}\right) + 5$. However, this equation holds only for Poisson-distributed vehicle arrivals and also leads to very long cycle times, thus, disregarding the requirements of pedestrians and cyclists. It can be used for isolated rural intersections with low pedestrian and bike demand.

Each signal group needs at least a green time as calculated according to Eq. (4.6) to satisfy the demand.

$$t_g = bC \tag{4.6}$$

Green times must never be less than 5 seconds. If Eq. (4.6) leads to a green time of less than 5 seconds, the green time becomes 5 seconds and independent from the cycle time, and thus has to be considered in the numerator in Eq. (4.5). Green times, like all other intervals, are commonly rounded to the nearest second. If the cycle time is longer than the sum of green times according to Eq. (4.6) and intergreen times, the surplus time can be assigned to any stage either relative to the respective flow rates b or by some other criteria.

The minimum green time for pedestrians is calculated in a way that pedestrians can cross at least half of the carriageway (in the case of center islands without separate signal groups for the two parts of the pedestrian crossing: the larger part and half of the shorter part). The pedestrian speed can be chosen between 1 m/s and 1.5 m/s, with 1.2 m/s being the standard value. If the green time is also transmitted by acoustic signals for seeing-impaired people, the minimum green time has to enable blind people to pass the complete crossing with the reduced speed of 1.0 m/s. Thus acoustic signals are commonly traffic-actuated with a separate signal group which can account for longer minimum green and intergreen.

4.4.5 Signal timing plan

The signal timing plan is constructed in the following order:

1. Green times of (motor) vehicles with consideration of the respective intergreen times in between;
2. Pedestrian green times are fitted into the program in a way that they are maximized under observation of intergreen times;
3. Green times are adjusted, if necessary
   a. pedestrians in a stage with permitted turning movements need a head start, if necessary by shifting vehicle green
   b. if green times of pedestrians are insufficient (cf. minimum green for pedestrians), green times have to be extended;
4. If additional green times are needed (see last point), cycle time and vehicle green might have to be recalculated;
5. Finally, the transition times for vehicles (red and yellow as transition from red to green, 1 s (max. 2 s); yellow as transition from green to red, 3, 4 or 5 seconds, depending on maximum speed), which are part of the intergreen times, are added at the end and beginning of vehicle red times.

An example signal timing plan is provided in Section 4.4.5.

4.4.6 Quality check

The last part of the timing process consists of a detailed quality check according to the German Highway Capacity Manual (HBS):

- Average delay of motor vehicles;
- Maximum waiting times of pedestrians and cyclists;
- Delay of public transport, if applicable;
- Queue length, if relevant for adjacent intersections or other facilities (e.g., railway crossings);
- Coordination with adjacent intersections, if applicable (performance index).

The average delay calculation of vehicles considers effective green time (assumed to be 1 seconds longer than signaled green time), permitted turning movements that have to wait for gaps in oncoming traffic or in pedestrian platoons. An *average* delay for vehicles of up to 70 seconds is considered to be satisfactory, i.e. level of service (LOS) D on a scale from A to F. For pedestrians and cyclists a *maximum* delay of 70 seconds is considered satisfactory.

4.5 Example

4.5.1 Intersection layout and signal groups

As explained earlier, intersections should be as compact as possible and accommodate all travelers. The example intersection will be designed for an urban situation with significant numbers of pedestrians and cyclists. Vehicle lanes are all set to be

3 m wide. The number of vehicle lanes is reduced to the minimum number which ensures safety and acceptable delay for all modes. In all approaches the traffic volume of left-turning and conflicting through traffic is significant. Hence, left-turning traffic should have a separate signal group and, thus, lane. All approaches and exits are equipped with a 2 m wide bike lane and advance waiting areas (5 m). Additional facilities for left-turning cyclists, as explained in Section 4.3.3, are not shown, but should be considered. Pedestrian crossings are at least 4 m wide. Through and right-turning vehicles are combined, but particular care has to be taken for conflicts between right-turning vehicles and cyclists and pedestrians. As the right-turning volumes are not particularly high and heavy vehicle share is low, a permitted green seems acceptable. Parallel pedestrians have to be on the crossing before permitted right-turning vehicles, which is ensured by starting their green time at least at the same time as vehicle green. One or two right-turning vehicles can wait in the intersection to yield to pedestrians without blocking through traffic, though they might block cyclists. The simplified intersection layout is shown in Fig. 4.11.

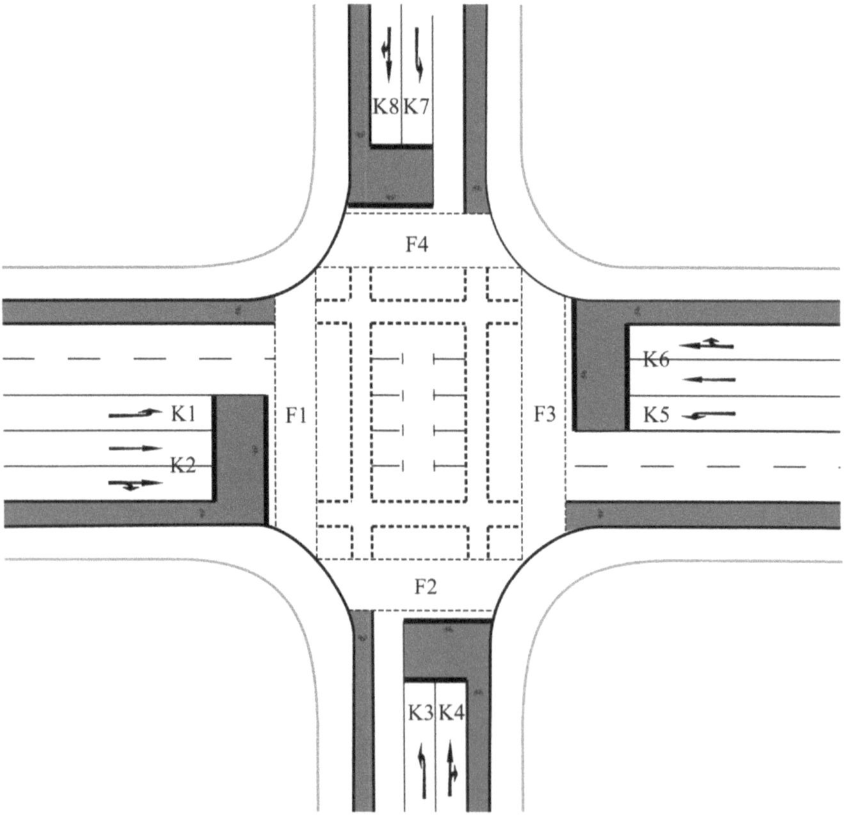

Figure 4.11 Example intersection with street design and signal groups.

Table 4.2 Intergreen matrix for example intersection

Clearing signal groups	Entering signal groups											
	K1	**K2**	**K3**	**K4**	**K5**	**K6**	**K7**	**K8**	**F1**	**F2**	**F3**	**F4**
K1		-	3	4	-	6	5	4	4	-	-	8
K2	-		5	5	5	-	4	3	5	-	7	-
K3	6	4		-	4	6	-	7	8	4	-	-
K4	5	4	-		5	7	4	-	-	5	-	8
K5	-	7	5	4		-	3	4	-	8	4	-
K6	5	-	4	3	-		5	6	7	-	5	-
K7	4	6	-	7	6	4		-	-	-	8	4
K8	5	8	4	-	5	4	-		-	8	-	5
F1	18	18	15	-	-	15	-	-		-	-	-
F2	-	-	13	13	10	-	-	10	-		-	-
F3	-	15	-	-	18	18	15	-	-	-		-
F4	10	-	-	10	-	-	13	13	-	-	-	

For protected left-turning and mixed permitted right-turning/through movements eight signal groups for vehicles (designated "K") and four signal groups for pedestrians (designated "F") are required. They are numbered counterclockwise (NB: cities might have deviating conventions to designate signal groups).

4.5.2 Intergreen times

Intergreen times are calculated for all conflicts as explained in Section 4.4.2. This results in the intergreen matrix given in Table 4.2. Empty cells show nonconflicting signal group sequences. Signal groups for vehicles consider cyclists (with advanced stop lines).

4.5.3 Stage sequence, cycle time, and green split

Protected left-turning traffic for all approaches requires at least a four-stage program. Because opposing through traffic has similar volumes and can be combined, separate left-turning stages are more efficient than a green-split program. Also, the opposing left-turning volumes are similar, making permitted-protected left-turning lead/lag times unnecessary. Stage sequence is usually determined such that the sum of intergreen times becomes minimal. The shortest sum of intergreen times for the four stages results from the stage sequence shown in Fig. 4.12.

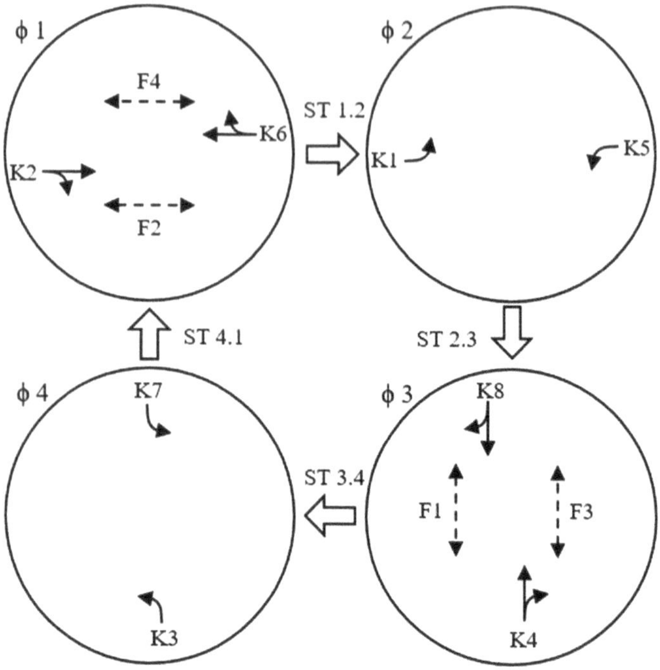

Figure 4.12 Stage sequence diagram for example intersections.

Cycle time and green splits are computed from the decisive (i.e., largest) flow ratios per stage, which is the largest flow ratio of all signal groups of the respective stage. The flow ratio for multilane signal groups (e.g., through and right-turning traffic on the East and West approach, i.e., K2 and K6) is assumed to be equal on both lanes, which requires proportional fitting of lane volumes for the two lanes with differing saturation flows. The saturation flow rate for mixed lanes is calculated as the harmonic mean of through and turning movements.

The cycle time is calculated from the sum of intergreen ($5 + 4 + 4 + 6 = 19$ seconds) and the sum of flow ratios. According to the German Highway Capacity Manual, the saturation flow rate for left-turning vehicles would be usually slightly higher than that for right-turning vehicles due to the smaller radius (cf. Section 4.4.1). Here, a saturation flow rate for turning traffic of 1800 and of 2000 vehicles/h for through traffic is assumed.

The flow ratios and intergreen times for all stages and the cycle time are given in Table 4.3. The cycle time is rounded up to 5 seconds, resulting in 70 seconds. Minimum pedestrian green time is calculated as half the width of the respective crossing divided by 1.2 m/s walking speed (cf. Section 4.4.4). The last but one column contains the critical path (maximum of the sum of green and intergreen times for vehicles and pedestrians, respectively). One second of surplus time remains, which is added to stage 1. The last column shows the final green split.

Table 4.3 Cycle time and minimum green times for example intersection

Stage	Flow ratio	Vehicle intergreen to next stage (s)	Pedestrian intergreen to next stage (s)	Cycle time (s)	Vehicle green time (s)	Minimum pedestrian green time (s)	Max green + intergreen (s)	Used vehicle green time (s)
$\Phi 1$	0.220	5	10		15	6	20	21
$\Phi 2$	0.133	4	4		9		13	11
$\Phi 3$	0.279	4	15		20	9	24	27
$\Phi 4$	0.086	6	4		6		12	7
	$\sum = 0.718$	$\sum = 19$		67			$\sum = 69$	
		Used cycle time:		70		Surplus time:	1	

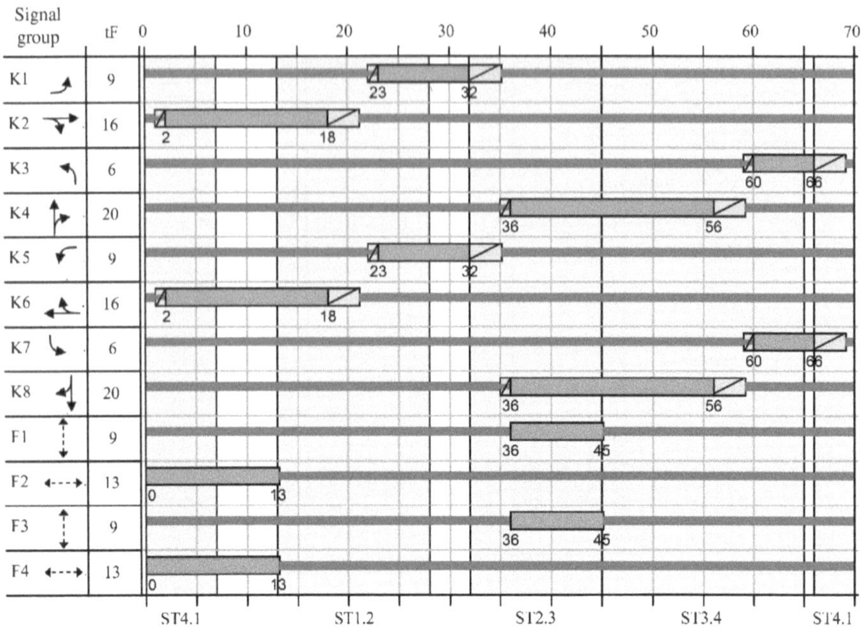

Figure 4.13 Signal timing plan for example intersection.

4.5.4 Signal timing plan

According to the principles described in Section 4.4.5 the signal timing plan is drawn (Fig. 4.13). The stage transitions including the minimum green times (5 seconds for vehicles and 6/9 seconds for pedestrians) are highlighted. It can be seen that stage 3 could not be shortened even if vehicle demand would be lower due to long pedestrian intergreen times (resulting from wide crossings).

4.5.5 Quality

The quality of traffic flow is checked according to the German Highway Capacity Manual with average delay for motor vehicles and maximum delay for cyclists and pedestrians as the indicator (stationarity during peak hour assumed). Table 4.4 contains the relevant data and level of service for all lanes and pedestrian crossings. It also provides maximum queue lengths. All major signal groups (i.e., through and right-turning traffic) achieve LOS D, which is considered acceptable. LOS E means that delay is considered long and a queue at the end of green is possible. However, a longer cycle time which would improve the quality for left-turning traffic would result in LOS E for pedestrians.

Table 4.4 Level of service for the example intersection

Lane	Signal group	Effective green time (s)	Traffic volume (vehicles/h)	Saturation flow rate (vehicles/h)	Flow ratio	Waiting time (s)	Queue length (m)	LOS
WL	K1	10	239	1800	0.132	88	83	E
WM	K2	17	436	2000	0.218	54	107	D
WR	K2	17	419	1916	0.219	55	105	D
SL	K3	7	137	1800	0.076	60	45	D
SR	K4	21	541	1938	0.279	57	133	D
EL	K5	10	169	1800	0.093	42	88	C
EM	K6	17	391	1949	0.201	42	90	C
ER	K6	17	402	2000	0.201	41	45	C
NL	K7	7	154	1800	0.087	45	119	C
NR	K8	21	534	2000	0.267	88	59	E
W	F1					61		D
S	F2					57		D
E	F3					61		D
N	F4					57		D

LOS, level of service.

Bibliography

Boltze, M., Busch, F., Friedrich, B., Friedrich, M., Kohoutek, S., Löhner, H., et al., 2011. AMONES: Anwendung und Analyse modellbasierter Netzsteuerungsverfahren in städtischen Straßennetzen (AMONES: Application and analysis of adaptive traffic signal control in urban road networks), Part 1–3, Straßenverkehrstechnik (55), Issue No. 5&6&7.

Boltze, M., Dinter, M., 1994. The project FRUIT—a goal-oriented approach to traffic management in Frankfurt am Main and the Rhein-Main Region. Traffic Eng. Control 35 (7/8), 437–444.

Boltze, M., Reusswig, A., 2005. Qualitätsmanagement fuer Lichtsignalanlagen. Sicherheitsueberpruefung vorhandener Lichtsignalanlagen und Anpassung der Steuerung an die heutige Verkehrssituation (Quality management for traffic signals), Berichte der Bundesanstalt für Straßenwesen, Unterreihe Verkehrstechnik, No. 128 Wirtschaftsverlag NW.

Boltze, M., Jiang, W., 2017. Berücksichtigung verschiedener Verkehrsteilnehmergruppen und Kriterien bei der Optimierung von Lichtsignalanlagen. Ein Vorschlag zum verkehrspolitischen Rahmen für die Gestaltung der Lichtsignalsteuerung in Städten. (Consideration of Different Road User Groups and Criteria in Traffic Signal Optimisation—A Proposed Policy Framework for the Design of Urban Traffic Signal Control). Straßenverkehrstechnik (61), Bonn (Germany), September 2017, pp. 615–621.

FGSV, 2007. Richtlinien fuer die Anlage von Stadtstrassen (RASt). FGSV Verlag, Edition 2006.

FGSV, 2010. Richtlinien fuer Lichtsignalanlagen: RiLSA – Lichtzeichenanlagen fuer den Strassenverkehr. FGSV-Verlage, Edition 2015.

FGSV, 2012. Empfehlungen für Verkehrserhebungen (EVE). FGSV-Verlag, Edition 2012.

FGSV, 2013. Richtlinien fuer die Anlage von Landstrassen (RAL). FGSV Verlag, Edition 2012.

FGSV, 2014. Hinweise zum Qualitätsmanagement an Lichtsignalanlagen: H QML, (Advice on Quality Management at Traffic Signals). FGSV Verlag.

FGSV, 2015. Richtlinien für Lichtsignalanlagen – RiLSA (German Standard for Traffic Signals). FGSV-Verlag, Cologne (Germany), Edition 2015.

FGSV, 2018. Richtlinien für das Sicherheitsaudit von Straßen RSAS (Guidelines for the Safety Audit of Roads), in final approval procedure.

FSV, 1998a. RVS 05.04.31 Einsatzkriterien (Deployment criteria), Oktober 1998.

FSV, 1998b. RVS 05.04.32 Planen von Verkehrslichtsignalanlagen (Planning of Traffic Signals), Oktober 1998.

FSV, 1998c. RVS 05.04.33 Ausführung, Abnahme, Betrieb, Instandhaltung (Implementation, Approval, Operation, Maintenance), Oktober 1998.

FSV, 1998d. RVS 05.04.34 Abnahme- und Prüfprotokoll (Approval and Inspection Protocol), Oktober 1998.

FSV, 2007. RVS 05.04.36 VLSA Plansymbole (Sympbols for Signalized Intersection Plans), November 2007.

FSV, 2013. RVS 05.04.35 Evaluierung von Verkehrslichtsignalanlagen (Evaluation of Traffic Signals), Februar 2013.

Hebenstreit, C., Fellendorf, M., Blokpoel, 2013. COLOMBO – Cooperative Self-Organizing System for Low Carbon Mobility at Low Penetration Rates, Policy Definition of Signal Control, EU FP7-ICT-2011-8, Deliverable 2.1

Hunter, B., Wolfermann, A., Boltze, M., 2011. Multimodal assessment of signalized intersections considering the number of travellers. In: TRB 90th Annual Meeting Compendium of Papers DVD, Washington, DC, 23–27 January 2011.

Keller, H., Friedrich, B., Wulffius, H., 2002. Verkehrsadaptive Netzsteuerung von Lichtsignalanlagen (Traffic Adaptive Network Control of Traffic Signals), Straßenverkehrstechnik (46), No. 7, pp. 333–338.

Kroen, A., 2013. Standardisierungsprozess für offene Systeme der Straßenverkehrstelematik (Standardization Process for Open ITS), Berichte der Bundesanstalt für Straßenwesen, F91, Bergisch-Gladbach.

Maier, R., Scholz, Th., Enke, M., 2007. Untersuchung zur Wirkung von unterschiedlichen Betriebszeiten der Lichtsignalanlagen in Städten (Schlussbericht) (Impacts of Operation Times of Urban Signalized Intersections, Final Report). Technische Universität Dresden.

Maier, R., Hantschel, S., Ortlepp, J., Butterwegge, P., 2015. Sicherheit von Grünpfeilen (Safety Aspects of Static Green Arrows/Right-Turn on Red at Signalized Intersections), Gesamtverband der Deutschen Versicherungswirtschaft e.V., Unfallforschung der Versicherer, Forschungsbericht Nr. 31.

Mensik, K., Beyer, F., 2013. Koordinierung von Lichtsignalanlagen fuer den Radverkehr (Coordination of traffic lights for bicycle traffic). Straßenverkehrstechnik 57 (10), 621–627.

Ortlepp, J., 2010. Betriebszeiten und Sicherheit von Lichtsignalanlagen (Operating times and safety of traffic lights). Straßenverkehrstechnik 54 (12).

Scholz, Th., Ortlepp, J., 2010. Verkehrstechnische Auswirkungen der Sonderphase für Linksabbieger an Knotenpunkten mit Lichtsignalanlage (Impacts of protected left-turns on traffic flow at signalized intersections). Contributed by Technische Universität Dresden, Fakultät Verkehrswissenschaften, Dresden. Published by Unfallforschung der Versicherer (UDV) im Gesamtverband der Deutschen Versicherungswirtschaft e. V. Berlin. <www.udv.de>.

Stephan, R., 2009. Vor 80 Jahren: Erfindung der Heuer-Ampel (80 years ago: Invention of the Heuer-Ampel), Straßenverkehrstechnik 2009, Heft 04, S. 241–245.

Wirth, W., Mechenbier, Ch., Pötzsch, J., 2009. Flächenbedarf und Befahrbarkeit kleiner Kreisverkehre (Space requirements and trafficability of roundabouts), Straßenverkehrstechnik 2009, Heft 04, S. 216–219.

Wolfermann, A., 2009. Influence of intergreen times on the capacity of signalised intersections, Schriftenreihe des Instituts für Verkehr, Fachgebiet Verkehrsplanung und Verkehrstechnik, Heft V 24, Technische Universität Darmstadt, ISSN 1613 8317.

Wolfermann, A., 2010. Improving the capacity of signalized intersections based on the assessment of signal change intervals. In: Transportation Research Board 89th Annual Meeting, Transportation Research Board, 2010, p. 17.

United Kingdom

5

Andy H.F. Chow[1,2] *and Konstantinos Ampountolas*[3]
[1]Systems Engineering and Engineering Management, City University of Hong Kong, Kowloon Tong, Hong Kong, [2]Centre for Transport Studies, University College London, London, United Kingdom, [3]School of Engineering, University of Glasgow, Glasgow, United Kingdom

5.1 History

Traffic signals in the United Kingdom were first invented by an engineer named John Peake Knight who was a railway manager (BBC, 2009). The concept of traffic signals was first applied to management of rail traffic in Great Britain. It was not until December 1868 when the world's first traffic lights appeared outside the Houses of Parliament in London to control traffic flow around the new Parliament Square. Unfortunately, despite their success at directing traffic, the first lights did not last long. Malfunction of a component leaded to explosion of the traffic lights and killing of an police operator (HistoryHit, 2015). Following the successful implementations in the United States in 1914 (Cleveland, Ohio) and 1918 (New York City), traffic lights arrived in London again in 1925, located at the junction of St James's Street and Piccadilly Circus.

Vehicle actuation techniques were experimented with in the 1930s in order to make traffic signals responsive to the prevailing traffic conditions. With the advent of computerized systems in the mid-1960s, many cities in the United Kingdom embarked on deploying centrally controlled and monitored traffic signal systems (Shepherd, 1994), which later led to the development of the Split Cycle Offset Optimization Technique (SCOOT) Urban Traffic Control System. In the 1970s, inductive loops were used associated with the signal control systems in order to improve their efficiency with respect to the prevailing conditions.

There have been 1000 new sets of traffic lights installed every year. The total number of traffic signals in the United Kingdom is now over 8000. By 2015, there were around 3000 signalized road junctions in Greater London. These are roughly a 50–50 split between inner and outer London (Greater London Authority, 2017).

Global Practices on Road Traffic Signal Control. DOI: https://doi.org/10.1016/B978-0-12-815302-4.00003-0

5.2 State-of-the-art

In the United Kingdom, the most widely used technique for calculating isolated and network-wide fixed-time signal plans is the transport research laboratory (TRL) TRANSYT method (Robertson, 1969, 1997; Vincent et al., 1980). This uses a mesoscopic simulation model of traffic to estimate performance in conjunction with a robust direct search optimization method to find a signal plan that performs well with a specified set of traffic flows.

The traffic model predicts a performance index (PI) for a network based on a fixed-time signal plan and set of average (representative) traffic flows. The PI is a measure of the overall cost associated with congestion and is a weighted combination of total vehicle delay and stops experienced by traffic within the modeled network. The signal optimization component within TRANSYT modifies signal timings and assesses whether those adjustments have reduced the PI. This iterative process is illustrated in Fig. 5.1.

The TRANSYT traffic model includes platoon dispersion to represent the effects of variability of vehicular speeds on flow propagation. The TRANSYT platoon dispersion model gives link outflow $q^L(t)$ at time t as

$$q^L(t) = F\,q(t - t_F) + (1 - F)q^L(t - \Delta t) \tag{5.1}$$

where t_0 is the free-flow travel time of the link, measured in steps Δt; $t_F = \beta t_0$ is a proportion of the free-flow travel time t_0; and $F = 1/(1 + \alpha t_F)$ is a smoothing factor.

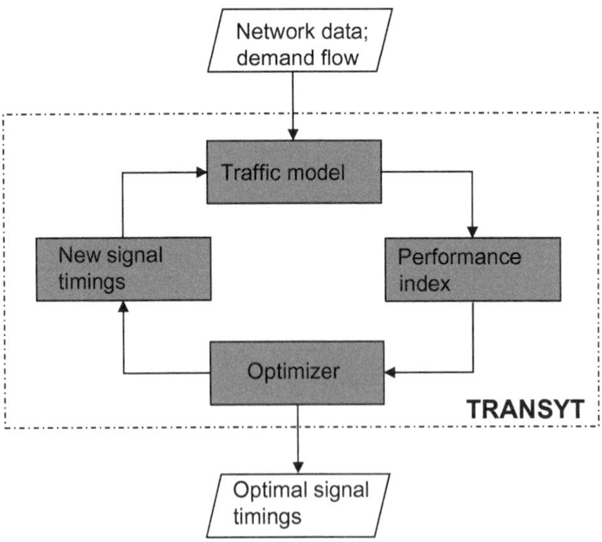

Figure 5.1 Illustration of the TRANSYT optimization process.

The expected uniform delay d_u is calculated from this link outflow in conjunction with the signal timings at the downstream junction using a macroscopic queueing simulation model.

A random delay term d_r is calculated according to the capacity of a link and the total flow during each cycle to correct for oversimplification in the simulation; this is calculated according to the method described by Kimber and Hollis (1979). Thus

$$d = d_u + d_r \tag{5.2}$$

The optimizer manipulates the stage durations and the offset at each junction in order to find improved performance. However, the optimizer requires that the cycle time be specified exogenously. This method provides a high standard against which other systems can be compared.

The latest versions of TRANSYT (after version 13) contain a number of changes to previous versions to enable more accurate modeling of platoon dispersion and queueing behavior at a stop line. The most significant of which is the addition of the celebrated cell transmission model of highway traffic (Daganzo, 1994). Field implementations of TRANSYT in the United Kingdom indicated savings of some 10%–15% of the average travel time through the network.

Other common modeling software for designing and optimizing signalized junctions in the United Kingdom include LINSIG (Simmonite, 1985; subsequently maintained and developed by JCT Consultancy[1]) and OSCADY PRO. LINSIG and OSCADY PRO are particularly suitable for preliminary design and for optimizing isolated signalized intersections with only minimal input data such as geometric characteristics, traffic demands, and maximum (or saturation) flows (Transport for London, 2010). Interaction between pedestrians and general traffic at crossing facilities and signalized junctions is of particular importance and necessary data (e.g., pedestrian demands) can be taken into account in the design and optimization process. Changes to traffic signal settings can have significant influence on the safety of road users, including pedestrians, cyclists, and general vehicular traffic. LINSIG can be used for detailed junction design and assessment of alternative models and signal settings for checking against junction controller specifications. LINSIG optimizes signal timings and provides an estimation of junction performance. Similarly, OSCADY PRO can be used to optimize alternative signal sequences at a junction by offering different potential configurations. The resulting design can be exported into a format compatible with other TRL products such as TRANSYT for use within a network-wide control framework. Both software tools require input of all junction information including geometric details and traffic flows (from historical data). A number of traffic signal plans are usually developed and assessed to represent the complete set of traffic conditions to be found on the

[1] JCT Consultancy: http://www.jctconsultancy.co.uk.

junction level at different times (morning peak, evening peak, and off-peak times) and days of the week.

Regarding applications to real-time, the SCOOT system (Hunt et al., 1981; Bretherton et al., 1998, 2003, 2004) is a centralized adaptive traffic control system developed and widely used in the United Kingdom (with over 200 installations of the systems in the United Kingdom and elsewhere). SCOOT operates upon the TRANSYT model, and uses online detector information to develop a signal control plan in real time. The central computer holds a complete signal plan for the road network, and modifies this plan gradually by making relatively small changes cycle by cycle in order to match it to measured conditions. The optimizer can vary stage durations (known as *splits*), the cycle time duration, and the offset at each junction according to model-based estimates of performance.

5.3 Signal timing procedure

The overall design process of a signalized junction follows the procedure specified in Design Manual for Roads and Bridges (Highways England, 1997; see also Department for Transport, 2006). The procedure generally consists of the following three stages:

- Stage 1: a predesign inspection of the proposed site to identify possible problems that would have a bearing on the design of the scheme;
- Stage 2: a design/safety check of the design carried out by the traffic infrastructure (TI) checking engineer;
- Stage 3: a site inspection carried out at commissioning to check that any problems identified in Stages 1 and 2 have been overcome and no other problems have occurred. If the site has been installed by a third party, TI must attend the inspection.

A scale plan of the site will be prepared with kerb and building lines. For schemes that entail no significant alteration to road kerb lines, base ordnance survey data may provide sufficient topographical information but care should be exercised to ensure that the details are both current and accurate. Limited surveys to check certain critical dimensions and to check the location of any physical objects, such as statutory undertaker's plant, manhole covers, drainage, lighting cables, lamp columns, road signs, trees, etc., may be all that is required. For more significant changes in layout, a full three-dimensional model of the existing topography may be required. For major changes early consultation with statutory undertakers, who may be affected, is recommended. These consultations may lead to a requirement to commission trial excavations to locate precisely existing plant in relation to the revised kerb alignments.

Hourly classified traffic counts, converted to pcu/h for each arm of the junction with separate turning traffic figures should be prepared. These counts should cover the morning and evening peaks, daytime off-peak, and any other significant event time, for example, Saturday shopping period. If the existing situation has queues, their length, in vehicles, should be measured every quarter of an hour. The flows should be in the form of a graphical summary if possible. If major changes to the road network, developments, or other factors are likely to result in changed traffic flows, predicted flows from traffic models should be provided.

The saturation flow of each arm of the junction should be assessed, either practically by on-site measurement using the TRL saturation flow program and a portable computer, or theoretically using the TRL method described in TRL research report (RR) 67 (Kimber et al., 1986). Saturation flows should wherever possible be measured practically on site but it is essential at critical junctions. The prediction method may be used as an alternative where site measurement is impractical. Cruise speed refers to the 85 percentile free-flow traffic platoon running speed for each approach, where the junction forms part of a linked network the following is also required.

5.4 Signal phasing

In the United Kingdom, traffic signals operate by showing the aspects cyclically: red, red-yellow, green, yellow, and red cyclically. The red-yellow and yellow durations are typically 2 and 3 seconds, respectively. The following summarizes some basic terminology that is used for traffic signals design in the United Kingdom:

- *Phase*—A phase is a group of one or more traffic (or pedestrian) links which always receive identical signal light indications.
- *Stage*—A stage is a part of the cycle during which a particular set of compatible streams (i.e., streams can proceed simultaneously) receives green. The duration of a stage typically will be at least 5–10 seconds. For a given cycle time and sequence of stages, the durations of the stages are also known collectively as the *split*, reflecting the division of the cycle into stages.
- *Intergreen*—Intergreen (or lost time) is the time lost to traffic while changing from one stage to the other. Some intergreen is used for clearing traffic in the previous stage for safety reasons. Intergreen for traffic clearance will be around 5–10 seconds in practice, depending on the size and the junction. Some intergreen may also be due to the time given to the pedestrian stage(s).

The sum of the interstage times over the entire signal cycle is sometimes called the *lost time*.

Signal timings at a junction are typically described in a stage-based approach in which the timings are described in terms of the sequence of stages, their durations, and the times during each of the interstages at which signals change. Stages cannot usually be eliminated from the sequence automatically because of the effect that this would have on the interstages. However, the choice of stage sequence can have a profound effect on the operation of the junction, and no rules have yet been devised for selection of the most appropriate one other than to calculate signal timings for a range of possibilities. Fig. 5.2 depicts an example junction with six phases (including pedestrian crossing phases). Fig. 5.3 illustrates the corresponding phase diagram in three stages. The British definition of intergreen is: "the period between the end of the green signal giving right of way for one phase, and the beginning of the green signal giving right of way for the next phase." The normal

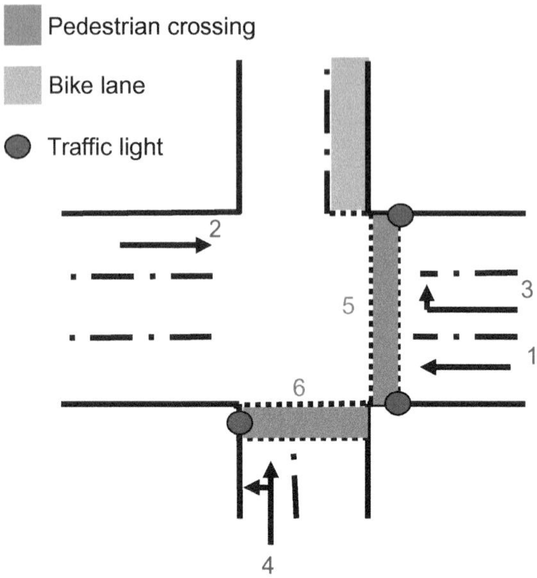

Figure 5.2 A junction with six phases.

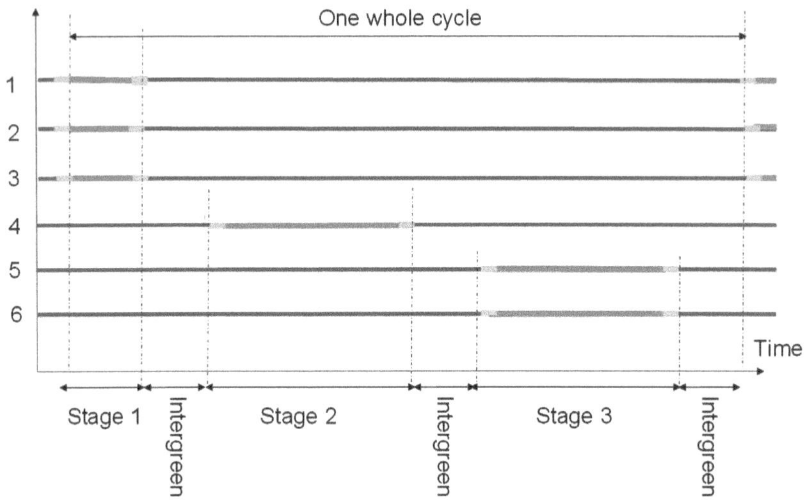

Figure 5.3 Example of a phase diagram of a simplified four-approach crossroads with six phases and three stages.

minimum is 5 seconds, made up of 3 seconds yellow after one green and 2 seconds before the next. The current method for calculating these in Britain can be found in the guideline in Department for Transport (2006). The intergreen period can be extended but never curtailed by external control.

5.4.1 Treatment of turning movements

Drivers drive on the left-hand side in the United Kingdom, and hence it should be right-turn to be protected instead of left-turn. All turning movements permitted: This is a common junction operation which forms the basis of signaling techniques. Vehicles on opposite approaches have a green signal, whilst those on the other two approaches have a red. When an all-round pedestrian stage is introduced, all vehicular movements are stopped. Each approach may have one or more lanes, a shared stop line, and simultaneous discharge. Right-turning traffic may impede vehicles wishing to proceed over the junction if an offside lane is shared, or even worse there is only a single lane.

With right-turn prohibited: Where there is a relatively minor right-turn flow the capacity of the junction is reduced by the road space occupied by the vehicle waiting to turn right and by the time which has to be provided to this movement in the cycle. If the right-turn maneuver is removed, then reduced delay and improved capacity can be expected. Where one exists, an alternative route can be indicated to traffic before the junction is reached.

The main alternatives are referred to Fig. 5.4:

- turn left before the junction, make two right-turns to appear at the junction on the left-hand arm (known as a "p" turn).
- pass through the junction, turn left and make two further left-turns to appear at the junction on the left arm (known as a "q" turn).

In the latter case the diverted traffic will pass through the junction twice and may adversely affect the expected improvements. If the diversion routes are residential the additional through traffic may be unacceptable.

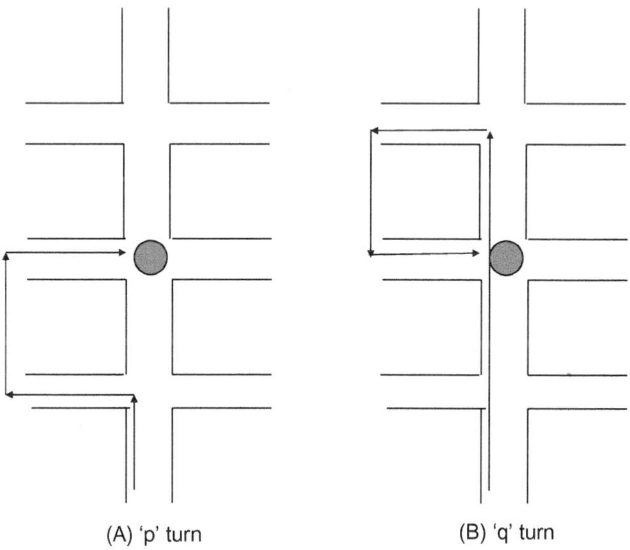

Figure 5.4 (A) "p" turn and (B) "q" turn.

With both left-turn and right-turn prohibited—This results in exclusive movements. If both opposing arms operate this way, the opportunity can be taken to run parallel pedestrian phases (walk with traffic).

Right-turn in the United Kingdom—Without special provision, some right-turning vehicles will turn in gaps in the opposing flow but if this is not possible they will complete their movement in the intergreen period. A typical junction will have space for two or three vehicles beyond the stop line. With, say, a 70-second cycle time this means that between 100 and 150 vehicles an hour will turn right in the intergreen. With a longer cycle time this "free" right-turn capacity decreases.

If the number of right-turners exceeds the number that can clear the junction in gaps, or during the intergreen, special provision will be probably be required. Note that if right-turners are left in front of the stop line after a change to another stage, not only will the driver feel in a vulnerable position but the presence of the vehicle may obstruct pedestrians wishing to cross.

There are three basic choices. If only one approach has a right-turn flow justifying special provision, typically an early cutoff can be used. However, if the right-turning movements from opposing directions both justify a right-turning stage then a staging with concurrent exclusive right-turning movements may be appropriate. The third option is to run the opposite arms separately but this normally has serious capacity problems.

5.4.2 Treatment of bicycles and pedestrians

The staging arrangement for a junction will have implications for how pedestrian and cycle facilities are provided. An exclusive pedestrian stage (all red for vehicular traffic) will typically require 20 seconds of the cycle time and will increase the lost time in a signal cycle disproportionately. The lost time for a simple junction could increase from, say, 12 $[=(2\times 7)-2]$ to 32 seconds and the cycle time would increase proportionately. This preliminary analysis will indicate whether an exclusive pedestrian stage is feasible, or desirable, for either vehicular or pedestrian traffic. Longer cycle times will increase waiting times for pedestrians as well as vehicular traffic. In this case the staging and layout of the junction will require walk with traffic pedestrian facilities where required. Ideally, pedestrian waiting times should not exceed 80 seconds.

Separated bicycle facilities would be considered in the same way. Staging schemes are advantageous to cycling if they help cyclists to maintain a steady speed and a direct course without interruption or obstruction from a position where they can be seen by drivers and pedestrians. Cycle safety may be improved through the use of advanced stop lines, widened carriageways, or dedicated cycle-lanes. In schemes where specialist provisions are proposed it is important to model the impact that these will have on all road users including public transport.

5.4.3 Treatment of transit priority

Transit vehicles (e.g., buses, trams) differ from other traffic in that they have different travel characteristics. In the United Kingdom, Robertson and Vincent (1975) showed how in cases such as this, buses could be modeled within the framework of the TRANSYT model as a distinct segment of traffic. They used separate links in a road network to represent the movements of buses, on which the travel times were calculated to incorporate the mean time spent by buses at stops on that link. This leads to the calculation of signal offsets that represent a compromise between those that would be ideal for the buses and for the other traffic, with an advantage to the buses and a consequent reduction in total passenger delay. This approach has been found to reduce bus journey times by about 8%, which is roughly equivalent to a 16% reduction in delay (DETR, 1997).

Selective vehicle detection (SVD) bus priority (BP) is a facility that distinguishes buses from general traffic and provides them with priority at traffic signals in London. The overall aim of SVD BP is to reduce delays and increase journey time reliability for timetabled buses. The use of SVD BP at a junction does not prevent it from altering timings to alleviate general traffic congestion where other dynamic traffic control systems are in operation. Traffic signals controlled by SCOOT should initially make provision for SVD BP. An assessment of SVD suitability can be made using the SCOOT BP guidelines to better understand the potential network impact (Transport for London, 2006).

The iBus system (Siemens) is an automatic vehicle location system to improve London's SVD BP. The system tracks all of London's 8200 buses to provide passengers with audiovisual announcements, improved information on bus arrivals, and to trigger priority at traffic junctions.

5.5 Parameter determination

Manual methods are still used in practice in the United Kingdom. Webster's (1958) method is a *stage-based* approach, and has been used widely. Webster's method requires that a suitable sequence of stages and the associated interstage structures be specified in advance. The method proceeds by identifying a *representative stream* for each stage from those that receive a green indication in it as the one which has the greatest ratio of mean arrival to saturation departure rates, or *flow ratio* $y_i = q_i/s_i$.

In the United Kingdom, typically cycle times in urban areas range from 60 to 120 seconds. The optimum cycle time c_0 (in seconds) is calculated from the Webster's formula

$$c_0 = \frac{1.5L + 5}{1 - Y} \tag{5.3}$$

where L is the total lost time per cycle; $Y = \sum_{j=1}^{K} y_j$ is the sum of critical demand–saturation ratio for all stages j; $y_j = \max\{q_i/s_i\}$, for all phases i in stage j; q_i and s_i are the total demand and total saturation flow, respectively, for phase i; K is the number of stages at the junction.

The time available in the cycle resulting from this calculation is then split into stages according to the formula

$$g_j = (c - L)\frac{y_j}{Y} \tag{5.4}$$

for $1 \le j \le K$, where g_j is the effective duration of stage j.

This allocation of green time to stages results in equal degrees of saturation for the representative streams, as

$$x_i = \frac{q_i c_0}{s_i g_i} = \frac{Y c_0}{c_0 - L} \tag{5.5}$$

This is because the representative streams are the most heavily loaded ones, all other streams having lower degrees of saturation. Thus the allocation of green time maximizes the reserve capacity of the junction at the given cycle time c_0. Webster showed that this method approximately minimizes the sum of the rates of delay in the representative streams. Whilst this is a reasonable proxy for the total rate of delay at the junction, it does ignore an arbitrarily large rate of delay incurred by traffic in streams other than the representative ones.

As a manual method, Webster's one works reasonably well. However, it is subject to a number of limitations that restrict its practical application to junctions that are relatively simple. In particular, the identification of representative streams can be difficult or even impossible when some streams have right of way in more than one stage. Furthermore, fine details of signal controlled junctions, such as importance weights, short lanes, gap acceptance for offside turners, and late starts and early cutoffs are difficult, if not impossible, to incorporate.

The saturation departure rate is best measured by site observation using standard methods (Road Research Laboratory, 1963). It is noticed that the mean rate at which vehicles cross the stop line during the effective green will depend upon the composition of the traffic, and in particular the proportion of large vehicles. Corrections can be made for this to convert observations to values in terms of standard passenger cars performing straight ahead maneuvers. The factors used to correct for the presence of different vehicle types are known as *passenger car equivalents*. Standard values are given in Table 5.1. The factors used to correct for the different maneuvers which are performed are known as *through car equivalents*.

In cases where a junction has not yet been built or where changes to the physical design are being considered, it is not possible to make field observations to measure the saturation departure rate. Instead, it can be estimated from geometric

Table 5.1 Passenger car equivalents

Vehicle type	Passenger car equivalents
Light vehicles	1.0
Light commercial	1.5
Heavy commercial	2.3
Buses and coaches	2.0
Motorcycles	0.4
Pedal cycles	0.2

Source: Kimber, R.M., McDonald, M., Hounsell, N.B., 1986. The prediction of saturation flows for road junctions controlled by traffic signals. Transport and Road Research Laboratory, Report RR 67. TRL, Crowthorne.

design data. Kimber et al. (1986) have established empirical relationships between the geometric characteristics of a stream of traffic and the saturation departure rate for it. For streams that do not contain opposed turning traffic, their estimate of saturation departure rate s in through car units (tcu) for a single lane is (Kimber et al., 1986)

$$s \approx 2080 - 140\delta_n - 42\Theta_G G + 100(w_l - 3.25) \tag{5.6}$$

where $\delta_n = 1$ for a nearside lane, 0 otherwise; G is gradient in percent; $\Theta_G = 1$ for positive (uphill) gradients, 0 otherwise; and w_l is the width of the lane (m).

The correction that was established for the presence of turning traffic is to divide by $(1 + 1.5f/r)$, where f is the proportion of traffic that is turning, and r is the radius of turn (m). The standard residual of the fitted model was 117 tcu/h, which illustrates a typical error of estimation.

It is assumed that traffic passes the junction only during a part of the cycle called the *effective green*. There is a difference between the effective green and the displayed (actual or design) green in practice, which is due to the acceleration and deceleration of vehicles. The difference between them can be measured on site, and such a difference is independent of the duration of green. As a rule of thumb, effective green typically starts 2 seconds later and ends 2 seconds later than displayed green, and is thus 1 second longer (Fig. 5.5).

5.6 Working example

We now demonstrate the UK stage-based signal timing procedure using the common example in this book (see Chapter 2, Principles of road traffic signal control). Based upon the configuration of the traffic movements (or phases), we arrange them into the stage structure and sequence illustrated in Table 5.2.

Taking that the lost time per stage is 4 seconds; calculate the cycle time by using Webster's formula, and the associated stage durations.

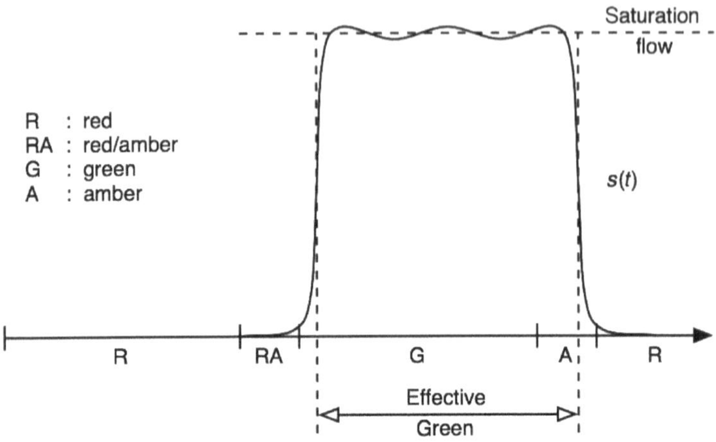

Figure 5.5 Effective green time and saturation flow.

Table 5.2 Scenario data of working example

Stage	Phases	Movement	Volume (pcu/h/L)
1	4	Westbound through (WBT)	695
	8	Eastbound through (EBT)	679
2	3	Eastbound right-turn (EBR)[a]	239
	7	Westbound right-turn (WBR)	169
3	2	Southbound through (SBT)	408
	6	Northbound through (NBT)	376
	1	Northbound right-turn (NBR)	137
	5	Southbound right-turn (SBR)	154

[a]Drivers drive on the left-hand side in the United Kingdom, and hence it should be right-turn to be protected instead of left-turn.

The critical y_j for each stage j is calculated as:

$$\text{Stage } 1{:}y_1 = \max\{q_4/s_4, q_8/s_8\} = \max\{695/1800, 679/1800\} = 0.386$$

$$\text{Stage } 2{:}y_2 = \max\{q_3/s_3, q_7/s_7\} = \max\{239/1800, 169/1800\} = 0.133$$

$$\begin{aligned}\text{Stage } 3{:}y_3 &= \max\{q_1/s_1, q_2/s_2, q_5/s_5, q_6/s_6\} \\ &= \max\{137/1800, 408/1800, 154/1800, 376/1800\} = 0.227\end{aligned}$$

Hence, $Y = \sum_{j=1}^{K} y_j = 0.386 + 0.133 + 0.227 = 0.746$.

Total lost time, $L = 3 \times 4 = 12$ seconds.

Each cycle will need an additional 20 seconds all red duration for pedestrian crossings. Hence, with Webster's formula we have the cycle time (rounding up to nearest integer):

$$c_0 = \frac{1.5L + 5}{1 - Y} = 111 \text{ second}$$

Consequently, the stage durations are:

$$\text{Stage } 1{:}g_1 = (c_0 - L - 20)\frac{y_1}{Y} = 41 \text{ seconds}$$

$$\text{Stage } 2{:}g_2 = (c_0 - L - 20)\frac{y_2}{Y} = 14 \text{ seconds}$$

$$\text{Stage } 3{:}g_3 = (c_0 - L - 20)\frac{y_3}{Y} = 24 \text{ seconds}$$

and

Stage 4 (pedestrian): 20 seconds

Bibliography

BBC Home, 2009. Local history—the man who gave us traffic lights (accessed 19.07.18).

Bretherton, D., Bowen, G., Wood, K., 2003. Effective urban traffic management and control—recent developments in SCOOT. In: Proceedings of the 82nd TRB Annual Meeting, Washington, DC.

Bretherton, D., Bodger, M., Baber, N., 2004. SCOOT—the future. In: Proceedings of the 83rd TRB Annual Meeting. Washington, DC.

Bretherton, R., Wood, K., Bowen, G.T., 1998. SCOOT version 4. In: Proceedings of the 9th International Conference on Road Transport Information and Control. Institution of Electrical Engineers, London, UK.

Daganzo, C.F., 1994. The cell transmission model: a dynamic representation of highway traffic consistent with the hydrodynamic theory. Transp. Res. B 28 (4), 269–287.

Department of the Environment, Transport and the Regions (DETR, 1997) Keeping buses moving, 1997. Local Transport Note 1/97. HMSO, London.

Department for Transport, 2006. General Principles of Traffic Control by Light Signals. Traffic Advisory Leaflet, 1/06, Parts 1 & 4. Department for Transport Traffic Advisory Unit, London.

Greater London Authority, 2017. Economic Impact of Traffic Signals. GLAEconomics, London, UK.

Highways England (formerly Highways Agency), 1997. Design Manual for Roads and Bridges. The Stationery Office (TSO), London, UK.

HistoryHit, 2015. 1868: First traffic light. <www.historyhit.com/1868-first-traffic-light/> (accessed 19.07.08).

Hunt, P.B., Robertson, R.D., Bretherton, R.D., Winton, R.I., 1981. SCOOT—a traffic responsive method of co-ordinating signals. Technical Report 1014, Transport and Road Research Laboratory.
Kimber, R.M., Hollis E.M., 1979. Traffic queues and delays at road junctions. Transport and Road Research Laboratory, Report LR 909. TRL, Crowthorne.
Kimber, R.M., McDonald, M., Hounsell, N.B., 1986. The prediction of saturation flows for road junctions controlled by traffic signals. Transport and Road Research Laboratory, Report RR 67. TRL, Crowthorne.
Road Research Laboratory, 1963. A Method of Measuring Saturation Flow at Traffic Signals. Road Note, 34. HMSO, London.
Robertson, D.I., 1969. TRANSYT method for area traffic control. Traffic Eng. Control 10, 276–281.
Robertson, D.I., 1997. The TRANSYT method of coordinating traffic signals. Traffic Eng. Control 38 (2), 76–77.
Robertson, D.I., Vincent, R.A., 1975. Bus priority in a network of fixed-time signals. Transport and Road Research Laboratory Report: LR 666. TRL, Crowthorne.
Shepherd, S., 1994. Traffic control in oversaturated conditions. Transp. Rev. 14 (1), 13–43.
Simmonite, B.F., 1985. LINSIG: a computer program to aid signal design and assessment. Traffic Eng. Control 26 (6), 310–315.
Transport for London, 2006. Bus priority at traffic signals keeps London's buses moving: Selective Vehicle Detection (SVD).
Transport for London, 2010. Traffic Modelling Guidelines: TFL Traffic Manager and Network Performance Best Practice. Version 3.0.
Vincent, R.A., Mitchell, A.I., Robertson, D.I., 1980. User guide to TRANSYT version 8. Transport and Road Research Laboratory Report: LR 888. TRL, Crowthorne.
Webster, F.V., 1958. Traffic Signal Settings. Road Research Technical Paper 39. HMSO, London.

France

6

Biao Yin[1], *Nicolas Chiabaut*[2] *and Xiaoyan Xie*[1]
[1]LVMT, UMR-T 9403, École des Ponts, IFSTTAR, UPEM, UPE, Champs-sur-Marne, France,
[2]Université de Lyon, IFSTTAR/ENTPE, LICIT, Rue Maurice Audin, Vaulx-en-Velin, France

6.1 Introduction

As an interacting area of several roads, the intersection has a key impact on passing traffic, and often faces the complexities of traffic information (traffic volumes, modes, etc.) and multidimensional road space (occupied zones, dedicated lanes, etc.). To avoid the conflict of different directional movements, one regular way to manage traffic is to make spatiotemporal usage of the intersection. In the early 20th century, modern signalization at intersections was invented to address the conflict. France has a long history of traffic signal developments. Dating back to 1912, the first traffic kiosk-signal device was placed in Paris. In 1923, the first single-colored traffic light device with ringing was deployed at an intersection. In 1936, a fixed-time control signal at an intersection was first implemented, followed by tricolor traffic lights in 1937 (MEGE Paris, 2017). Nowadays, there are about 30,000 signalized intersections in France (5–12 intersections per 10,000 inhabitants). This number is still significantly increasing in most major cities, especially since the time when trams and the BHNS (Bus à Haut Niveau de Service, namely Bus Rapid Transit System) were created (Damas, 2016). Traffic signal control at intersections is critical to improve mobility, traffic safety, and the city environment. From a survey, however, traffic congestion cost the average French household €677 in 2013; and drivers in Paris wasted an average of 69 hours in traffic congestion in 2017 (INRIX, 2017). Over recent decades, transportation sectors and research institutes have engaged in improving road traffic conditions, with the goal of constructing a more efficient, safer, and sustainable road environment. Besides improving control systems, some new and distinctive configurations of intersections have been promoted, such as roundabouts, intersections with a central island, and intersections integrated with tramway systems. Accordingly, the intersections with signalization are implemented by control systems with specific strategies (Certu, 1999, 2008, 2010; Le Cocq, 1973; Sétra, 1970, 1998). For example, a channelized intersection with a central island was invented, so that traffic flows are directed by signalization at a signalized roundabout. According to a report, this kind of intersection can successfully solve delays due to high volumes of left-turn traffic (Certu, 1999). On the other hand, traffic signal control systems were explored to enhance

Global Practices on Road Traffic Signal Control. DOI: https://doi.org/10.1016/B978-0-12-815302-4.00006-6

capacity, efficiency, and security by incorporating with other traffic management strategies, such as transit signal priority (Certu, 2001) and connected vehicle technology.

This chapter aims to investigate the developments, practices, and contributions of road traffic signal control at intersections in France. We give detailed insights into fixed-time control strategies at isolated intersections, specifically under the conditions of pedestrian crossings and left-turn traffic (France is a right-driving country). Further, a signalized intersection with a central island is studied as an example to investigate its operation performances, such as capacity and traffic delay. Moreover, the roundabout as a distinctive type of intersection concerning traffic safety is discussed.

This chapter is organized as follows. Section 6.2 describes the state-of-the-art of signal control at intersections in the country. Intersection categories, signal implantations and main signal control systems are introduced. Section 6.3 gives an analysis of fixed-time control strategies at isolated intersections. The two most used intersections, including conventional intersections and the signalized intersections with a central island, are considered. An example of the latter case is given. Then, we introduce the special consideration of roundabouts. In the final section, conclusions are drawn.

6.2 The state-of-the-art of signal control at intersections

6.2.1 Introduction to infrastructures

6.2.1.1 Intersection categories

In France, intersections are divided into three main categories: (1) ordinary intersections, (2) roundabouts, and (3) the intersections with a central island (Certu, 1999, 2008, 2010; Le Cocq, 1973; Sétra, 1970, 1998). In the ordinary intersections, a four-leg intersection (or crossroads) is the most common type (see Fig. 6.1A). This type of intersection only needs a limited space and is able to regulate high traffic

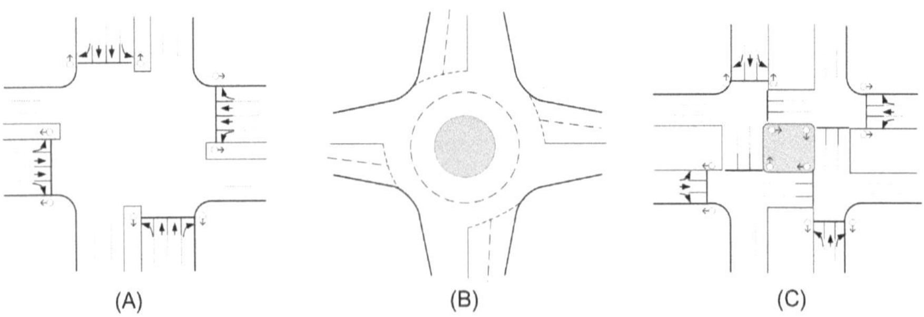

Figure 6.1 Types of intersections. (A) Four-leg intersection (crossroads). (B) Roundabout. (C) Signalized intersection with a central island.

demands (Certu, 1999; Le Cocq, 1973). Other types, like three-way intersections ("T" or "Y" junctions) and more than four-leg intersections, are fewer in urban cities. The second category is the roundabout (see Fig. 6.1B). This type of intersection needs sufficient space and satisfies low traffic flow (Alonzo, 2004; Brilon, 2015; Sétra, 1998; The Economist, 2016), which can be seen more in periurban environments. Most importantly, the roundabout has been proven as a safer solution than ordinary intersections in France. Based on the previous two types of intersections, signalized intersections with a central island have been invented (Certu, 2008) (see Fig. 6.1C). They are designed with greater capacity and safety when using island refuges for vehicles and pedestrians (Certu, 2008).

The above flat intersections represent the majority of cases in France, except those on uneven surfaces with underground passages.

6.2.1.2 Traffic lights

With a rapid growth of vehicles in the 20th century, traffic lights are deployed at road intersections to separate vehicular flows in time splits for safety purposes by avoiding conflict. With increasing traffic demands, signal timing of traffic lights is required to reduce traffic delay. Such a requirement is not only to maximize the throughput of the intersection, but also to coordinate neighboring intersections at a network level.

In practice, some intersections are installed with traffic lights, but some are not. The installation of traffic lights depends on traffic conditions, locations, and costs of infrastructures, etc. To clarify this requirement, it is possible to propose general criteria, which serve as a standard for light installation, especially when external factors are largely out of date and intersections are rescaled. The related quantitative criteria of traffic light implantation are summarized in Table 6.1 (Le Cocq, 1973).

6.2.2 Signal control software and systems

Signal control at intersections relies on rules and control algorithms to make traffic regulation safe and efficient. Constantly increasing traffic flows allow us to recognize that traffic lights can gain safety, but also lead to increasing traffic delays due to periodically stopping vehicles at intersections. Signal control algorithms are then required to make operations more efficient. The most recommended strategy for fixed-time control at signalized intersections is Webster's method. Based on this method, two traffic regulation programs were widely used in France: PROTEE for three-leg intersections, and HERMES for four-leg intersections (Le Cocq, 1973; Sétra, 1970). On the other hand, actuated or adaptive control methods were appearing. These methods use real-time traffic information for parameter adjustments. In France, CRONOS (Boillot et al., 1992) and PRODYN (Henry and Farges, 1990) were two of the most well-known adaptive control systems.

From the mid-1970s, many precursors argued the disadvantages of vehicular dominant approaches for traffic control. They recommended considering the needs

Table 6.1 Implantation quantitative criteria for traffic lights

N°	**Reasons**	**Quantitative criteria**
a.	Heavy vehicle flows	$Q1 > 600$ veh/h and $Q2 > 200$ veh/h[a]
b.	Long waiting on the lane of the conflicting phase	$Q1 > 800$ veh/h and $Q2 > 100$ veh/h
c.	Heavy pedestrian flows	$P > 250$ pedestrian/h, and $Q > 600$ veh/h without refuge island or 1000 veh/h with refuge island
d.	Management of a complex intersection	Intersection with more than six legs supporting a total traffic of more than 800 veh/h
e.	Intersection located on a coordinated axis	Intersection located on a coordinated axis with more than 50 veh/h on the second route, if it benefits the priority
f.	High number of accidents	More than five bodily injury accidents per year
g.	Combination of the previous reasons	Two of the criteria **a** to **f** verified at 80%, or three of the criteria **a** to **f** verified at 70%

[a]veh/h = uvpd/h (vehicles per hour) in Le Cocq (1973). P = pedestrian flow, Q = vehicle flow, $Q1$ = total flow of one phase (two directions), and $Q2$ = total flow of another phase (perpendicular street in the case of a four-leg intersection. For an intersection with more than four legs, the flows of the two busiest phases are taken). The flow criteria must be checked for 8 hours/day.

of pedestrians and cyclists, and good service quality from transportation systems. A robust traffic management system should include several integrated subsystems with multiple functions. For example, a systematic method called GLOCALE was designed from the 1980s, considering the overall perspectives of multimodal, travel behaviors, network deployments, and the analysis of service quality (Certu, 1999).

One classical expert system for traffic management is CLAIRE, which has been developed by IFSTTAR since the 1980s (Gabard et al., 1987). It is capable of detecting congestion, recording congested origins, generating real-time strategies based on predefined settings, and implementing suitable solutions. In other words, it can compute fixed-time signal plans on line. The system has been implemented successfully in many cities in France, such as Paris, Toulouse, and Rennes, and coupled with SCOOT to reduce road congestion in some UK cities, such as London and Leeds. Later, an advanced version named CLAIRE-SITI was developed to manage multimodal road traffic (e.g., cars, buses, and tramways) (Olivier, 2002; Scemama and Tendjaoui, 2013). Presently, CLAIRE-SITI is implemented for about 600 signalized intersections (including 100 intersections with a tramway system) in the city of Toulouse. Nowadays, many real-time intelligent systems, such as GERFAUT II (CEREMA, 2012) and GERTRUDE (Gertrude SAEM, 2018), are designed to take into account public transport priorities, measurements of traffic condition, and capacity. For example, the systems for tramway regulation at signalized intersections have been implemented, with two schedule systems for regulating traffic and tramways and a video analyzer to recognize the occupancy of an area.

There are a total of 13 regions in metropolitan France. The regions or urban areas use different traffic signal control systems in different times. For example, the Paris region used the first control system SURF1000 from 1979 to 1981; realized the system extension and modernization from the 1980s to the 1990s; and developed the system SURF3 for multimodal traffic since 2012 (Charansonney, 2014). Another important feature of signal controllers is the implementation of controller standardization, which was first published by AFNOR in 1990 and updated in following years (AFNOR, 2016, 2018). Many French manufacturers produce signal controllers with the declared standards. Some typical manufacturers (>80% market) with their products (presented in the format of manufacturer—system) are listed as follows: FARECO—GALLERY and GARBARINI, AXIMUM—M@ESTRO, LACROIX TRAFIC—TRAFFY 2/3 and SOFREL, and POLYVELEC—AZUR II (Sétra, 2017).

6.3 Strategies of fixed-time control at intersections

The analysis of fixed-time control strategies at isolated intersections is introduced in this section. Two types of signalized intersections are considered: conventional intersections and the signalized intersections with a central island.

6.3.1 Signal control at conventional intersections

In most urban areas, traffic management occurs at ordinary signalized intersections. An intersection getting rid of the influences from neighboring intersections is called an isolated intersection. Signal control at an isolated intersection lays the foundation for studying multiintersections, which are at least two adjacent intersections sharing traffic information and coordinating signal actions in a global view.

Traffic signal control refers to the adjustment of parameters, such as signal phase, phase duration, and cycle length, to ensure the safety of motorists and pedestrians, while gaining maximum throughput. Fixed-time control is a traditional and early method to regulate traffic flow. Regarding its control parameters, cycle length, phase number, and phase split are predetermined and fixed. According to the official road signalization instruction (MEDDE, 2012), minimum green time (or flashing yellow time) is 6 seconds, fixed yellow time is 3−5 seconds, and all-red clearance time is 3 seconds. Fixed-time control systems, such as TRANSYT, were implemented early worldwide. They have satisfying control performances for intersections where traffic volumes are predictable, stable, and constant. The main drawback of fixed-time control is that it cannot handle unexpected traffic conditions. Some fixed-time control systems have been developed with more functions, such as coordination of adjacent intersections through adjustments of signal timing. In this chapter, we only consider fixed-time control at an isolated intersection.

A fixed-time signal control problem can be described from two aspects: (1) the physical constraints to be respected by signal timings, and (2) the performance measures to be optimized. Here, we only focus on the first aspect for the purpose of knowing about the signal timing procedure. To calculate signal timing parameters, such as optimum cycle length and green split, precalculations as follows are often considered:

- Entrance capacity of intersection: parking, bus stop, and taxi station;
- Exit of intersection: pedestrian crossing and right-turn traffic;
- Left-turn traffic volume;
- Signal control referring to traffic conditions, at peaks (morning peak, evening peak), off peaks, night, etc.

We mainly discuss the impacts of pedestrian crossing and left-turn traffic flow on signal timings regarding traffic engineering and operation in France.

Firstly, the choice of the number of phases is important. The operation of traffic lights involves the lost time, including all-red clearance time between two phases, the time to keep security distance, and the lost green time for vehicle acceleration and deceleration (4–8 seconds per cycle). Therefore, it is imperative to make the switching signal phases as few as possible to limit the lost time. It is suggested to consider two phases initially to arrange conflicted flow groups.

Then, cycle length is calculated to regulate traffic periodically. The length limits are determined by traffic conditions. In France, at an urban intersection, the cycle length is set in the range of 40–90 seconds; whereas, at a suburban intersection or one with large capacity, the upper limit can be adjusted to a maximum of 120 seconds. According to Webster's method, the optimum cycle length to achieve minimum delay is adopted by:

$$C_o = \frac{1.5L + 5}{1 - X} \tag{6.1}$$

where L is the total lost time per cycle (i.e., the sum of all-red intervals ρ and lost time at predominate entry t_L for all phases), and X is the sum of X_i, which is the maximum of the ratios of traffic entry volumes to saturation flow rate per phase (i.e., the ratio on predominate entry per phase; in the later context, we call X_i the charge of phase). In fact, the adopted cycle length C is a little larger than C_o.

The last procedure is about the determination of the duration of green light appearance that gives right-of-way to directional flows by phases. The available green time g_i' is obtained by the proportionally available cycle length, according to charge ratio per phase. That is,

$$g_i' = \frac{X_i}{X}(C - L) \tag{6.2}$$

Therefore, the real green time per phase is finally confirmed based on g_i' added to lost time t_L and minus yellow time Y, namely $g_i = g_i' + t_L - Y$, often with $t_L = 5.5$ seconds and $Y = 3$ seconds.

6.3.1.1 Is cycle length impacted by pedestrian crossings?

Assume the pedestrians at a simple two-phase intersection with crossing time θ, and if $\theta > C - g_i'$, the cycle length is modified by following procedures.

1. First, the red time of the phase r^* should be equal to pedestrian crossing time θ. Thus, for the first phase we have $r_1^* = \theta$.
2. Then, the effective green time of the second phase is calculated as $g_2^* = r_1^* - (\rho_1 + \rho_2) - t_{L,2}$. ρ_1 and ρ_2 are the all-red intervals in phases 1 and 2, respectively.
3. The effective green time of the first phase is obtained proportionally. That is, $g_1^* = g_2^* g_1' / g_2'$.
4. Finally, the new cycle length is given by $C^* = g_1^* + g_2^* + L$.

If the new calculated cycle is superior to the upper-limited cycle length, it means this method is valid and should give other solutions for pedestrian crossings, such as designing a two-stage crosswalk with a splitter island.

6.3.1.2 Is cycle length impacted by left-turn traffic?

After each phase, there may be three situations related to left-turn flows: (1) two entries without vehicles; (2) only one entry exists with vehicles; (3) two entries exist with vehicles. If all phases are in situation (1), there is no problem for the previous cycle setting. If one or several phases are in the situations (2) and (3), problems should be addressed as follows.

- If one phase is in situation (2), the problem can be addressed in two ways: whether reducing the green time of opposite flow or increasing cycle length and permitting a dedicated left-turn phase.
- If one phase is in situation (3), the problem can be solved by decomposing the phase into two distinct phases using two alternative ways: (a) two-way synchronous left-turn phase; (b) one-way left-turn jointed with straight and right-turn phase.

As for situation (2), the green time of opposite flow reduces τf, where τ adopts 1.5–2.5 seconds according to exit lanes and f is the stocked left-turn vehicles. If, adding a dedicated left-turn phase, the new charge of a considered phase i is recalculated as

$$X_i' = \max\{X_s, (X_l + X_{l'})\} \tag{6.3}$$

$$X_s = \frac{q_s}{S_s} \tag{6.4}$$

$$X_l = \frac{\tau f}{C} \tag{6.5}$$

$$X_{l'} = \frac{q_{l'}}{S_{l'}} \tag{6.6}$$

where $q_{s(l)}$ and $S_{s(l)}$ are the traffic demand and saturation rate for straight (left-turn) flow, respectively.

As for situation 3(a), the new charge of phase i is $X'_i = X'_s + X'_l$, where X'_s and X'_l are the maximum ratios of the two compatible group flows over the saturation rates regarding straight and left-turn traffic, respectively. In situation 3(b), the new charge of phase is calculated by the sum of the two charges related to the ingress traffic of two opposite approaches. In summary, according to the above charge of phase affected by left-turn flows, the new cycle length, which is functional with the charge of phase, can finally be obtained by Eq. (6.1).

6.3.2 Signalized intersection with a central island

Some road managers want to install traffic lights at existing roundabouts for traffic management. This request often happens in situations where high traffic demand meets the limited capacity of a roundabout. It is suggested to consider a very large size of intersections with a central island.

At a large signalized intersection, the central island permits canalizing of turning movements, especially left-turn movements. To optimize the throughput of the intersection, it requires that waiting vehicles cannot spill back over the storage zone of the intersection. The number of vehicles turning left in the storage zone per signal cycle is defined in Table 6.2 (Certu, 2008).

We introduce a two-phase control strategy for a signalized intersection with a central island, as shown in Fig. 6.2. For *green indications (phases 1 and 2)*, the straightforward flows traverse the intersection without conflicts; the right-turn flows

Table 6.2 Number of vehicles in storage for left-turn traffic volume

Cycle length, C (s)	Left-turn traffic volume G (vehicles/h)						
	50	100	150	200	300	400	500
50	1	2	3	3	5	6	7
60	1	2	3	4	5	7	9
75	1	3	4	5	7	9	11
90	2	3	4	5	8	10	12

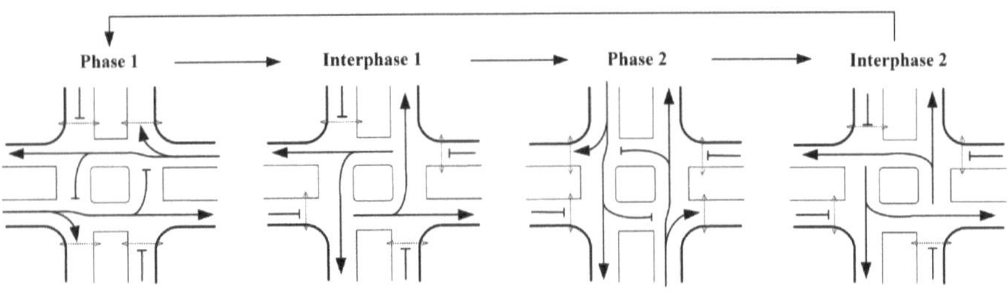

Figure 6.2 Example of two-phase control scheme.

respect the priority of the pedestrian crossing and pass the exits of approaches; the left-turn flows move to waiting zones. For *intergreen phases (interphases 1 and 2)*, all entries at approaches are closed by red signals; the left-turn vehicles on waiting zones pass freely.

The security time represented by a red lag for interphase changes is very important when setting signal timings for successive and inverse signal lanes. Three general principles for setting the security gap are: (1) do not stop straightforward vehicles at circular during the gap of two entry phases; (2) the gap permits allowing left-turn vehicles at circular to move before the phase of inverse traffic that vehicles are waiting at entries; (3) the lights for pedestrian crossings start simultaneously at the beginning of each phase; to minimize the conflicts between pedestrians and right-turn vehicles, the crosswalk lights close early when current phase still works.

6.3.3 An example

An example of a fixed-time controlled intersection with a central island is illustrated in Fig. 6.3. The symbolic intersection and its one possible signal plan are shown in Fig. 6.3A and B, respectively. The control scheme follows the diagram in Fig. 6.2.

In the signal plan, the cycle time C is set to 60 seconds. The intergreen time for two phases of entry lanes (e.g., A1 and C1) is 5 seconds. Vehicles at entry lanes can only move when stopping 1 second later than the lights of circular lanes turning red. On the contrary, vehicles at circular lanes should wait for 2 seconds after the lights of entry lanes turning red (e.g., A1 and C2). The green time at circular lanes for current movements lasts 4 seconds more than the setting of entry lanes (e.g., A1 and A2). No gap is needed between the lanes for vehicles at the same entries and the passages for pedestrians (e.g., A1 and PA1). The yellow time is 3 seconds.

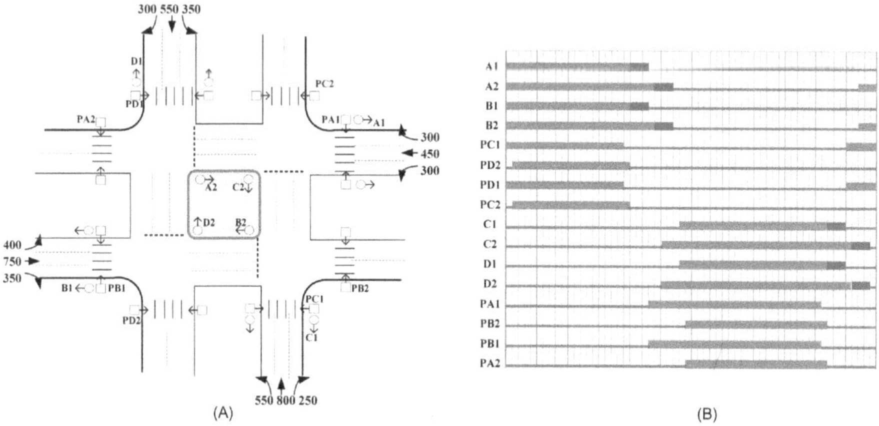

Figure 6.3 (A) An intersection with a central island, and (B) a two-phase signal plan.

Signal control performances in terms of capacity, reserve level, maximum queue length, and delay are evaluated as follows. The calculation of capacity of signalized intersection is $c = S(C - L)/C$, where S is the saturation flow (veh/h), C is the cycle length (in seconds), and L is the total lost time (simply the sum of yellow time and intergreen time). We can use $c_R = (c - q)/c$ to describe the capacity reserve of the intersection. The queue length is $l_{max} = q(C - g)l_v$, where g is the green duration, l_v is the average length of vehicles with 5 m, q is demand in veh/s (i.e., veh/h/3600). The average delay per vehicle when the lights are for vehicle lanes, is defined as

$$d_v = \frac{(C-g)^2}{2C(1 - q/S)} \tag{6.7}$$

and the average delay per vehicle when the lights are for pedestrian crossings, is defined as

$$d_p = \frac{(C-g)^2}{2C} \tag{6.8}$$

According to the signal timing plan mentioned earlier, as well as the formulas of indicators, the performances of a signal plan can be evaluated for traffic lanes at entries and circulars, as well as pedestrian crosswalks. Detailed results of tested traffic volumes are summarized in Table 6.3.

Table 6.3 Performance of fixed-time control at intersection for traffic and crossings

	Demand	**Green time**	**Capacity**	**Reserve**	l_{max} **(m)**	**Delay (s/veh)**
A1	350	20	600	42%	19.4	16.6
C2	583	31	930	37%	23.5	10.4
PA1	–	28	–	–	–	8.5
PC2	–	19	–	–	–	14.0
B1	500	20	600	17%	27.8	18.5
D2	400	31	930	57%	16.1	9.0
PB1	–	28	–	–	–	8.5
PD2	–	19	–	–	–	14.0
C1	533	24	720	26%	26.7	15.3
B2	500	27	810	38%	22.9	12.6
PC1	–	24	–	–	–	10.8
PB2	–	23	–	–	–	11.4
D1	400	24	720	44%	20.0	13.9
A2	433	27	810	47%	19.8	11.9
PD1	–	24	–	–	–	10.8
PA2	–	23	–	–	–	11.4

6.4 Special considerations

Modern roundabouts are very popular for traffic control around the world, and especially in France, Australia, and Spain. A modern roundabout requires entering traffic to give priority to traffic already within the roundabout, which is called "yield-at-entry" control. The roundabout may include signal installations, pedestrian islands, and cycling lanes. This type of intersection shows the advantages of good conditions of safety, fluidity, speed reduction, and the opportunity to simplify complex crossroads. French roundabouts first appeared in the French Highway Code in 1983. Since then, they have been greatly promoted, and there are currently estimated to be more than 30,000. According to the survey of Environment Rating Scales Institute (ERSI) in 2014, France has the highest roundabout ratio, that is, 1 roundabout in every 45 intersections. Nearly 80% of these roundabouts are in towns or suburbs (Thai Van and Balmefrezol, 2000). Official institutes estimated that the country adds on average more than 1000 roundabouts to its road network each year. Safety is the main reason driving this growth. Roundabouts can make motorists change their trajectories, with lower permissible speed when traveling on circular roads (generally, less than 30 km/h). Compared to ordinary intersections, roundabouts have fewer conflicts within the crossroads (see Fig. 6.4). Safety improvements make roundabouts more and more attractive. The number of crashes causing injuries at intersections fell on average between 57% and 78% following the construction of modern roundabouts. The number in the United States was 51% (EuroRAP, 2015). Small, single-lane roundabouts have proved to be extremely successful, while very large roundabouts with multiple lanes cause problems.

In many cases, roundabouts are safer, less congested, and more environmentally friendly (Chevallier, 2008), but harder to control even when they are signalized. In France, there are signalized and nonsignalized roundabouts. Single-lane roundabouts or "miniroundabouts" that are mostly deployed in towns or suburbs, operate without traffic lights. Some multilane or special roundabouts install traffic lights at the entries, even on circular roadways. Roundabouts with traffic lights can make

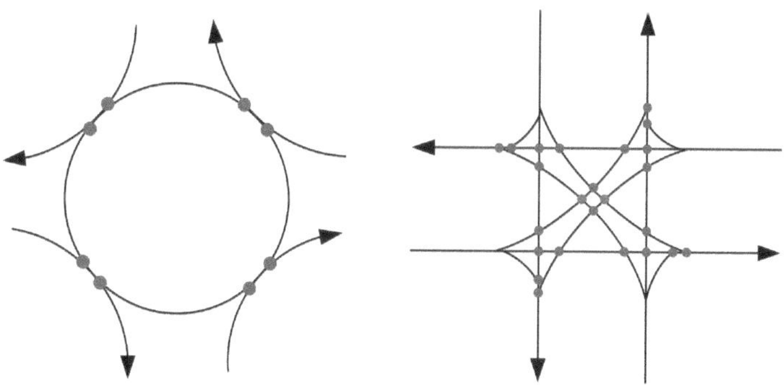

Figure 6.4 Comparison of possible conflicts on a roundabout versus an ordinary intersection.

traffic more balanced and coordinate nearby intersections. Moreover, the establishment and rules of roundabouts consider drivers, pedestrians, cyclists, and even public transit priority (Azhar and Svante, 2011; Baranowski, 2004). For example, cyclists can choose to ride through a roundabout like a car or to walk it like a pedestrian by using the crosswalk of the roundabout. We introduce the following cases to recognize only the importance of roundabout signal control, regarding pedestrian crossings, augmentation of capacity, prevention of queue overflow, and giving transit signal priority (e.g., tramways), since nonsignalized roundabouts are beyond the scope of this book.

6.4.1 Pedestrian crossings

Pedestrian signals for crossing are installed on the two sides of a street. For the purpose of security, pedestrian crossings are located at least 15 m from roundabouts. The green signals start immediately when pedestrians press the crossing buttons. An interval of 30 seconds is necessary between two cycles of demands. The green time for pedestrian crossings reduces to zero, according to the crossing distance (e.g., 1 second per 1 m).

6.4.2 Augmentation of capacity

There is an argument about whether or not traffic lights should be set permanently at roundabouts. It was said that in low-volume traffic situations, nonsignalized methods have better performances; whereas, in high-volume traffic situations, traffic lights are always implemented (Al-Madani, 2003; Azhar and Svante, 2011). Ebrahim Fouladvand et al. (2004) explored the question about under what circumstances people should control an intersection with traffic lights. In their study, through comparing quantitatively the overall delays regarding roundabouts with or without signalization, the threshold of traffic volume was finally obtained, over which the alternated scheme operates optimally. For the same issue in France, it is not encouraged to implement permanent traffic lights for small or medium-sized roundabouts. A central island (with optimal radius ≥ 30 m) should be designed at a very large-scale intersection with traffic lights for organizing left-turn flows. In this case, the control mechanism is to treat the entries as an ordinary intersection, allowing traffic lights to be installed at entries and at circular roads. The capacity of the intersection can be calculated easily by the software—Girabase (Cerema, 2018).

6.4.3 Prevention of queue overflow

A proper setting of traffic lights regulates traffic flows in balance. Using signals for metering at roundabouts, queues can be handled and prevented from blocking nearby intersections or occupying rapid roadways and tram lanes. It is recommended that traffic lights are set at roundabouts for frequent sharp traffic flows.

When meeting saturated traffic conditions, the pedestrians' demand for crossing the street is invalid to avoid perturbing the signal cycle.

6.4.4 Public transport priority

It is also possible to give a signal priority to public transport at roundabouts. In France, roundabouts are the main type of intersections for new lines of trams under traffic light management. Traffic lights are set at a roundabout to avoid the conflict between trams and vehicles, and trams have a priority to traverse the roundabout without a stop. When a tram approaches (the detection of the tram is at 80–140 m away from a roundabout) the roundabout, traffic lights are switched to red. Detection of the tram at the exit of the roundabout orders the extinction of lights, and resends a priority to traffic on the roundabout (Guichet, 2005). Many French cities, such as Nantes, Lyon, Montpellier, and Bordeaux, successfully use this type of roundabout. Other types of public transport, such as buses, are suggested to circle the roundabout rather than traversing it.

6.5 Conclusions and perspectives

In this chapter, we have introduced an essential procedure of traffic signal timing for fixed-time control at isolated intersections and on-site system implementations in France. Specifically, a fixed-time signal plan is designed for a two-phase signalized intersection with a central island as an example. Moreover, we have discussed the traffic signal control systems that are operated with the incorporation of different kinds of intersection configurations (e.g., roundabouts), as well as the related strategies for public transport priorities (such as pedestrian crossings and tramway systems).

Security and efficiency are two principles in many traffic management systems. In France, with the guarantees of traffic safety, transport sectors and institutes have continued to explore new solutions, such as travel mode shifts, for future challenges of increasing traffic demand. To do so, public transport priority and connected vehicle technology for intersection control are two useful methods. In the early 1980s, France was the first country to introduce Transit Signal Priority. Public transport priority at intersections attempts to modify the operation of intersections (extending or reallocating green time, implementing a dedicated phase) to minimize or eliminate bus waiting times at traffic signals. Moreover, people use different transportation modes in urban cities. It is important to design efficient traffic control systems in mixed traffic contexts. On the other hand, connected vehicle technology is progressing very rapidly. A factor to improve traffic control efficiency is due to two-way wireless communication of intervehicles and vehicle-intersection, which could support real-time decision making and improve control efficiency at intersections.

Bibliography

AFNOR, 2016. Régulation du trafic routier-Signaux lumineux de circulation routière-Caractéristiques techniques-R25. Norm. Fr. NF P99-200/A1 Janvier 2016 (in French).

AFNOR, 2018. Régulation du trafic routier-Carrefour à feux-Principes de maintenance. Norm. Fr. NF P99-050 Mars 2018 (in French).

Al-Madani, H.M.N., 2003. Dynamic vehicular delay comparison between a police-controlled roundabout and a traffic signal. Trans. Res. A: Policy Pract. 37 (8), 681–688. Available from: https://doi.org/10.1016/S0965-8564(03)00024-7.

Alonzo, E., 2004. Du rond-point au giratoire. Parenthèses Editions, pp. 1–162 (in French).

Azhar, A.M., Svante, B., 2011. Signal control of roundabouts. Proced. – Soc. Behav. Sci. 16, 729–738. Available from: https://doi.org/10.1016/j.sbspro.2011.04.492.

Baranowski, B., 2004. Pedestrian crosswalk signals at roundabouts: where are they applicable? Pedestrian Crosswalk Signals at Roundabouts: ITE District 6 Annual Meeting June 2004.

Boillot, F., Blosseville, J.M., Lesort, J.B., Motyka, V., Papageorgiou, M., Sellam, S., 1992. Optimal signal control of urban traffic networks. In: Proceedings of the Sixth IEE International Conference on Road Traffic Monitoring and Control, pp. 75–79.

Brilon, W., 2015. Roundabouts: A state of the art in Germany. National Roundabout Conference Location, Vail Colorado, United States.

CEREMA, 2012. Etude fonctionnelle de Gerfaut II. Retrieved from <http://www.its-actif.org/spip.php?article80>

Cerema, 2018. Girabase (logiciel version 4.0). Retrieved March 15, 2018, from https//www.cerema.fr/fr/centre-ressources/boutique/girabase-logiciel-version-40.

Certu, 1999. Carrefours Urbains. Guide, pp. 1–239, in French.

Certu, 2001. Priorité aux feux: Pour les véhicules de Transport en Commun, Aménagement Exploit. la voirie, coll. CERTU Dossiers n°108, 3, 1–230, in French.

Certu, 2008. Carrefours à feux avec îlot central. Centre d'études sur les réseaux, les transports, l'urbanisme et les constructions publiques, Ouvrage, Publications du Certu, pp. 1–55 (in French).

Certu, 2010. Guide de conception des carrefours à feux, Ouvrage, Publications du Certu, pp. 1–77 (in French).

Charansonney, L., 2014. Paris road traffic management system. Retrieved March 15, 2018, from http//www.urba2000.com/PREDIM-16042014/Luc-CHARANSONNEY-16042014-V1.pdf.

Chevallier, E., 2008. Modélisation dynamique du tra c et du bruit au niveau des carrefours giratoires. Thesis of Ph.D., No d'ordre 2008-ISAL-0081, Lyon, October, pp. 1–138 (in French)

Damas, C., 2016. Les feux tricolores, où en est - on? TRANSFLASH N° 404 (MARS – AVRIL 2016), CEREMA 11–12, in French.

EuroRAP, 2015. Circle of influence: how France has used roundabouts to cut casualties. Retrieved January 5, 2018, from <http://www.eurorap.org/wp-content/uploads/2015/04/iRAP-case-study-5_France-roundabouts.pdf> (in French).

Fouladvand, et al., 2004. Characteristics of vehicular traffic flow at a roundabout. Phys. Rev. E - Stat. Physics, Plasmas, Fluids, Relat. Interdiscip. Top 70, 4. Available from: https://doi.org/10.1103/PhysRevE.70.046132.

Gabard, J.F., Henry, J.J., Abours, S., Lesort, J.B., 1987. CALIFE: on-line calculation of fixed time plans. Control in Transportation Systems. Issyles-Moulineaux, pp. 233–237.

Gertrude SAEM, 2018. Intelligent transport service. Retrieved January 8, 2018, from <http://www.gertrude.fr/nos-missions/systeme-de-transport-intelligent/>.

Guichet, B., 2005. Evolution of roundabouts in France and new uses. Transportation Research E-Circular 1–7. Retrieved from <http://trid.trb.org/view.aspx?id = 775433>.

Henry, J.J., Farges, J.L., Tuffal, J., 1984. The PRODYN real time traffic algorithm. In: Proceedings of the 4th IFAC-IFIP-IFORS Conference on Control in Transportation Systems, June, pp. 307–311. doi:10.1016/B978-0-08-029365-3.50048-1.

Henry, J.-J., Farges, J.-L., 1990. PRODYN. Control, Computers, Communications in Transportation. Elsevier, pp. 253–255.

INRIX, 2017. Los Angeles Tops INRIX Global Congestion Ranking. Releases, Press 1–14.

Le Cocq, J.P., 1973. Carrefours à feux. Sétra, France, pp. 1–117 (in French).

MEDDE, 2012. Instruction interministérielle sur la signalisation routière – 6ème partie – Feux de circulation permanents. VC20160215. Retrieved March 15, 2018, from http//www.equipementsdelaroute.equipement.gouv.fr/IMG/pdf/IISR_6ePARTIE_vc20120402_cle573dda.pdf, in French.

MEGE Paris, 2017. Histoire de la Signalisation Routière: Chronologie de l'histoire de la signalisation routière à Paris. <http://Www.Megedoudeau.Jmbertho.Odns.Fr/Index.Php/Eclairage-Public/Signalisation-Routiere/Histoire-de-La-Signalisation> (in French).

Olivier, C., 2002. Modèle conceptuel de données générique pour la représentation des réseaux intermodaux. Rapport Inrets, Projet STIP – VP (in French).

Scemama, G., Tendjaoui, M., 2013. CLAIRE-SITI Système pour la supervision intelligente, intermodale, intégrée des réseaux de transport de surface. TEC Transport Environnement Circulation, pp. 1–11 (in French).

Sétra, 1970. Methode de calcul des carrefours à feux. SETRA-D.C.E. (in French).

Sétra, 1998. Aménagement des carrefours interurbains. Guide Technique, pp. 1–130 (in French).

Sétra, 2017. Signal controllers in declared standard. Retrieved January 8, 2018, from <http://www.equipementsdelaroute.developpement-durable.gouv.fr/IMG/pdf/certificats_de_conformite_controleurs_de_feux_maj_2017-12-20_cle759573-1.pdf>.

Thai Van, M., Balmefrezol, P., 2000. Design of roundabouts in France. Transp. Res. Rec. 1737, 92–97.

The Economist, 2016. Roundabouts: French revolution. Retrieved January 1, 2018, from <https://www.economist.com/news/europe/21697235-invented-britain-modern-traffic-circles-invade-frances-hexagon-french-revolution>.

Switzerland

7

Thomas Riedel[1] *and Monica Menendez*[2]
[1]Verkehrs-Systeme AG, Winterthur, Switzerland, [2]Division of Engineering, New York University Abu Dhabi (NYUAD), Abu Dhabi, United Arab Emirates

7.1 Control strategy

7.1.1 Statistics and environment

Swiss cities and county areas are relatively small: 100−150 signalized intersections per city is a typical number, 500 intersections in a county is about the maximum. The average number of inhabitants per signalized intersection in cities ranges between 1000 and 2000. The percentage of total intersections that are coordinated for motorized traffic ranges between 10% and 65% within cities, and is less than 5% when considering whole counties.

Software for purely developing fixed signal plans is not used. Coordination algorithms are usually manually developed as no coordination optimization tools for the development of traffic-dependent intersections are known. Software systems in use (in descending order of popularity) are: VS-PLUS, FESA, TS2000, and a proprietary Modula-based programming environment owned by the largest city in Switzerland—Zurich.

Hardware is mainly provided by the two Swiss companies, VRAG and Bergauer AG, by the German company Siemens, and by the Austrian-German company Swarco. Hardware manufacturers usually provide their own software tools for signal control. Initial installations are done by the hardware companies including safety-relevant unchangeable programming (e.g., conflict matrix, intergreen matrix, minimum and maximum green and red times) with a warranty of 3−5 years. Changes can be made by the user itself (city road authority, engineering company) or need an intervention by the hardware company for reprogramming, depending on the software system used.

7.1.2 General remarks

Swiss intersections in larger cities tend to be complex, with more than four legs and many individually controlled movements, for cars and public transport, but also for bicycles and pedestrians. Public transport may share the car lanes, and/or have extra lanes in the middle of the street or at any other place. It may share a turning lane even though the public transport itself is not turning. Public transport stops may be

Global Practices on Road Traffic Signal Control. DOI: https://doi.org/10.1016/B978-0-12-815302-4.00007-8

nearside or farside (i.e., just before or just after the intersection). As a result of these complexities, there are easily several dozens of individually controlled movements to be handled. Additionally, intersection layout is often rather compact, as space is typically very limited in downtown areas.

As for the signal control algorithms, these vary across cities. In places like Zurich, one of the main objectives of signal control is to provide full priority to public transport—besides the legal requirement of increasing safety at the intersection (Stadelmann et al., 1995; Loder et al., 2018). Zurich has developed all traffic control in-house; other cities might rely on external services and software as described in Section 7.1.1.

Last but not least, we should note that many intersections in Switzerland, even in the large cities, are completely unsignalized (without signal control nor any type of sign), as long as safety considerations and the load balance permit it (Guler and Menendez, 2016; Menendez et al., 2015).

7.1.3 *Use of fixed-time signal control*

In Switzerland, most signalized intersections are traffic-actuated. Cases where intersection control is not actuated include:

- New intersections that are not yet equipped with detection means;
- Intersections where the detection means are switched off due to temporary construction;
- Intersections temporarily affected by roadworks;
- Intersections with too many conflicting public transport streams, where it is not feasible to provide priority to any of them with actuated control.

Oftentimes, the actuation of signal control is driven by public transport (Riedel and Brunner, 1994; Guler and Menendez, 2014), not necessarily by cars.

Fixed signal timing plans are often used as a foundation for developing traffic-actuated intersection control. Therefore, and given the limitations discussed above, the main focus of this chapter is not on the development of actually used fixed signal timing plans, but on those fixed-time signal plans used as a foundation for other types of intersection control in Switzerland.

As a matter of fact, on each traffic-actuated controller there must be at least one fixed-time signal plan, called a backup signal plan. This signal plan is used in case of hardware failures, in order to still guarantee safe traffic signal operations.

7.2 Intersection layout

7.2.1 *Single intersection*

In Switzerland, traffic lights are often switched to yellow blinking when the demands are low and the lights are not necessary for traffic safety. This is the case mostly at night. Therefore, all intersections are designed so that they remain safe for crossing even when the traffic lights are not in operation. As such, the following characteristics are typical for the layout of Swiss signalized intersections:

- For all vehicle and bicycle lanes, there is a right of way sign provided in case of nonoperating signals (yield, stop, priority).
- Crossing for pedestrians should always be safe. Pedestrian crossings that span more than three vehicle lanes without a splitter island are usually not accepted.
- A splitter island shall be present between the two driving directions in the case that one of the directions consists of two or more lanes.
- For a pedestrian crossing, both crossing directions shall be shown green (or red) at the same time.
- For safety reasons, some cities do not allow two different pedestrian phases (two-step crossing), one for each side of the splitter island.
- Some cities try to avoid stopped pedestrians on the splitter island due to space concerns.
- Two-step crossings for bicycles are rare.

In general, intersections should be compact in order to avoid long crossing times, especially for pedestrians. Ideally an intersection is controlled by three to four stages only. As such, it typically permits short cycle times, and therefore, short waiting times for pedestrians and bicycles.

7.2.2 Multiple intersections

Many cities in Switzerland (e.g., Zurich) may design green waves primarily for public transport and not for private vehicles (Dönier and Riedel, 2015; Loder et al., 2017). Such green waves are, additionally, more common when using short cycle times. Green wave design includes optimum public transport stop placement and optimum routing (e.g., minimizing left turns).

With a high number of public transport vehicles and short cycle times, there are typically enough public transport windows available to give public transport the necessary priority, even without actuation: there might be a maximum waiting time of around 30 seconds before the coordinated area. Afterwards, the vehicle is taken by the coordinated flow. In case of an unexpected longer dwell time at the bus stop, another waiting time of 30 seconds might be necessary in order to catch the next coordinated window.

When coordinating for private vehicles, the number of consecutive coordinated intersections is normally limited to a value between three and five. For a higher number of intersections, capacity might decrease too much due to platoon dispersion.

7.3 Signal heads and Swiss signal specifications

Traffic lights are normally placed at or above the stop line, and typically, they are vertically aligned.

7.3.1 Public transport signals

Public transport signals are under local legislation. Most contain four white dots. Sometimes, a fifth dot is used for priority-call feedback. Fig. 7.1A shows the lamp combination for "red." Fig. 7.1B–D shows the lamp combination for "green." They specify the green turning direction, that is, turning either left or right, turning only right, and turning only left, respectively. Fig. 7.1E shows "red" with an additional feedback lamp on the top that indicates whether a priority call has been registered. Sometimes this lamp is blue, and sometimes this lamp might blink corresponding to a "red-yellow" state, that is, giving some preparation time to the public transport vehicle.

7.3.2 Pedestrian signals

Pedestrian signals are also under local legislation. There exist signals with red and green only, or with green flashing before red. There exist also pedestrian signals with three colors and yellow before red.

7.3.3 Signals for nonpublic transport vehicles

Vehicle signals use red-yellow before green, and yellow before red to provide preparation time.

A *green circle signal* gives the right of way, but unprotected traffic may interrupt the flow under the following conditions:

- When turning left, vehicles or pedestrians parallel to the through movement can interrupt the flow.
- When turning right, pedestrians or bicycles parallel to the through movement can interrupt the flow.

  A *green arrow signal* gives full right of way. No unprotected traffic may interrupt the flow. The arrow may be on the signal head frame when the green signal does not permit an arrow picture, for example, for special bicycle signals or for public transport signals.

  A *green arrow signal with a blinking yellow circle signal* is equivalent to a full circle green signal with respect to the right of way. Additional blinkers may be placed at the conflict points.

  If the green circle signal is replaced by a *yellow blinking circle signal*, the right of way is limited to crossing the stop line. Any conflicting traffic may occur afterwards. Such lights are often used at roadworks.

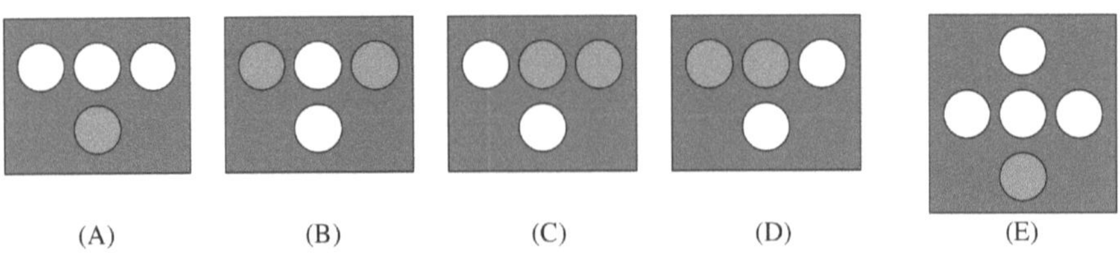

Figure 7.1 Possible signal combinations for public transport.

Figure 7.2 Examples of multiple traffic signals.

Fig. 7.2 shows a collection of signals, explained below from left to right:

- Public transport signal for straight traffic sharing a motorized vehicle right turn lane; there is also public transport turning right, but it has to respect the vehicle right turn signal. This is indicated by the black arrows on the signal head frames;
- Vehicle right turn signal with full right of way (green arrow signal and repeated arrow on signal head frame);
- Bicycle lane through signal;
- Bicycle lane right turn signal, with full right of way because the blinker is off. When the blinker is on, there is an unprotected pedestrian movement that is green as well.

7.4 Signal timing inputs

Once the intersection layout has been defined, the creation of an initial fixed-time signal plan is based on three important inputs:

- Traffic load on each lane;
- Intergreen times for conflicting movements;
- Cycle length.

7.4.1 Traffic load

If possible, the critical traffic load is measured by traffic count values every 15 minutes. If no count values are available, the peak hour values are calculated as a percentage of the daily traffic count values. The traffic load per lane is expressed in vehicle units. It is calculated as the weighted sum of vehicle counts from different vehicle classes (cars, trucks with and without trailers, bicycles).

For control design purposes, the 15-minutes values are usually summed up to 1-hour values. The peak hour values are used for the final design with some spare capacity in order to cover the peak 15 minutes as well. Usually, the peak hour counts should not exceed 85% of the capacity offered at an intersection in order to guarantee proper functioning.

7.4.2 Intergreen times

Intergreen times (also known as clearance times) define the minimum time between a signal switching to red and another signal switching to green. Intergreen times exist for all conflicting traffic relations and their corresponding signals. Therefore, they can be seen as part of a matrix, the intergreen matrix. Intergreen times are usually greater than or equal to 0 seconds. Negative intergreen times are not allowed, as they cannot guarantee safe operation of the intersection.

When timing the signals in Switzerland, the considered states are not "red" and "green," but "closed" and "open" states with the necessary constant transitions between them. "Open" may be represented as "green" or "yellow blinking." Transition times are constant for red-yellow before green (1 seconds) or depend on the permitted speed (yellow before red: 3 seconds until 50 km/h, 4 seconds for 60 km/h; 5 seconds for 70 km/h and above).

All yellow or green blinking time (for pedestrians) belongs to the intergreen times.

Intergreens are calculated according to a set of rules (RiLSA) based on geometry, assumed speed, acceleration and deceleration per vehicle class (or pedestrian), safety factors or offsets, and other local settings. Primarily intergreen times depend on the sojourn time of a car (or pedestrian, or bicycle) between the stop line and the geometric conflict point for entering and clearing traffic.

There are some basic rules across the whole country, and then the individual cantons (states) and cities add their own restrictions to those basic rules.

If the intergreen times are not known yet, they are assumed to be 5 seconds in all cases. For every intersection it is necessary to compute the critical intergreen times. These are the intergreens in the critical stage sequence path. The critical intergreen times are the longest intergreens in a stage transition. The critical stage sequence path is the stage sequence path with the longest stage transition times and the longest stage durations given by the assumed traffic load. They need to be inspected for plausibility even when the 5 seconds assumption is used. They also need special attention in case of a clearing distance of more than 25 m, with bicycles being the critical traffic type.

Evidently, the shorter the intergreens are, the higher the capacity of the intersection will be. Short intergreens are only possible with compact intersection design.

7.4.3 Cycle length

Cycle times in Switzerland tend to be very short, to easily accommodate pedestrians and bicycles. A cycle time between 60 and 90 seconds is generally recommended. Cycle times greater than 120 seconds are to be avoided, as they lead to long waiting times for nonmotorized modes, as well as long reds and potentially cyclic spillbacks (as blocks tend to be rather short in many Swiss cities).

Shorter than 60 seconds cycle times are also used, for example, 45 seconds in Zurich, but the intersection topology must be such that a maximum of three stages can be used and the intergreens should be short due to short trips over the shared areas. These very short cycle lengths are normally used in many Swiss cities to avoid long waiting times for pedestrians.

If no cycle time is specified, usually a 90-seconds cycle time is assumed.

7.4.4 Further restrictions

There may be offset restrictions imposed by county or city law, as pedestrian green offsets when green is in parallel to through and turning traffic: pedestrians need to get green before the traffic stream whose turning traffic is unprotected.

Also, minimum green and red times may be given by law, as well as the intergreen calculation method.

There may be flashers required at unprotected conflict points. There might be pedestrian greens required simultaneously to the parallel motorized traffic. There may be induced pedestrian phases by the arrival of a public transport vehicle.

7.5 Signal timing procedure

7.5.1 Optimum green time

The optimum green time per access partially depends on the discharge rate of vehicles from the intersection. Such a discharge rate is especially affected by the headway of the first cars that leave the intersection after being stopped at the signal. These initial headways are highly dependent on local driving behavior, and the yellow-before-green signal aims at reducing it. In Switzerland, saturation headways between cars range between 1.5 and 2.6 seconds. Local driving behavior in Zurich leads normally to two optimum green lengths. An optimum green length is defined as the time period after which the capacity lost by switching to another phase and back again is compensated by the gains in headways:

- When switching to another phase and back again, intergreen time losses occur and thus capacity is lost.
- Time headways in car platoons are usually short at the head of the platoon and become longer with a later position in the platoon. This decreases the lane capacity.

Optimum green lengths in Zurich are 6 seconds (with an equivalent of three cars) and 25 seconds (for 11 cars). Short green lengths are normally used for secondary approaches only.

7.5.2 Pedestrian phasing

The pedestrians are expected to cross with a mean speed of 1.2 m/s. The minimum green time is calculated assuming that a pedestrian starting to cross at the beginning of the green should be able to cover two-thirds of the distance to the opposite curb or the splitter island while the signal is green. During the remaining third of the distance the signal may be yellow or green blinking.

Some cities require pedestrians to cross within one single green phase without the need of a stop on any splitter islands, as long as the pedestrians start to cross right at the beginning of their green and keep a walking speed of at least 1.2 m/s.

While applying such a restriction, some Swiss cities allow shortening the pedestrian green duration when the intersection has a critical load and when a splitter island is present. In such cases, the minimum green time for pedestrian crossing can be shortened to the time needed to cross the wider side of the street, the splitter island, and half of the narrower side of the street.

7.5.3 Signal timing plan

When searching for the optimum signal timing plan, we need to find the optimum stages and stage sequence to handle the given traffic loads, while respecting the intergreen times and the chosen cycle length. The list below shows the steps followed in this process.

1. Identify the movement with the highest load. This might be a straight movement or a turning movement.
2. Add the nonconflicting movements with the highest loads in descending order, until no more nonconflicting movement is available. This is done to form the highest capacity stage.
3. Identify the next movement with the highest load that has not been considered in the previous stage(s).
4. Repeat steps 1–3 until all movements have been considered.
5. Once all movements have been considered, add (noncritical) pedestrian and bicycle movements to the created stages.
6. If necessary, check the use of inner offsets in case of multiple partial intersections, coordinations, mixed traffic movements, and substages.
7. Check whether all critical movements are in conflict with each other.
8. Determine the optimum stage sequence by minimizing the sum of intergreen times.
9. Look at the chain of the critical movements within the stages. They form a closed chain. The green time distribution on these movements is based on a saturation load of 1800 pcu/h, that is, 2 seconds headway.
10. Fill the remaining movements around the chain of critical movements in such a way that they do not violate any intergreen time. If required, respect offsets at the start and end of the green time. Such offsets may arise because, for example, unprotected pedestrian movements must always be given green at least 2 seconds before the parallel and turning conflicting car movements start.
11. In case of public transport movements on separate lanes, green windows need to be defined for these movements as well.

Minimum green times differ from city to city and from canton (state) to canton. Usually, cars have a minimum green time of 4 seconds, while pedestrians get the time required to cross two-thirds of the pedestrian crossing. Pedestrian minimum green times vary frequently by project.

It is common in Switzerland to develop for each intersection four programs, corresponding to four different time periods: morning peak, evening peak, off-peak, and night.

7.5.4 Further uses of fixed-time traffic signal plans

An optimum fixed-time signal plan can be used also for:

- Calculating the characteristic values for the intersection, for example, percentage of used green time or expected delay time;
- Setting up a microsimulation;
- Adding traffic actuation to the intersection based on the available detection means and the degrees of freedom allowed by the different possible movements.

7.6 Example

7.6.1 Layout remarks

The example intersection does not correspond to any Swiss intersection. The intersection size is unusually large and the layout does not meet the Swiss safety rules. Normally in Switzerland an access cannot include more than three lanes.

Pedestrian crossings with a length of more than 20 m are not allowed due to safety regulations, unless they include splitter islands. Long pedestrian crossings lead to long clearance times and hence long intergreen times. Long intergreen times decrease the intersection capacity. The Swiss safety regulations require that the intersection must also be safe for pedestrians when the traffic lights are switched off.

According to the current Swiss norms for pedestrian crossings a splitter island is needed on a signalized pedestrian crossing after a maximum number of three vehicle lanes. Splitter islands should be placed where the direction of driving reverses. The islands should be large enough to accommodate "stranded" pedestrians, that is, pedestrians who were not able to cross the whole pedestrian crossing at once. Usually there are pedestrian push buttons on the islands to call for another pedestrian phase.

Fig. 7.3 shows the modified intersection layout in order to comply with the Swiss rules, with one splitter island in the middle of the intersection and an additional splitter island in case of four access lanes.

7.6.2 Traffic load and bicycle safety

The current Swiss norms (SN 640023A) assume an ideal saturation load of 1800 pcu/h, that is, 2 seconds headway per vehicle. In dense city areas with frequent commuters (like Zurich and Basel), even 2000 pcu/h can be observed when cycle times are short. In the example we assume 2000 pcu/h for the straight movements and 1800 pcu/h for all other movements.

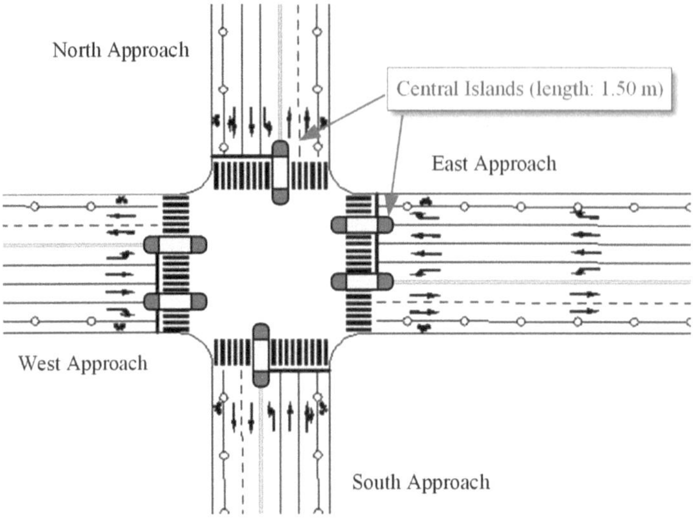

Figure 7.3 Example intersection layout adapted to the Swiss norms.

In the example, bicycles can only turn right or drive straight. Left turning is only possible indirectly, that is, by using the straight movement of the traffic coming from the right in a second step. Allowing bicycle left turns is common practice in Switzerland even though the city of Berne has decided in favor of indirect left turns for future intersection layouts.

The bicycle lane layout as shown in the west and east approaches is undesirable in Switzerland for safety reasons, as the right turn car movement is on a separate lane and is in conflict with the bicycle through movement. If the car movement has a separate lane, the lane will be shared by cars and bicycles. Hence, it should be located at the very right of the access lanes.

7.6.3 Numbering of movements

Usually, the movements for motorized traffic are numbered first in a clockwise direction. Then the public transport movements are numbered, followed by bicycle movements, and finally the pedestrian movements. The order of the movement types can vary between different cantons (states).

Fig. 7.4 shows the numbering and naming used for the example:

- "V" stands for motorized vehicles;
- "P" stands for pedestrians;
- "Bic" stands for bicycles;
- The numbering is independent from the preceding letter(s) and corresponds to the channel number of the signal group.

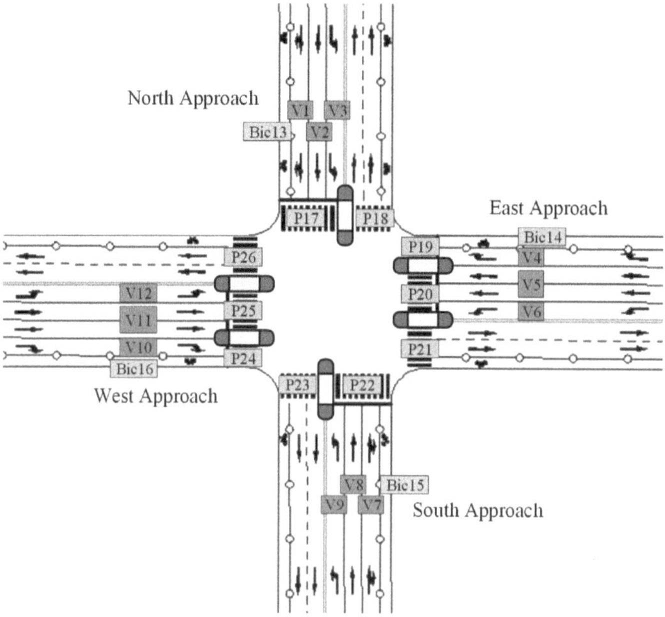

Figure 7.4 Example intersection numbering scheme.

7.6.4 Basic assumptions for the signal timing plan

The following three input values are used to calculate the signal timing plan:

- Critical traffic load per lane, given in the example by pcu/h.
  In case of two parallel lanes, the right lane takes 60% of the traffic and the left lane 40%. A traffic allocation of 50%/50% can be useful for first feasibility studies, but it can be seen from field studies that both lanes are evenly used only when the system is under complete saturation, that is, at its capacity limit. The allocation of 60%/40% is far more frequently observed. Often, the distribution depends also on the signalization regarding lane directionality, that is, if it is already indicated that the lanes will go to different directions on a downstream intersection.
- Intergreen time calculation or estimation.
  When the intergreen times are not known, a general 5 seconds intergreen time is used wherever a conflict between two movements is identified. As the pedestrian crossings in the example are quite long, the intergreen times for pedestrians have been verified.
- Cycle time.
  We assume a cycle time of 90 seconds, which is rather long for Switzerland, but not so long at a global level.

7.6.5 Conflict matrix

Before developing the stages and the stage sequence, the conflict matrix needs to be determined. Fig. 7.5 shows the conflict matrix. An "X" stands for a conflict. A conflict is a logical conflict, that is, a physical conflict between a right turn

c.SG	e. SG	1	2	3	4	5	6	7	8	9	10	11	12	13	14	15	16	17	18	19	20	21	22	23	24	25	26
		V1	V2	V3	V4	V5	V6	V7	V8	V9	V10	V11	V12	Bic13	Bic14	Bic15	Bic16	P17	P18	P19	P20	P21	P22	P23	P24	P25	P26
1	V1					X	X			X	X	X	X	X	X		X	X						X			
2	V2					X	X			X	X	X	X		X		X	X						X			
3	V3					X	X	X	X			X	X		X	X	X	X				X					
4	V4							X	X				X		X	X			X	X							
5	V5	X	X	X				X	X	X			X	X		X					X						X
6	V6	X	X	X				X	X	X	X	X		X		X	X				X			X			
7	V7			X	X	X	X					X	X		X		X		X				X				
8	V8			X	X	X	X					X	X		X		X		X				X				
9	V9	X	X			X	X					X	X	X	X		X						X				X
10	V10	X	X				X							X			X							X	X		
11	V11	X	X	X			X	X	X	X				X		X						X				X	
12	V12	X	X	X	X	X		X	X	X				X	X	X			X							X	
13	Bic13	X				X	X			X	X	X	X					X						X			
14	Bic14	X	X	X	X			X	X	X			X							X							X
15	Bic15			X	X	X	X					X	X						X				X				
16	Bic16	X	X	X			X	X	X	X	X											X			X		
17	P17	X	X	X										X													
18	P18				X			X	X				X			X											
19	P19				X										X												
20	P20					X	X																				
21	P21			X								X					X										
22	P22							X	X	X						X											
23	P23	X	X				X				X			X													
24	P24										X						X										
25	P25											X	X														
26	P26					X				X					X												

c.SG = clearing signal group, e. SG = entering signal group

Figure 7.5 Example intersection conflict matrix.

movement and a parallel bicycle or pedestrian movement can be overruled by not setting an X. Such overruling creates the so-called conditional conflicts.

Overruling a physical conflict is only allowed in Switzerland for movements with up to 150 pcu/h.

For each conflict there must be an intergreen time. All intergreens together form the intergreen matrix. For the signal group numbering and names see Fig. 7.4.

7.6.6 Stages and stage sequence

In the following, we compute the states and their sequence based on the steps described in Section 7.5.3.

1. Identify the motorized traffic movement with the highest load. This is the movement for signal group V5 in the example.
2. Add the nonconflicting movements with the highest loads in descending order, until no more nonconflicting movement is available. Create the highest capacity stage.
   This leads to stage A.
3. Identify the next movement with the highest load that has not been considered in the previous stage. This is the movement for signal group V7.
   This leads to stage B. Thus stages A and B handle the straight (i.e., through) traffic.
4. Repeat steps 1–3 until all movements have been considered.
   This leads to stages C and D which handle separate left and right turns.
5. The (noncritical) pedestrian and bicycle movements have already been added to the created stages.
   In Fig. 7.6 the critical movements are marked with a circle. These movements can be used in further steps to calculate the intersection efficiency and other characteristics, such as the 95% likelihood spillback lengths, and the average expected waiting times.
6. If necessary, check the use of inner offsets in case of multiple partial intersections, coordinations, mixed traffic movements, substages.
7. Check whether all critical movements are in conflict with each other.

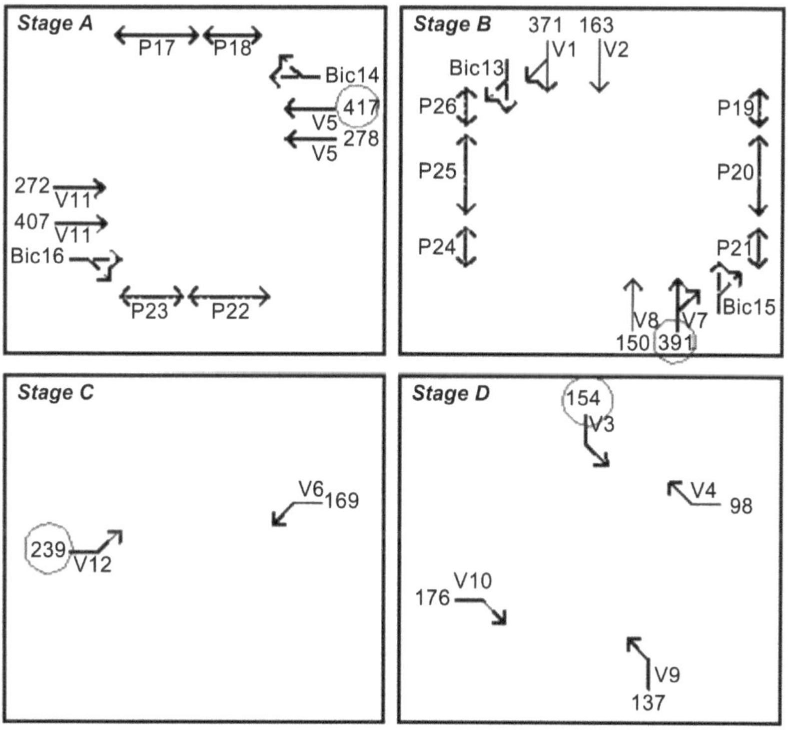

Figure 7.6 Example intersection stages.

Inner offsets are not applicable in the present example. For point 7 it shall be noted that V10 of stage D has the highest load but is not in conflict with the critical movement V12 of stage C. Therefore, it could already be integrated into a partial stage C without movement V6. Thus, V10 is not the critical movement of stage D, but V3 (the next smaller in traffic load) is.

Fig. 7.6 shows the resulting stages. For the signal group numbering and names see Fig. 7.4.

7.6.7 Signal timing plan

Based on a cycle length of 90 seconds, the given traffic loads per movement, and the calculated conflicts and intergreen times, the signal timing plan can be calculated by using stages A–D in alphabetic order. The maximum speed is assumed to be 50 km/h.

Fig. 7.7 shows the compute signal timing plan.

- TOB stands for t open (green) "begin", TOE for "end", TOD for "duration".
- RES stands for "reserve until 100% load."
- C stands for capacity in pcu/h.
- X stands for expected used green time.
- For the signal group numbering and names see Fig. 7.4.

The signal timing plan has the following particularities:

- Two consecutive splitter islands are unusual in Switzerland but legal.
- The pedestrian signal groups use an offset of 0 seconds for begin and end of green; therefore they show green at the same time for the whole pedestrian crossing.
- The pedestrian crossings on the west and east approaches got an offset of 1 seconds, that is, an early start of 1 seconds with reference to the parallel motorized traffic (on the north and south approaches). This is for safety reasons to make sure that the pedestrians are already on the pedestrian crossing when the motorized traffic gets green. Some cities declare such offsets as compulsory.
- The turning movements that are in conflict with the parallel pedestrian and bicycle movements show enough space to cope with stopping traffic due to pedestrian and bicycle priority. If this was not the case, the maximum flow for these movements would have to be decreased.

7.6.8 Next steps

Based on the signal timing plan, traffic actuation can also be implemented. To do so, detectors need to be placed and their roles need to be defined, for example, calling green, extending green (dilemma zone), or shortening green (outflow spillback). By adding traffic actuation, green times can be optimized, and hence waiting times for all types of traffic, including pedestrians, can be decreased.

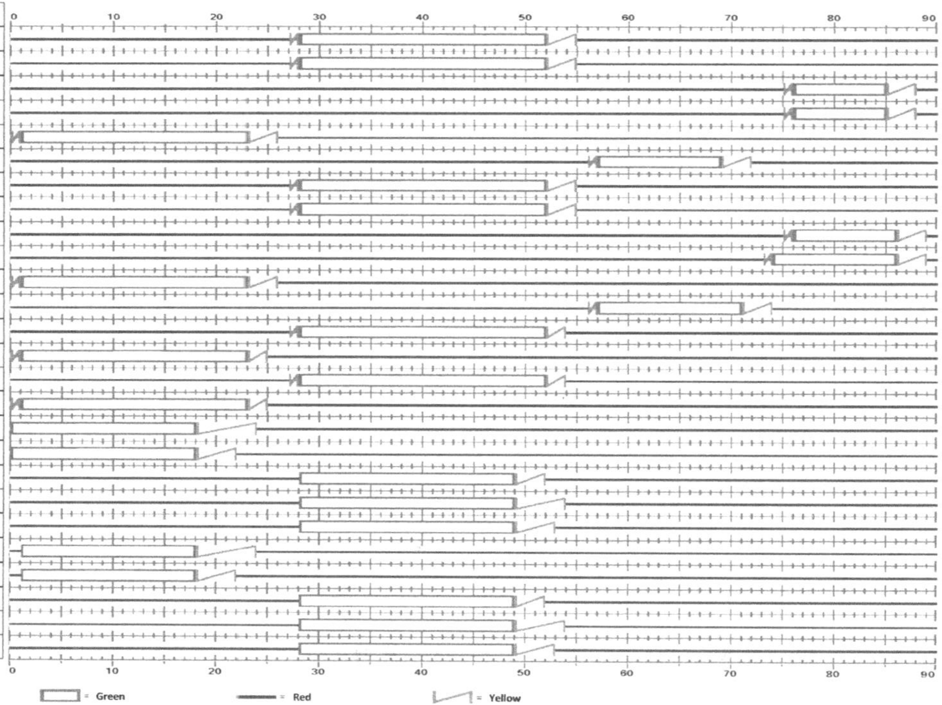

Signal group	TOB1	TOE1	TOD1	RES	C	x
V1	28	52	24	5	480	0.77
V2	28	52	24	17	533	0.31
V3	76	85	9	1	180	0.86
V4	76	85	9	4	180	0.54
V5	1	23	22	3	978	0.85
V6	57	69	12	4	240	0.70
V7	28	52	24	4	480	0.81
V8	28	52	24	17	533	0.28
V9	76	86	10	3	200	0.68
V10	74	86	12	3	240	0.73
V11	1	23	22	4	978	0.83
V12	57	71	14	2	280	0.85
Bic13	28	52	24	20	0	0.00
Bic14	1	23	22	18	0	0.00
Bic15	28	52	24	20	0	0.00
Bic16	1	23	22	18	0	0.00
P17	90	18	18	1	0	0.00
P18	90	18	18	1	0	0.00
P19	28	49	21	0	0	0.00
P20	28	49	21	0	0	0.00
P21	28	49	21	0	0	0.00
P22	1	18	17	0	0	0.00
P23	1	18	17	0	0	0.00
P24	28	49	21	0	0	0.00
P25	28	49	21	0	0	0.00
P26	28	49	21	0	0	0.00

Figure 7.7 Example intersection signal timing plan.

Acknowledgments

The authors wish to thank the engineering office Rudolf Keller & Partner in Muttenz, Switzerland, and particularly André Zerbin for sharing his experience with us.

Bibliography

Dönier, G., Riedel, Th, 2015. Regelungsgrundsätze in der Stadt Zürich; Entwicklung eines Leitfadens (Traffic Control Principles in the City of Zurich; Development of a Guideline). DAV Department, City of Zurich (in German with partial English translation).

FGSV, 2015. Richtlinien für Lichtsignalanlagen – RiLSA (German Standard for Traffic Signals). FGSV-Verlag, Cologne, Germany, Edition 2015 (in German with partial English translation).

FGSV, RiLSA; Richtlinien für Lichtsignalanlagen (Guidelines for Traffic Signals), German Road and Transportation Research Association, Steering Committee Traffic Control and Traffic Safety, Edition 1992 (with minor modifications), Translation 2003, Forschungsgesellschaft für Straßen-und Verkehrswesen e.V., Köln.

Guler, I., Menendez, M., 2014. Analytical formulation and empirical evaluation of pre-signals for bus priority. Transp. Res. B: Methodol. 64 (2014), 41–53.

Guler, I., Menendez, M., 2016. Methodology for estimating capacity and vehicle delays at unsignalized multimodal intersections. Int. J. Transp. Sci. Technol. 5 (4), 257–267.

Loder, A., Ambühl, L., Menendez, M., Axhausen, K.W., 2017. Empirics of multi-modal traffic networks: using the 3D macroscopic fundamental diagram. Transp. Res. C: Emerg. Technol. 82 (2017), 88–101.

Loder, A., Ambühl, L., Menendez, M., Axhausen, K.W., 2018. A Case Study of Zurich's Two-Layered Perimeter Control. Transportation Research Arena (TRA), Vienna, April 2018.

Menendez, M., Guler, I., Puffe, E., 2015. Traffic flow at uncontrolled urban intersections with attention to different modes of traffic; determination of representative standard values and interrelations. Technical Report VSS 2011/308, Prepared for the Research Commission for Road Transport in Switzerland (VSS), November 2015.

Open Traffic Systems City Association e.V. OCA: statistical data about Switzerland and Germany, see <https://www.oca-ev.info/> (in German).

Riedel, Th, Brunner, U., 1994. Traffic control using graph theory. Control Engineering Practice. Pergamon Press, UK, pp. 397–404, 2(3).

Swiss norms (SN). Research Commission for Road Transport in Switzerland (VSS): Swiss Norms (in German).

Swiss norm (SN). 640838, 1992, 2018. Lichtsignalanlagen; Zwischenzeiten (Traffic lights; intergreen times).

Swiss norm (SN). 640837, 1992, 2015. Lichtsignalanlagen; Übergangszeiten und Mindestzeiten (Traffic lights; transition times and minimum times).

Swiss norm (SN). 640023A, 2008. Leistungsfähigkeit, Verkehrsqualität, Belastbarkeit; Knoten mit Lichtsignalanlagen (Performance, traffic quality, capacity; intersections with traffic lights).

Swiss norm (SN). 640832, 1992 and 12368-E, 2017. Lichtsignalanlagen; Kopfnorm (Traffic control equipment – signal heads).

Swiss norm (SN). 640835, 1995. Lichtsignalanlagen; Abschätzen der Leistungsfähigkeit (Traffic signals; performance estimation).
Swiss norm (SN). 640840, 1971. Lichtsignalanlagen; Koordination in Strassenzügen mit der Methode der Teilpunktreserven (Traffic signals; Coordination by the method of partial point reserve).
Stadelmann, H., Riedel, Th, Vollenweider, M., 1995. Introduction to the Traffic Control Concept of the City of Zurich. DAV Department, City of Zurich (in German, English translation available).

Turkey

8

Yetis Sazi Murat[1] *and Fatih Gundogan*[2]
[1]Faculty of Engineering, Civil Engineering Department, Pamukkale University, Denizli, Turkey, [2]ISBAK Inc., Istanbul, Turkey

8.1 Background

8.1.1 History

The establishment of traffic signal facilities around the world coincides with the end of the 1800s, long before motorized vehicles were on the roads. In Turkey, despite the uncertainty, the first traffic signal installation was in 1929, according to newspaper reports. The first traffic signal, formerly known as an "automatic beacon," was implemented in Istanbul. The Municipality of Istanbul officially declared this first practice in an edition of the Cumhuriyet newspaper (Cumhuriyet, 1929). It was then used in other large cities such as Ankara, Izmir, and Bursa. Similar to other countries, intersection control has been managed by traffic police due to an increase in traffic accidents (Sezen and Apaydın, 2012).

8.1.2 Guidelines and handbooks

Although some national handbooks, guidelines, and standards do exist in Turkey, traffic signal design and programming mostly follow international references. The historical background and influences from both European countries and the United States have created a great deal of confusion in terminology and implementation. Additionally, due to the lack of generally accepted guidelines, different cities apply different methods for traffic signal control and design. The following are some guidelines and handbooks used in Turkey (TS EN 12368, 2015; EN 11937):

Highway Design Handbook: Published by Turkish Highway Directorate and includes design principles of signalized and nonsignalized intersections.

Road Traffic Regulation: General rules and regulations are described and documented in this national guide (The Official Gazette, 1983).

TS 11937: Standard that describes the design principles of signalized intersections. Content includes definitions such as location of signaling units, mast height, width of pedestrian crossings, and distance of the stop line.

TS 7635: Standard that describes the design principles of pedestrian crossings to be applied in urban roads. Includes guidelines for walkway types to be selected depending on vehicle and pedestrian demand.

TS 7769: Standard for defining intersection-related terms.

Global Practices on Road Traffic Signal Control. DOI: https://doi.org/10.1016/B978-0-12-815302-4.00008-X

Apart from these sources, other international handbooks and guidelines, such as Highway Capacity Manual (HCM) (TRB, 2000), German Guideline for Traffic Signal (RILSA) (FGSV, 2010), and Manual on Uniform Traffic Control Devices (MUTCD) (FHWA, 2009), are also used in traffic signal design.

8.1.3 Number of signalized intersections

In general, there are two bodies providing the management and maintenance of signalized intersections in Turkey:

- Municipalities;
- General Directorate of Highways.

The municipalities provide installation, maintenance, and operation of signalized intersections within the provincial borders. The General Directorate of Highways provides installation, maintenance, and operation of signalized intersections located on intercity roads. In some cases, although the responsibility of intercity roads belongs to the General Directorate of Highways, the responsibility of signals can be transferred to municipalities with a protocol.

Some surveys about signalized intersections were conducted by ISBAK (ISBAK, 2017). Table 8.1 shows the types of control systems and the number of intersections per 1000 people in Turkey. In the last column of the table, I represents the number of intersections in the city and P represents the population of the city (per 1000 people).

In 2017, the total number of signalized intersections in 28 cities was 7089 and the total population was 54,430,904. The number of intersections per 1000 people was 0.129. The proportion of each signal control types is given here (ISBAK, 2017):

- Fixed time: 62%;
- Semiactuated: 26%;
- Fully actuated: 7%;
- Adaptive: 4%;
- Flash: 1%.

As can be seen, the majority of the signals are under fixed-time mode. The General Directorate of Highways, which is mostly responsible for rural roads, operates about 2094 traffic signals, of which only 30 are under semiactuated mode.

Signal controllers are one of the significant components in the market. Controllers produced by several signal controller manufacturers are being used, and the rough market share is indicated here (ISBAK, 2017):

- ISBAK: 69%;
- Teknotel: 12%;
- Siemens: 15%;
- Rayennur: 4%;
- ISSD: 0.1%;
- Tankes:0.6%.

Table 8.1 Some statistics about the number of signalized intersections (ISBAK, 2017)

City	Population	Number of intersections	Fixed time	Semiactuated	Fully actuated	Adaptive	I/P
İstanbul	14,804,116	2209	40	1695	400	74	0.149
Ankara	5,346,518	856	817	39	–	–	0.160
İzmir	4,223,545	503	253	150	–	100	0.119
Bursa	2,901,396	358	46	150	162	–	0123
K,Maraş	1,112,634	124	121	2	–	1	0.111
G,Antep	1,974,244	268	225	43	–	–	0.136
Çorum	288,578	37	28	–	–	9	0.128
Muğla	923,773	91	76	15	–	–	0.099
Kastamonu	146,103	19	12	7	–	–	0.130
Sakarya	976,948	120					0.123
Kocaeli	1,830,772	213	16	86	103	–	0.116
Elazığ	431,294	32	32	–	–	–	0.074
Mersin	1,773,852	66	62	4	–	–	0.037
Sivas	365,135	39	39	–	–	–	0.107
Edirne	173,037	6	6	–	–	–	0.035
Van	1,100,190	27	27	–	–	–	0.025
Konya	2,161,303	388	320	–	–	68	0.180
Şanlıurfa	1,940,627	63	–	63	–	–	0.032
Rize	145,739	9	9	–	–	–	0.062
Tekirdağ	972,875	90	37	5	–	48	0.093
Antalya	2,328,555	476	401	70	–	5	0.204
Adana	2,201,670	315	–	–	–	–	0.143
Kayseri	1,358,980	205	155	50	–	–	0.093
Hatay	1,555,165	172	172	–	–	–	0.111
Eskişehir	844,842	136	110	9	–	17	0.161
Denizli	1,005,687	125	110	15		–	0.124
Erzurum	762,021	45	45	–	–	–	0.059
Malatya	781,305	47	34	13	–	–	0.060

Figure 8.1 Sample signal units from the city of Ankara.

8.1.4 Signal units and signal display sequence

For signal units, the Turkish standard (i.e., prepared by the Turkish National Standard Institute, such as TS 11937) should be used if available, otherwise, European standards (i.e., approved by the Council of European Standards) are preferred. In general, round glass signal lamps should be used in diameters of 200 or 300 mm. One hundred millimeter signal units are found in some cities but primarily for bicycles (https://asyatraffic.com). In 2007, Istanbul started using 5 mm LED lamps. Since then, very few cities use halogen and/or glow filament signal units.

Arrow signal heads are used for left or right turns according to the traffic volumes (Fig. 8.1). Auxiliary (or exclusive) lanes are usually provided at the intersection for left- or right-turning traffic. Arrow signals are also used for cases where auxiliary lanes are not installed for turning flows. In the city of Ankara, arrow signal heads are also used for through (straight) flows (http://www.ankara.bel.tr/en). Fig. 8.1 shows the signal units used in the city of Ankara.

8.2 Control strategies selection

8.2.1 Signal system installation principles

In Turkey, there is a lack of guidelines for signal system installations. Intersections near schools and hospitals are mostly signalized due to safety concerns. Also, the nine signal warrants in the MUTCD (FHWA, 2009) are accepted. The following steps are followed for signal installations within Istanbul Metropolitan Municipality's responsibility areas.

1. The planned location for signal system installation will be evaluated and decided upon by the Transportation and Traffic Regulation Authority (TTRA) of the city.

2. The TTRA decision is forwarded to the Road Maintenance Department if a geometrical change is needed at the intersection. If the location is geometrically ready, the decision is forwarded to the Traffic Directorate.
3. The TTRA decision location is examined and evaluated before the implementation phase starts, and if there is no obstacle to the establishment of the system, the necessary institutions are informed before production and permissions are obtained and a license is issued.
4. It is implemented by the Traffic Directorate in accordance with the intersection infrastructure and the superstructure project.
5. Following the completion of the manufacturing phase, the signaling system is activated by the traffic engineer and signal programming officer.
6. Apart from the operating mode (semiactuated, fully actuated, fixed time), applications such as green-wave coordination and public transportation priority are also evaluated in the decision phase.

8.2.2 Selection of operating modes

The operation mode of signalized intersections is predominantly fixed time in Turkey. In particular, the intersections under the responsibility of the General Directorate of Highways are established on a fixed-time basis. In the municipalities other than Istanbul, the fixed-time mode is the preferred operation method. In cities such as Sakarya, Kocaeli, and Konya, due to the increasing population density and the existence of light rail systems, semiactuated systems and public transportation priority systems have been gaining popularity in recent years. Traffic volume is the most important factor when establishing an intersection with semiactuated or fully actuated control. It is essential to determine the presence/absence of vehicles on both side streets and main streets to minimize waiting times (Murat and Gedizlioglu, 2002).

If coordinated signal operation mode is selected, the distance between the intersections and the average speed in the arterial are essential. If it is desired to reduce the vehicle delays on an arterial, the intersections on the arterial should be coordinated.

8.3 Signal timing procedure

8.3.1 General timing design process

Before designing signal timing for isolated intersections, geometry and traffic flow data need to be gathered. For this purpose, the following data are obtained from field studies:

- Intersection geometry (number of approaches, lanes, etc.);
- Six-hour vehicle traffic volume data (including 2 h each for the morning and evening peak periods and the off-peak period);
- Pedestrian volume;
- Special turning lanes;
- Bus stop locations;
- Roadside parking;
- Crash statistics.

The data listed earlier are significant for a reliable signal timing design (Gedizlioglu, 1994). Next, brief explanations of these data and their relations to signal timing design are summarized.

Intersection geometry is one of the important components. The number of approaches, number of lanes on each approach, and the lane width must be specified. These parameters must be carefully evaluated before making any changes. Any necessary changes in geometry need to be planned. After these evaluations are done, signal timing design can start.

It is a known fact that traffic volumes are a major input for achieving an accurate and reliable signal timing design. Traffic volume fluctuations often require different signal timing plans. Thus, 6 h of traffic volume data, including traffic composition (i.e., heavy vehicles, pedestrians, and bicycles) should be collected. Using Istanbul as an example, about 500 intersection traffic volume counts are conducted annually. The traffic department also uses traffic analysis and simulation tools for geometric and signal timing design.

Besides traffic volumes, pedestrian volumes are also taken into consideration in signal timing design. Specifically, pedestrian volumes will determine what phase plan is needed. If an intersection has high pedestrian volumes, a fully protected phase plan is generally selected. This means that all of the pedestrians at the intersection have a special phase. Otherwise, the phase plan for pedestrians may be different.

Special turning lanes provide many advantages in signal timing design, and they should be considered if the geometry of an intersection is sufficient and the turning volumes need such turning lanes. However, due to settlements around intersections in most urban areas, there is generally not enough space for providing the turning lanes. Narrow streets are one of the major challenges in city centers in Turkey. Thus, turning lanes are mostly used at rural intersections or in newly settled urban areas. The existence and location of public transportation facilities affects signal timing design, mainly due to potential blockages to general vehicles if no turn bay is provided for buses.

The roadside parking has a similar blockage effect on traffic flows. Therefore, it is also taken into consideration in signal timing design. Parking hours, locations, and other information need to be collected.

In addition to these parameters, the crash statistics for the intersection are another factor to consider in the signal design process. Crash frequency and type need to be investigated before the design starts. Fig. 8.2 illustrates the signal timing design process by considering various factors.

8.3.2 Calculation process of a signal timing plan

In Turkey, Webster's critical movement method is used in signal timing design, as depicted in Fig. 8.2. The signal phasing and phase plan are determined in the first step. The signal phase for pedestrians and its required timing are decided in the second stage. Pedestrian signal timing is determined based on the intersection geometry and pedestrian walking speeds. A protected (separate) phase is generally assigned for all pedestrian flows at an intersection. The traffic volumes and saturation flow values are determined after the phase plan.

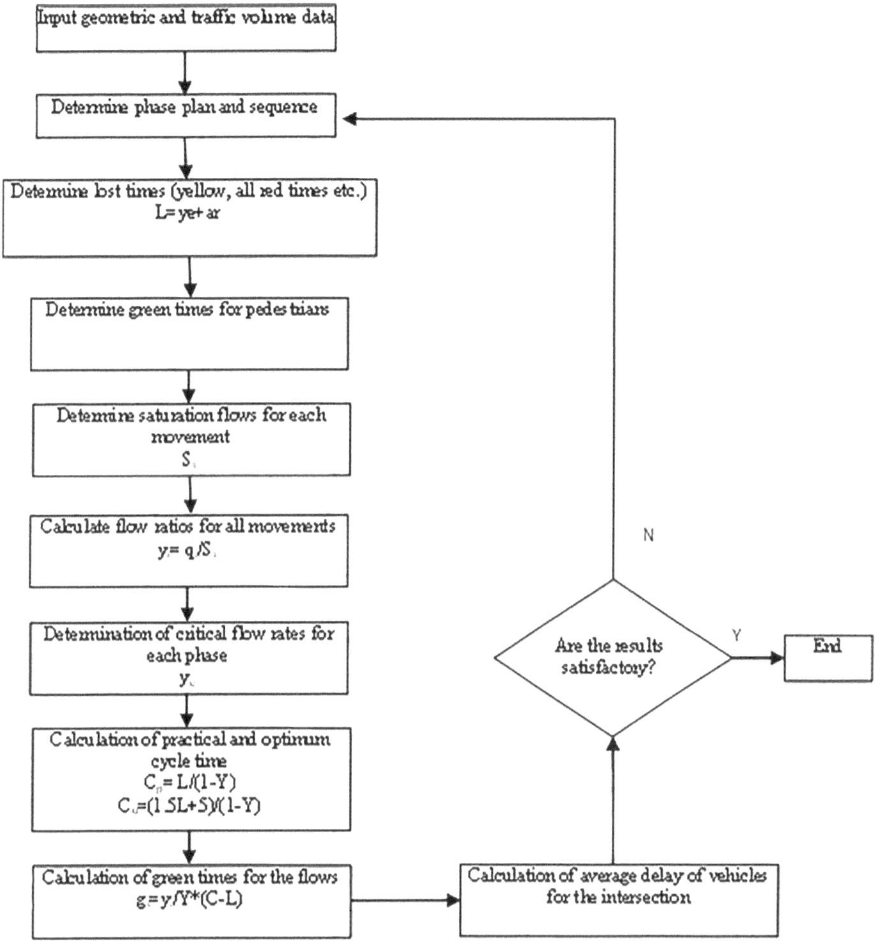

Figure 8.2 Flow chart of the signal timing design procedure in Turkey (Murat, 1996).

The flow rates (ratios) are calculated for traffic flows and phases. The biggest flow ratios for each phase are determined and assigned as the critical ratio for the phase. The critical ratio for intersection (*Y*) is then calculated by summing all of the critical ratios. Using this value and the lost time for the intersection (*L*), the signal cycle time can be calculated by Webster's formula. The green time of each phase is calculated using the flow ratio and cycle time values.

8.4 Signal phasing

Signal phase plans are determined based on both traffic and pedestrian flows at an intersection. Although it varies by intersection geometry and traffic conditions, three-phase control schemes are the most common, followed by four-phase control schemes (Ayfer, 1977). Some typical phasing designs are illustrated in Fig. 8.3 and more discussions on these phasing applications follow.

(A)
(B)
(C)

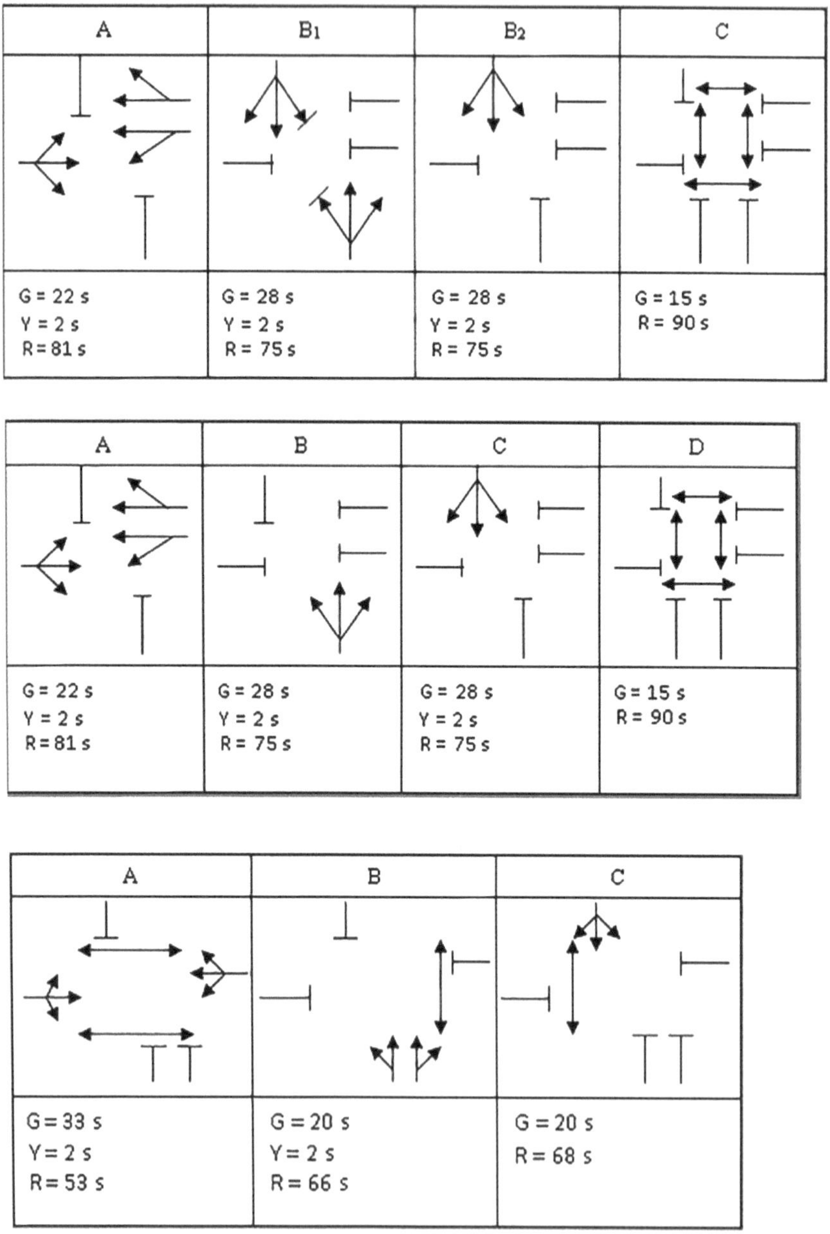

Figure 8.3 (A) Early cutoff (late opening) application; (B) protected pedestrian phase plan; (C) staged (permitted) pedestrian phase plan.

8.4.1 Signal phasing for left-turning and right-turning flows

If intersection geometry has sufficient space for providing a separate right-turn lane, continuous turning to the right is allowed. The right-turn phase is canceled in order to only allow pedestrian crossing.

For left turns, a separate phase is given if the left-turning volume in each phase is greater than two vehicles per cycle. If it is less than this value, it is preferable to apply a filter phase or early cutoff (late opening), as shown in Fig. 8.3A. As can be seen in Fig. 8.3A, traffic flow from the south direction is cutoff early in the B2 phase. The application of early cutoff or late opening is taken into account in the phase duration calculation, and the critical movements are determined accordingly.

8.4.2 Phase plans for pedestrian and bikes

The pedestrian phase is either protected (i.e., a separate phase) or staged (permitted) as shown in Fig. 8.3B and C. The treatment for bicycles is similar to pedestrians in signal phasing design (Ozdirim, 1993).

8.5 Parameter determination

8.5.1 Saturation flow rate

Saturation flow rate is used extensively in Turkey for signal timing calculation. Saturation flow rate is generally determined by field studies. However, in terms of practical solutions, it is suggested that the basic saturation flow rate value needs to be corrected according to some parameters. The basic saturation flow rate is considered to be 1800 vehicles/h per lane under ideal conditions. This base value is used in signal timing calculations and adjusted based on other parameters such as traffic composition and turning traffic volumes. Although studies on saturation flow rate have been carried out in Turkey, no generalized values have yet been obtained that can be used throughout the country. Table 8.2 lists some saturation flow rate values from various studies. Studies have found that the saturation flow rate in the central business district is dramatically reduced due to violations of minibus drivers, parking maneuvers, and jaywalkers (ISBAK, 2007; Gundogan, 2012).

Table 8.2 Sample saturation flow rates from Turkey

City	Saturation flow rate (vehicles/h)	References
Izmir	1720	Calışkanelli and Tanyel (2018)
Denizli, Izmir	1520	Cetin and Murat (2013)
Tokat, Yalova, Karabuk	1546–1827	Polat et al. (2015)
Istanbul	1827	Gundogan (2012)

8.5.2 Capacity calculations

The following formula is used for capacity calculation of signalized intersections.

$$C_{cap} = S\frac{g}{C} \tag{8.1}$$

where C_{cap} = capacity of the lane (vehicles/h); S = saturation flow (vehicles/h); g = green time (s); C = cycle time (s).

As seen in this formula, the capacity, C_{cap}, is directly related to the saturation flow rate, S value. Therefore, it is important to obtain accurate saturation flow values from the field.

8.5.3 Cycle length

Webster's formula (Webster, 1958) is used for determining the cycle length, with the minimum and optimal cycle lengths obtained from the following formulas:

$$C_{min} = \frac{L}{1 - Y} \tag{8.2}$$

$$C_{opt} = \frac{1.5L + 5}{1 - Y} \tag{8.3}$$

As can be seen from these formulas, the intersection lost time (L) and the total flow ratio (Y) values are needed for the cycle time calculations. Lost time is regarded as the time that cannot be fully used to serve the traffic. After the phase plan is determined, the lost times are determined. This is an approach widely accepted in Turkey. Additionally, the methods from HCM, RILSA, and Akçelik and related software are also used.

In Turkey, the minimum cycle time is 40 s, and the maximum cycle ranges between 120 s and 140 s. Cycle times of up to 250 s can be used under some extreme situations (Kutlu, 1993).

8.5.4 Green times

After the cycle length is determined, the green times of each phase are calculated. The following formula is used for calculating the green times.

$$g_i = \frac{y_i}{\sum y_i}(C - L) \tag{8.4}$$

where g_i = green time for phase i (s); y_i = flow rate (vehicles/s) for phase i; C = cycle time (s); L = total lost time for the intersection (s).

8.5.5 Pedestrian times

The calculation of pedestrian phase times are based on road width and pedestrian speed, as shown in Eq. (8.5). Using these parameters and the following relation, the signal time for the pedestrian is calculated.

$$G_p = \frac{R_w}{S_p} \tag{8.5}$$

where G_p = green time for pedestrians (s); R_w = width (length) of the road (m); S_p = average speed of pedestrians (m/s).

However, it has been observed in practice that constant pedestrian phase times ranging between 10 s and 15 s are also used.

8.5.6 Intergreen times

The intergreen times, composed of yellow and all-red intervals, are a signal design parameter for traffic safety considerations. The minimum yellow time in Turkey is 2 s. Higher values are used depending on the intersection geometry and traffic conditions. In Istanbul, the yellow interval is set as 4 s for major intersections (mostly with red light running monitoring) and 3 s for all other intersections.

The all-red time is a design parameter mainly for traffic safety. Its main purpose is to clear the vehicles out of the intersection before the conflicting signal phases start. According to Ayfer (1977), all-red time is lost time. The all-red time applied is usually 2 s. However, considering the high traffic volumes in the morning and evening peak hours, different all-red intervals are used and can vary between 3 and 8 s.

8.5.7 Effective green times and lost times

The effective green time is obtained by subtracting the total lost times from the calculated green times. It can vary by the number of phases and the calculated green times.

Lost time is the portion of the cycle that cannot be used to serve the traffic. Lost time occurs at start-up and during yellow and all-red intervals. Not the entire yellow is considered lost time, but the entire all-red is considered lost time.

8.6 Special considerations

This section discusses some custom applications and placement of poles and units. Custom applications include leading pedestrian interval, special pedestrian signals for right-turning movements, dual signal light, flashing right turn, approach flash, and signalized roundabouts. Placement of poles discusses the location of poles and adjustment of units in the field.

Figure 8.4 Sample application of leading pedestrian interval at right turn.

8.6.1 Custom applications

8.6.1.1 Leading pedestrian interval at right turn

This rule is applied in Istanbul with a reference to the article "Road Traffic Act." The purpose of implementing this rule is to determine the optimal phase number and to reduce the lost times and delays at the intersections without creating an extra phase for the pedestrian currently crossing with the right turn. Fig. 8.4 shows a sample with a leading pedestrian interval at right turn.

In the application, if there is no sign for right-turning vehicles, they are obliged to give the right-of-way to the crossing pedestrian(s) or bicycle(s) on the bicycle path, if any.

8.6.1.2 Special pedestrian signals for right-turning movements

This application was implemented according to the MUTCD. It is applied generally at the intersections where pedestrian traffic is high. In Ankara, the rules are as follows:

- If the red light is on, the vehicles need to stop.
- If the signal unit is off the vehicles can pass.

Fig. 8.5 shows some typical applications of special pedestrian signals for right-turning movements in Ankara. In this application, the right-turning vehicles should be stopped if the red light shown in the figure is on. The vehicles can only pass if the signal unit is dark.

Figure 8.5 Special pedestrian signals for right-turning movements from the city of Ankara.

Figure 8.6 Dual signal units and signal poles with LED from city of Adana.

8.6.1.3 Dual signal unit

In this application, there are two units in the signal lamps. The red lamp is in one unit and the green/yellow lamp is in the same unit and through the LED arrays. Fig. 8.6 shows a typical application of dual signal units and signal poles with LED in the city of Adana.

8.6.1.4 Flashing right turn

A flashing right-turn signal may be installed at signalized intersections where the right turn is channelized. The flashing color varies according to the speed of the vehicles. The signal flashes red at high speeds and flashes yellow at lower speeds.

Figure 8.7 Signalized roundabouts: (A) type A (left) and (B) type B (right).

8.6.1.5 Approach flash

Flashing signals are usually installed at high-speed roads outside of the city center, making a yellow flash at about 100 m before the intersection. Approach flashers are also installed to warn drivers while approaching a signalized intersection that is after a bend in the road. The flashing signal can be installed at a standard position or overhead.

8.6.1.6 Signalized roundabouts

Signalized roundabouts are common in Turkey and they are mostly operated with fixed-time operations. There are two types of geometries. Type A, which is typically seen in urban areas, has storage lanes around the center for all movements from all directions (Fig. 8.7). These storage lanes (areas) serve especially for left-turning vehicles from all directions. In type A, there are two stop lines for the movements; the first is located at each arm of the intersection and the second is on the storage area around the center. Type B is typically seen in rural areas only and has storage lanes for the main roads (i.e., storage lanes are used only for two ways). In this application, storage lanes are not allocated for traffic flows on the side roads. These two types of roundabouts are shown in Fig. 8.7.

A signal timing method with fixed-time control was developed by Cakici and Murat (2016). In this method, signal timing for the circulation area is determined separately and then merged with the timing calculated for the intersection approaches. The details of the method can be found in Cakici (2014) and Cakici and Murat (2016).

8.6.2 Placement of poles and units

8.6.2.1 Locations of poles and units

There is no standard for the location of signal poles. According to Ayfer (1977), the position of the signal poles addressing vehicles should not be closer than 2 m to the stop line. The distance between the end point of the sidewalk and the signal pole varies depending on the approach speed. Accordingly, if the approach speed is less than or equal to 60 km/h, site distance from the roadway should be 50 cm and if it is less than or equal to 70 km/h, this distance should increase to 75 cm (TS 11937, 1996).

According to TS 11937, the standard height of the units should be 2.1 m, and should be 4.50 m if they are installed overhead. The console width should be between 2.50 m and 4.50 m.

8.6.2.2 Number of units

Although no specific standard exists for the number of units to use, TS 11937 *1.1.2 —Number of signal poles* under the heading states (TS 11937):

> *This number should be sufficient to allow the signals to be seen by each of the vehicles when the vehicles entering the intersection restrict the views of each other. When there are two or more lanes in the intersection entrance:*
>
> 1. *If there is a median in the middle and the median width is ≥2.00 m, an additional signal post should be installed in the refuge.*
> 2. *If there is no refuge in the middle or the width is less than 2.00 m, an additional signal post must be installed on the left side of the road and on the side of the strip where the counter traffic is located.*
> 3. *It should be placed on both left-handed and right-handed additional and out-of-the-way signals as if it were on a high-speed road.*

8.7 Working example

A sample intersection shown in Fig. 8.8 is used to illustrate the major steps for the signal timing design. The figure includes all of the necessary information, including geometry, traffic volumes, and traffic composition.

Step 1: Assignment of movement numbers

The first step in the signal timing design is assigning movement numbers. The movement order is shown in Fig. 8.9.

Step 2: Determination of the phase plan

Based on the traffic volumes and turning ratios, the four-phased plan is selected and is shown in Fig. 8.10. As seen in the figure, a fully protected phase is assigned for pedestrians. This application is very common in Turkey.

After determining the phase plan, the lost times for the intersection are calculated. A 2-s yellow for each phase and 5-s all-red time for the intersection are selected. Additionally, a 2-s start-up lost time is also selected for each phase. Based on these values, the total lost time for the intersection is determined by Eq. (8.6).

$$L = y + ar + slt \tag{8.6}$$

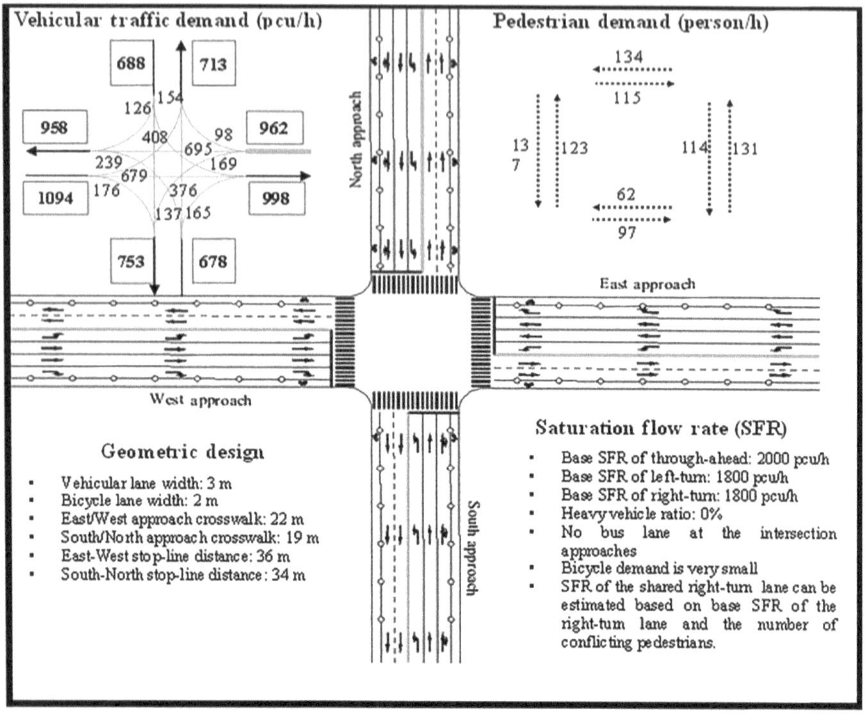

Figure 8.8 Intersection geometry and traffic volumes.

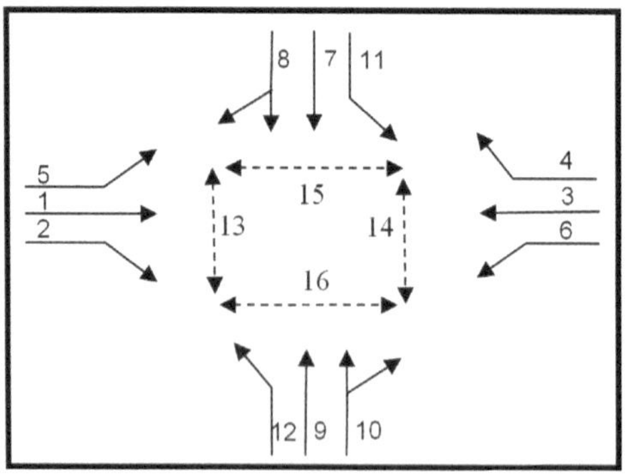

Figure 8.9 Movement numbers.

where L represents the total lost time for the intersection, y represents the yellow time for each phase, ar represents the all-red time for the intersection, and slt represents the start-up lost time for each phase. Thus, the total lost time for the intersection is calculated as 23 s.

$$L = 2 \times 5 + 5 + 2 \times 4 = 23\ s$$

Phase I	Phase II	Phase III	Phase IV	Phase V
1-2-3-4	5-6	7-8-9-10	11-12	13-14-15-16-B

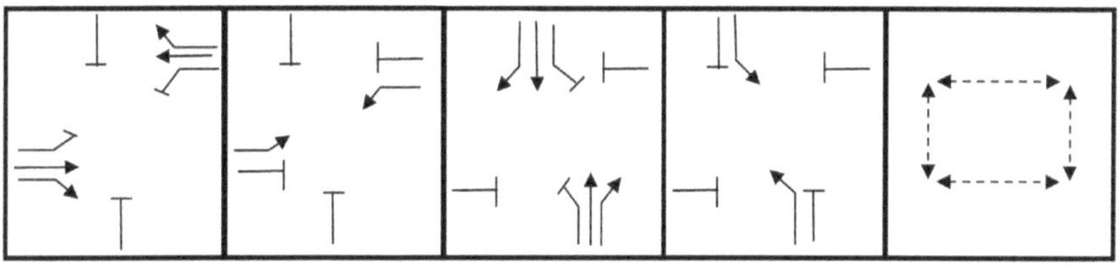

Figure 8.10 Signal phase plan.

Step 3: Determination of signal timing for pedestrians

Signal timing for pedestrians is determined considering the maximum crosswalk length and assumed pedestrian speed. In this example, the E-W crosswalk is calculated. The pedestrian speed is assumed to be 1.5 m/s (an average value).

Eq. (8.7) is used for calculating the pedestrian time.

$$Pst = CL/ASP \tag{8.7}$$

where Pst represents the pedestrian signal timing (s), CL represents crosswalk length (m), and ASP represents assumed pedestrian speed. Using the crosswalk length and pedestrian speed, 15 s is obtained for the pedestrian signal time.

$$Pss = 22/1.5 = 15\ s$$

Step 4: Calculation of flow ratios for each phase

Flow ratios or rates are one of the determinants in the signal design procedure. Below are the calculations of the flow rates for all of the movements according to the signal phase plan.

Phase I:

$$y_1 = \frac{q_1}{S_1} = \frac{679}{2 \times 2000} = 0.17 \quad y_3 = \frac{q_3}{S_3} = \frac{695}{2 \times 2000} = 0.17$$

$$y_2 = \frac{q_2}{S_2} = \frac{176}{1800} = 0.09 \quad y_4 = \frac{q_4}{S_4} = \frac{98}{1800} = 0.05$$

Phase II:

$$y_5 = \frac{q_5}{S_5} = \frac{239}{1800} = 0.13 \quad y_6 = \frac{q_6}{S_6} = \frac{169}{1800} = 0.09$$

Phase III:

$$y_7 = \frac{q_7}{S_7} = \frac{308}{2000} = 0.15 \quad y_9 = \frac{q_9}{S_9} = \frac{300}{2000} = 0.15$$

$$y_8 = \frac{q_8}{S_8} = \frac{226}{1800} = 0.13 \quad y_{10} = \frac{q_{10}}{S_{10}} = \frac{241}{1800} = 0.13$$

Phase IV:

$$y_{11} = \frac{q_{11}}{S_{11}} = \frac{154}{1800} = 0.09 \quad y_{12} = \frac{q_{12}}{S_{12}} = \frac{137}{1800} = 0.08$$

Step 5: Determination of critical movements

The critical movement for each phase is determined by comparing the flow ratios. Based on these calculations, the movements 1, 5, 7, and 11 are determined to be critical. The flow ratio for the intersection, Y, is determined by summing up the critical flow ratios.

$$Y = \sum_{i=1}^{n} y_i = 0.17 + 0.13 + 0.15 + 0.09 = 0.54$$

Step 6: Calculation of optimum cycle time

Webster's formula [defined previously in Eq. (8.3)] is used for obtaining the optimum cycle time. In this formula, the total lost time for the intersection and the flow ratio of the intersection are used.

$$C_{opt} = \frac{1.5L + 5}{1 - Y}$$

$L = 23$ s, $Y = 0.54$

$$C_{opt} = \frac{1.5 \times 23 + 5}{1 - 0.54} = 95 \text{ s.}$$

The signal timing for the pedestrians should be added to this value. Thus the cycle time is calculated as

$$C = 95 + 15 = 110 \text{ s}$$

Step 7: Determination of green times

Eq. (8.8) is used for determining green times.

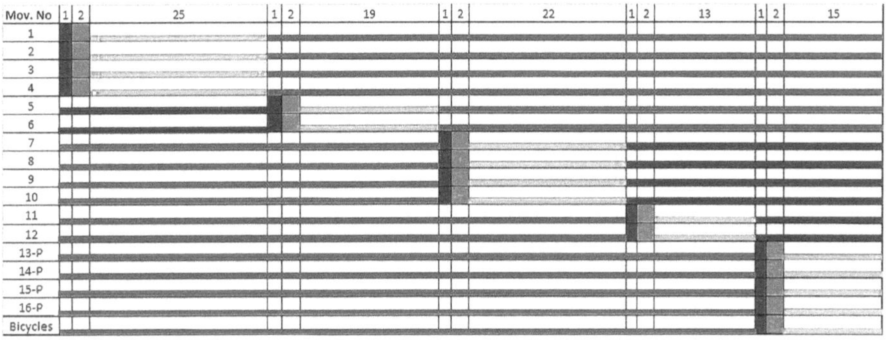

Figure 8.11 The signal cycle plan of the intersection.

$$g_i = \frac{y_i}{Y}(C - L) \tag{8.8}$$

Using Eq. (8.8), green times for each phase and the movements are determined. The signal cycle plan of the intersection is obtained considering green times and other signal timing values. Fig. 8.11 shows the signal cycle plan of the intersection.

$$g_1 = \frac{0.17}{0.54}(110 - 23) = 25 \text{ s} \quad g_3 = \frac{0.15}{0.54}(110 - 23) = 22 \text{ s}$$

$$g_2 = \frac{0.13}{0.54}(110 - 23) = 19 \text{ s} \quad g_4 = \frac{0.09}{0.54}(110 - 23) = 13 \text{ s}$$

Step 8: Check for degree of saturation values of movements

After calculating the green times, the degree of saturation for each movement is calculated using Eq. (8.9). In this equation, x_i represents the degree of saturation value of movement i, q_i represents the traffic volume of movement i, S_i represents the saturation flow value of movement i, and g_i represents the green time of movement i.

$$x_i = \frac{q_i C}{S_i g_i} \tag{8.9}$$

All of the degrees of saturation values of the movements are checked. The signal timing design needs to be revised if the degree of saturation value of any movement is higher than 1.0. A new phase plan may be used for the revision. Table 8.3 shows the degrees of saturation values calculated for the movements considered in the example.

Step 9: Determination of intersection delay

The delays of vehicles for each movement are calculated using the formula proposed in the HCM (TRB, 2000). The delay value for the intersection is obtained

Table 8.3 Degree of saturation values of the movements

Movement number	Degree of saturation value (X)
1	0.69
2	0.40
3	0.71
4	0.22
5	0.70
6	0.49
7	0.71
8	0.57
9	0.69
10	0.61
11	0.63
12	0.56

Table 8.4 Movement and intersection delays

Movement number	Traffic volumes (pcu/h)	Delay by HCM 2000 (s/veh)	Intersection delay and level of service
1	679	30.64	31.95 (C)
2	176	34.05	
3	695	32.86	
4	98	35.59	
5	239	30.29	
6	169	30.72	
7	308	29.92	
8	226	32.74	
9	300	32.05	
10	241	32.68	
11	154	32.29	
12	137	35.18	

by weighting the delays from each approach. Table 8.4 shows the calculated delay values for each movement and the intersection. As can be seen in the table, the intersection level of service is determined to be *C*. This level is considered acceptable from the design point of view, therefore the procedure is finalized.

Bibliography

Ayfer, M.O., 1977. Traffic Signal Control. Publication of General Directorate of Highways, Ankara, Turkey, p. 184.

Cakici, Z., 2014. Investigation of design procedure of roundabouts with signals. MSc Thesis. Pamukkale University Institute of Natural and Applied Sciences, Denizli, Turkey.

Cakici, Z., Murat, Y.S., 2016. A new calculation procedure for signalized roundabouts and performance analysis. Tech. J. Turkish Chamber Civil Eng. 27 (4), 7569–7592.

Calışkanelli, S.P., Tanyel, S., 2018. Investigation of saturation flow value at signalized intersections. Tech. J. 29 (1), 8225–8248.

Cetin, M., Murat, Y.S., 2013. A new method for determining saturation flow depending on driver behaviour. Tech. J. Cham. Civil Eng. 24 (2), 6399–6414.

Cumhuriyet. Gazetesi (Turkish Newspaper) issued on March 19, 1929.

FGSV, 2010. Forschungsgesellschaft für Strassen– und Verkehrswesen e.V., RİLSA, Richtlinien für Lichtsignalanlagen (Guidelines for Traffic Signals).

FHWA, 2009. Manual on Uniform Traffic Control Devices (MUTCD). FHWA, Edition 2009.

Gedizlioglu, E., 1994. Traffic Engineering Lecture Notes. Istanbul Technical University. Unpublished, 130 p.

Gundogan, F., 2012. Simplified traffic responsive signal control method for developing large cities. PhD Thesis. Graz University of Technology, Graz, Austria.

IETT, 2018. Photo Archives. IETT, Istanbul, Turkey.

ISBAK, 2007. Saturation Flow Study. ISBAK, Istanbul, Turkey, p. 110.

ISBAK, 2017. The Market Survey of Turkey. ISBAK, Istanbul, Turkey, 180 p.

Kutlu, K., 1993. Traffic Surveys and Counts. Publication of Istanbul Technical University, Istanbul, Turkey.

Murat, Y.S., 1996. Analysis of traffic flows at urban intersections of denizli by some softwares. MSc Thesis. Pamukkale University, Graduate School of Natural and Applied Sciences, Denizli, Turkey. 159 p.

Murat, Y.S., Gedizlioglu, E., 2002. A new approach for fuzzy traffic signal control. In: 13th Mini-EURO Conference on Artificial Intelligence in Transportation Systems and Science. Politecnico di Bari, June 10–13, 2002, Bari, Italy.

Ozdirim, M., 1993. Traffic Engineering, Vol. 1 and Vol. 2. Publication of the General Directorate of Highways, Ankara, Turkey, 835 p.

Polat, A., Sarisoy, G., Ogut, K.S., 2015. Determination of saturation flow values and departing lost times at signalized intersections. In: Proceedings of the 11th Transportation Congress, Chambers of Civil Engineers, May 27–29, 2015, Istanbul, Turkey, pp. 129–138.

Sezen, S., Apaydın, A., 2012. Transportation in Istanbul: From Coach to Marmaray. IETT, Istanbul, Turkey, 227 p.

TRB, 2000. Transportation Research Board, Highway Capacity Manual. National Research Council, Washington, DC, USA.

TS 11937, 1996. Turkish Standards.

TS EN 12368, 2015. Turkish Standards.

The Official Gazette, 1983. Turkish Road Traffic Act, Act No: 2918, 18.10.1983. no 18195, arrangement no 5, vol. 22, 687 p. Ankara, Turkey.

Webster, F.V., 1958. Traffic Signal Settings, Road Research Laboratory Technical Paper No. 39. HMSO, London, UK.

Australia and New Zealand

9

Chaitrali Shirke[1], *Ashish Bhaskar*[1] *and Edward Chung*[2]
[1]School of Civil Engineering and Built Environment, Queensland University of Technology, Brisbane, QLD, Australia, [2]Department of Electrical Engineering, Hong Kong Polytechnic University, Hong Kong, P.R. China

9.1 Background

This chapter is a synthesis of the general practices of signal control in Australia, with a focus on basic principles and methodology used for designing isolated fixed-time signals. The chapter also briefly describes the Australian standards and guidelines that specify standards of signal infrastructure such as signal heads and controllers. Most of the discussions are based on the traffic signal practice in Australia. Practice in New Zealand is very similar.

9.1.1 History and development of traffic signal control in Australia

The first signal operated by an electromechanical fixed-time controller was adopted in Melbourne in October 1928. By mid-1929 signals were synchronized at a number of Swanston Street intersections, Melbourne (May, 2005). However, these signals were not efficient and caused unnecessary delays. In October 1933, vehicle-actuated signals (imported from England) were installed in Sydney. Tenders were invited for the installation of further sets of signals in 1935, but due to excessive prices, no further signals were installed until 1937. Thereafter, the number of signals increased rapidly. The 1000th signal was installed in New South Wales (NSW) on April 8, 1974. As per NSW (2010), by August 1991, there were over 2120 traffic signals in operation at intersections throughout NSW and a further 280 at mid-block sites. By December 2009 there were 3791 traffic signals in service in NSW including mid-block sites. In the state of Queensland the first traffic signal was installed on January 21, 1936. To date, the number of signalized intersections in Australia is of the order of 10,000.

Not only has the number of signals increased, but the signal control technologies have also evolved over time, which has led to the development of adaptive and intelligent traffic control, such as Sydney Coordinated Adaptive Traffic System (SCATS). SCATS was developed in Sydney, Australia, in 1970. Such systems do not only help in efficient management of vehicular traffic but also effectively serve other road users such as pedestrians, cyclists, and public transport.

Global Practices on Road Traffic Signal Control. DOI: https://doi.org/10.1016/B978-0-12-815302-4.00009-1

9.2 The state-of-the-art traffic signal control

9.2.1 Current nation-wide guidelines

The transport department of the state and local government plays a leading role in the standardization of design, implementation, and operation through national or industrial manuals and guidelines. The Australian Road Research Board (ARRB) (https://www.arrb.com.au/) is recognized as one of the strong contributors to the applied research for Australian and New Zealand state road agencies. The theoretical base of techniques used by traffic engineers in Australia for analysis of the capacity and timing requirements at the signalized intersection can be found in the ARR 123 report. In addition, Austroads, the peak organization of Australasian (a region that comprises Australia, New Zealand, and neighboring islands in the Pacific Ocean) road transport and traffic agencies provides a series of guides which cover the design, construction, maintenance, and operation of the road network in Australia and New Zealand. All road agencies across Australasia (Australia, New Zealand, and neighboring islands) have adopted the Austroads Guides.

Out of these guides, The Austroads Guide to Traffic Management (AGTM) Part 9: Traffic Operations (Austroads, 2016a) provides guidance on the configuration and operation of traffic signal systems, both Intelligent Transport Systems (ITS) and manual. In addition, traffic theory, including queuing and platoon dispersion, are explained in AGTM Part 2 (Austroads, 2015a). Traffic studies and analyses performed to analyze the capacity and performance of signalized intersections are described in AGTM Part 3 (Austroads, 2017a). AGTM Part 6 (Austroads, 2017b) provides details about layout, lane management, and function design of traffic signals. AGTM 10 (Austroads, 2016b) provides guidance on the aspects of traffic signal which are visible to users, for example, signs, signal displays, etc. Moreover, safety considerations are covered in AGTM Part 13 (Austroads, 2015b).

In addition to the above guides state transport authorities such as VicRoads (Victoria government), TMR (Queensland Department of Transport and Main Roads), RMS (Road and Maritime Services, NSW), MRWA (Main Roads, Western Australia), and NZTA (New Zealand Transport Agency) use the Australian Standards (AS). For example, AS 1742.14−2014 Manual of Uniform Traffic Control Devices (MUTCD) is used to determine how to implement the practice/option referenced in the Austroads Guides. Furthermore, each authority publishes the supplements or complementary documents for each part of the AGTM and relevant Australian Standard that mentions any complementary guidelines which need to be considered.

9.2.2 Major signal control manufacturers and control systems

Traffic signal controllers used in Australia and New Zealand are as per standards defined in AGTM Part 10 by Austroads (2016b), along with Australian Standards specified in AS 2578−2009 that provides mechanical, electrical, and functional requirements for the control equipment used for the operation of road traffic signals.

There are two types of controllers, namely post-mounted controllers and ground-mounted controllers (RTA, 2008). A post-mounted controller is restricted to a maximum of six signal groups, which are normally mounted on a kerbside post. Ground-mounted controllers provide a maximum of 24 vehicle signal groups and 8 pedestrian signal groups, 24 vehicle detectors, and 8 pedestrian detectors. All ground-mounted controllers provide a maximum of seven phases.

Furthermore, controllers may operate in SCATS or non-SCATS conditions. As SCATS is the most commonly used software in Australia, the SCATS-compliant controller called TSC/4 is widely used in Australia. The major distributors of traffic controllers are Tyco Traffic and Transportation, Quick Turn Circuits, and Aldridge Traffic Controllers (RTA-NSW, 2013).

9.2.3 Traffic signal control software

Traffic signal software is used at different levels of control. For instance, some software work with a local controller, some help with central management, and some are used for signal timing development.

The well-known SCATS that aids at all three control levels and was first developed by the New South Wales Roads and Traffic Authority (RTA) is adopted by most states in Australia and New Zealand (Lowrie, 1990). SCATS is an adaptive urban traffic management system that constantly adjusts the signal timing for each movement and co-ordinates consecutive traffic signals to optimize traffic flow across a whole city, region, or corridor. It operates at three levels namely: Central Computer—Communications and Database Functions; Regional Computer—Strategic Control; and Local Traffic Controllers—Tactical Control. It can determine stage splits at intersections, alter the cycle time of intersections either individually or in groups, and introduce cycle or plan-dependent options.

In addition to SCATS, STREAMS, an ITS platform, is provided by Transmax, which helps to manage motorways as well as arterial traffic on a single platform (Transmax.com.au, 2018). Along with adaptive traffic control for arterials, it also includes a few more traffic management measures such as ramp metering, lane control, ramp queue clearance, and motorway speed management, etc. Transmax deploys STREAMS across Australia, with customers in Queensland, Victoria, South Australia, Western Australia, ACT, and Tasmania. Sometimes SCATS is used for adaptive traffic control of arterial traffic under the STREAMS platform.

Furthermore, SIDRA INTERSECTION software, developed in Australia in 1984 (Akcelik & Associates_Pty.Ltd., 1984), is widely used for the design and evaluation of signal timings at individual intersections and/or networks of intersections.

9.2.4 Signal heads and signal display

The signal heads and signal display follow the Australian Standards (AS 1742.14–2014, AS 2144–2014, AS 2339–2017). AS 2144–2014 specifies the design, construction, and performance of traffic signal lanterns that use light-emitting diode light sources and are intended to control the movement of vehicles, including bicycles and pedestrians. AS 1742.14–2014 specifies the type and layout of signals, aspect, and display to be used at intersections controlled by traffic signals. AS 2339–2017 specifies the form, dimensions, material, and finish of tubular steel posts and mast arms, and attachments used for the support of traffic signal lanterns, pedestrian push-button assemblies, and ancillary equipment.

Signal faces are made up of a number of signal aspects, generally in one or two columns. Circular aspects (disks) are used as a first preference. Arrow aspects are used to control particular movements at traffic signal installations, including U-turn movements. Aspects are arranged in columns with the red aspect upper-most, the yellow aspect central, and the green aspect at the bottom. The basic three-aspect signal face consists of red, yellow, and green circular aspects in a single column (refer to Fig. 9.1A), whereas a multicolumn signal faces can contain four, five, or six aspects (refer to Fig. 9.1B).

The location of signal faces with respect to approach is determined in accordance with AS 1742.14 (Standard-Australia, 2014a). For a given approach the primary signal faces whose main function is to warn approaching traffic of the state of the signals, and stop traffic at the correct position, are located on the left-hand side, adjacent to the stop line or another point where traffic is required to stop. Secondary signal faces, which indicate the start of a running phase to the traffic

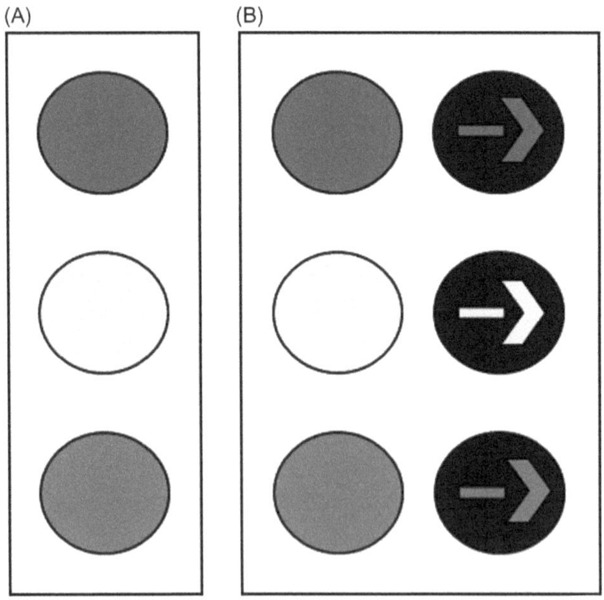

Figure 9.1 Examples of signal phase layout: (A) three-aspect signal face layout; (B) six-aspect multicolumn signal face layouts.
Adapted from Austroads, 2016b. Guide to traffic management part 10: traffic control and communication devices. Second edn, AGTM10-16, Austroads, Sydney, NSW.

that is stopped, are located to the right side of the roadway, beyond the point where traffic is required to stop. Tertiary signal faces are provided to back-up the secondary signals and these faces are located to the left side of the roadway beyond the point where traffic is required to stop. The location of signal faces for an undivided, divided, and T-junction intersection along with a pedestrian-actuated traffic signal are shown in Fig. 9.2.

For pedestrian signals, a red steady human figure is for the "DON'T WALK" aspect, and a green walking figure is the "WALK" aspect (see Fig. 9.3). The basic sequence of displays is steady red to green to flashing red to steady red. The flashing red symbolic steady pedestrian is displayed during the pedestrian clearance period and indicates that pedestrians on the crossing may complete their crossing, but pedestrians who have not yet started to cross are not to leave the refuge. The flashing red pedestrian symbol may be directly replaced with a yellow countdown timer display (within the same aspect) indicating the number of seconds left (stopping at 1) before a steady red standing pedestrian is displayed (refer to Fig. 9.3).

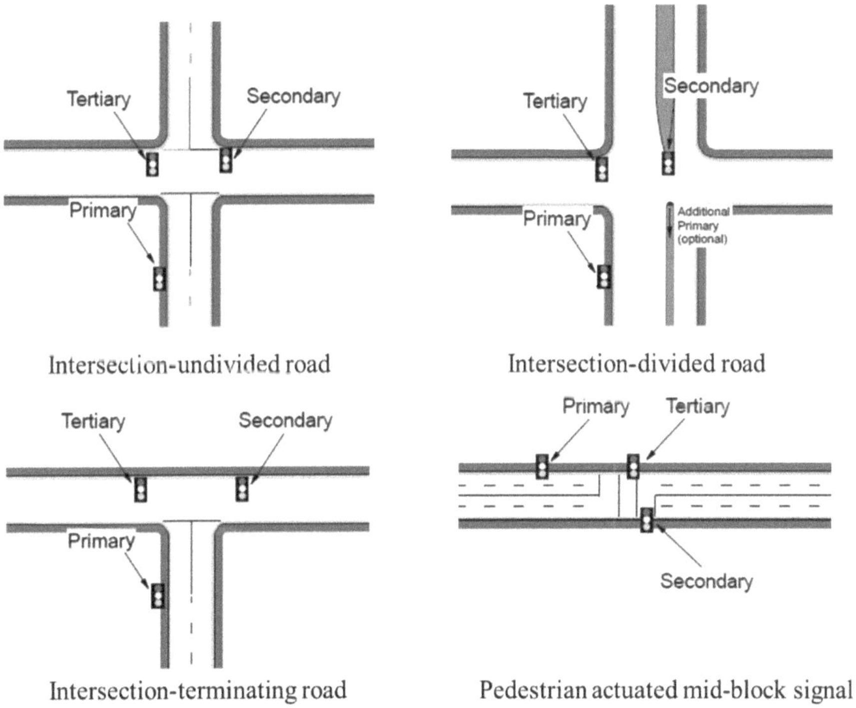

Figure 9.2 Location of signal heads.
Adapted from Austroads, 2016b. Guide to traffic management part 10: traffic control and communication devices. Second edn, AGTM10-16, Austroads, Sydney, NSW.

Figure 9.3 Pedestrian signal with a countdown.
Adapted from Austroads, 2016b. Guide to traffic management part 10: traffic control and communication devices. Second edn, AGTM10-16, Austroads, Sydney, NSW.

9.3 Selection of control strategies

9.3.1 Installation of traffic signals

The transport authorities of various states specify the warrants to determine the requirement of a traffic signal at a particular location. Generally, these warrants consider the following factors (RTA-NSW, 2008):

- Traffic flow;
- Traffic conflicts;
- Traffic accident statistics;
- Pedestrian requirement;
- Access to major roads;
- Cost of installation;
- Availability of funds;
- Maintenance cost;
- Practicality;
- Feasibility; and
- The posted speed limit is not more than 80 km/h.

9.3.2 Control strategies

The various types of signal control strategies available across Australia are as follows:

- *Fixed-time control*: fixed duration and sequence of signal phases;
- *Semiactuated control*: generally minor movements are actuated, which allows for indefinite extension of green time for major movements until minor phase demand is actuated;
- *Actuated traffic control*: all movements (phases) are actuated, which allows for variation of phase sequence and duration depending on traffic demands; and

- *SCATS master isolated (SMI) control*: intersection controller can be linked to the SCATS regional computer, which gives the option to control as fully actuated or SMI (Lowrie, 1990). SMI is a fully actuated control where the SCATS regional computer determines the maximum green time for each cycle as per varying demand.

Out of these control types, the best candidate is chosen for a particular site based on the following conditions:

- Facilities provided by a signal controller, such as fixed sequence, variable sequence, actuation, and master control;
- The capacity of the controller: this is determined by the number of signal groups which must be switched, and the number of detector and push-button input circuits;
- Operation and maintenance; and
- Cost and availability

Of all the control types adapted across Australia, this chapter describes the step-by-step procedure to design a fixed-time control at the isolated intersection.

9.4 Design steps

This section presents the signal timing procedure for an isolated and fixed-time control, recommended in ARR 123. Fig. 9.4 summarizes the major steps of signal timing design.

The input data required for timing and performance analyses is as follows:

- Detailed intersection layout: configuration, geometric details of each approach, for example, approach distance, number of lanes, lane width, lane discipline, and basic saturation flow;
- Demands for traffic movements are identified through vehicle detectors, pedestrian push-buttons, and, where relevant, externally supplied data from a master control computer; and
- Design volumes.

The signal timing design process described in Fig. 9.4 provides cycle time, displayed green times for phases, and effective green times for movements. Table 9.1 specifies symbols used for the remainder of this chapter.

The major steps of signal design specified in Fig. 9.4 are described in detail next.

9.4.1 Design period and volume

The very first step of signal design is to know the design period and design volume. If the signal timings need to be adjusted at an existing intersection, the current year traffic is used. Whereas, if the performance of existing intersections needs to be improved or new signal needs to be installed, then design calculations should be carried out for the flow expected at the time of the opening and the design should be checked for flows expected in the near future.

The average weekday peak hour traffic flows in the design year are used for signal designs. Usually, the signal timings are designed separately for morning and

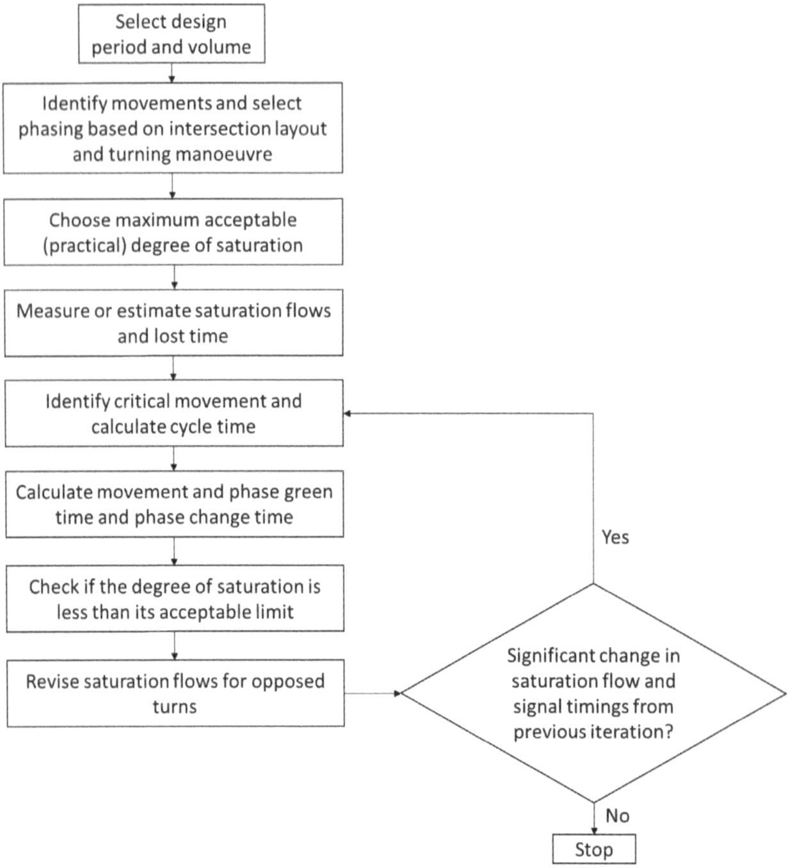

Figure 9.4 Flowchart for fixed-time signal timing design.

evening peak periods, especially for urban intersections because traffic pattern is different during these peaks due to the directional nature of traffic movement, that is, home to work and work to home.

9.4.2 Movements and phases

Each movement has characteristics which determine its saturation flow, lost time, and ultimately time to be allocated to meet its capacity. The two major points to consider while defining movements are right of way provisions and lane allocation or lane utilization. For example, traffic in an underutilized lane is a separate movement, whereas traffic in lanes with equal utilization is combined as a single movement. During each signal phase, one or more assigned movements receive the right of way. The movement which receives the right of the way during more than one phase is called overlapped movement.

In general, the design of signal phases aims to enhance efficiency (without compromising safety) by simultaneously running the nonconflicting movements.

Table 9.1 Symbols and their definition

Symbol	Definition	SI unit
A	Start loss	Seconds
B	End gain	Seconds
C	Cycle time	Seconds
c_m	Minimum cycle time	Seconds
c_p	Practical cycle time	Seconds
c_o	Optimum cycle time	Seconds
c_{max}	Maximum acceptable cycle time	Seconds
F	Phase change time	Seconds
g	Effective green time	Seconds
g_c	Critical movement green time	Seconds
g_m	Minimum effective green time	Seconds
g_u, g_d	Upstream and downstream effective green times	Seconds
G	Displayed green time	Seconds
G_{min}	Minimum displayed green time	Seconds
I	Intergreen time	Seconds
l	Movement lost time	Seconds
l_c	Critical movement lost time	Seconds
L	Intersection lost time	Seconds
N_d	Average stop-line queue at the start of green period on downstream approach	
O	Offset	Seconds
Q	Arrival flow	Veh/h
Q	Capacity	Veh/h
s	Saturation flow rate	Veh/h
t	Movement time	Seconds
t_m	Minimum required movement time	Seconds
T	Sum of various movement time	Seconds
u	Green time ratio for movement	
U	Intersection green time ratio	
x	Degree of saturation for movement	
X	Intersection degree of saturation	
x_p, X_p	Practical degree of saturation for individual movement and intersection, respectively	
y	Flow ratio for a movement	
Y	Intersection flow ratio	
W	Effective width of crosswalk or kerb to kerb crossing distance	Meters

Austroads (2016a) describes the major factors affecting phasing design as layout, alignment, and traffic flows. Layout includes the number of lanes and the length of left and right turn lanes available for each movement on the approach and departure of each intersecting road. (Alignment refers to the horizontal and vertical alignment in regard to the angle at which roads intersect and sight distance available to allow safe filtering of right turn movements.) Traffic flow is the amount of traffic, including the proportion of heavy vehicles, in each through or turning movement. Other factors to be considered for phase design include pedestrian facilities and coordination. The design procedure is started by considering trial phasing.

For the Australian method of signal timing design, movements are of particular importance, because instead of "phase-based" method, a "movement-based" method is adopted. Here, "phase" is the same as "stage" defined in Chapter 2, Principles of road traffic signal control.

In the "phase-based" method, the control parameters (such as intergreen time, minimum green time, and lost time) are specified for phases and apply to all movements receiving the right of way during that phase. The objective of this approach is to minimize the number of phases because it is believed that the intersection lost time can be decreased by decreasing the number of phases. However, this is true only when overlap movement is not present.

In the "movement-based" method, the aforementioned control parameters are calculated for movements. The objective of this approach is to maximize the number of overlaps, even though that will increase the number of phases because maximizing the number of overlaps will reduce the total time required to satisfy the capacity requirements of all critical movements at the intersection. This approach is more flexible and efficient as it can handle complicated phasing with multiple overlap movements.

Capacity and timing analysis is performed for each movement to determine the intersection capacity requirement. Interested readers should refer to section 2.3 of ARR 123 for details on movement saturation flow and lost time (difference in start loss and end gain). A movement-based approach facilitates a better understanding of the relationship between movement and signal phasing characteristics as well as allowing easier handling of complicated phasing systems.

9.4.3 Degree of saturation

Capacity (Q) for each movement is defined as the maximum sustainable rate at which vehicles can depart (saturation flow, s) multiplied by the proportion of cycle time (c), which is effectively green for that phase (green time ratio, $u = g/c$).

$$Q = s * \frac{g}{c} \tag{9.1}$$

The movement degree of saturation (x) is the ratio of arrival flow (q) to capacity (Q). The movement capacity is adequate if x is less than one (i.e., capacity is greater than arrival flow rate). The movement degree of saturation can be expressed as below, where y is defined as the flow ratio, that is, the ratio of arrival flow to saturation flow (q/s).

$$x = \frac{q}{Q} = \frac{qc}{sg} = \frac{y}{u} \tag{9.2}$$

For an intersection, the degree of saturation (X) is obtained as the ratio of $Y = \sum y$ to $U = \sum u$, where the summation is over critical movements only. The U

for an intersection can be expressed as the ratio of total available green time ($c-L$, where L is the intersection loss time) to the cycle time (c). Hence the intersection degree of saturation (X_p) is expressed as follows,

$$X_p = \frac{Y}{U} = \frac{Y \times c}{c - L} \tag{9.3}$$

To manage excessive delays and queues, practitioners specify the practical maximum degree of saturation sought for different movements (x_p) and intersection (X_p). The recommended range for a practical maximum degree of saturation is 0.8–0.9 (Akcelik, 1978). A different value can be chosen for different movements, which leads to an unequal degree of saturation method (Akcelik, 1979). In the case of an equal degree of saturation method, all movements and intersections have the same practical degree of saturation.

9.4.4 Movement lost time

The movement lost time (l) is the difference between the start loss (a) and end gain (b) times. Here, start loss is the sum of intergreen time (I) and loss caused by the driver's reaction time and acceleration of vehicles. The intergreen time (yellow plus all red) of movement is the time gap between the start of the display green for the movement to the end of the display green for the previous phase. Yellow time and all-red time can be calculated using Eqs. (9.4) and (9.5), respectively (Austroads, 2016a).

$$t_y = t_r + 0.5\frac{v_D/3.6}{a_d + 9.8G_r} \text{ subject to } t_y \geq 3.0 \tag{9.4}$$

$$t_{ar} = 3.6L_c/v_D \text{ subject to } t_{ar} \geq 1.0 \tag{9.5}$$

where t_y = yellow time (s); t_r = reaction time (s), usually between 1.0 and 1.5 seconds; v_D = design speed (km/h); a_d = the deceleration acceptable to the majority of drivers (m/s^2), commonly 3.0 m/s^2; G_r – average approach grade over the stopping distance; t_{ar} = all-red time (s); L_c = clearance distance between the stop line and the furthest point of potential conflict with vehicles or pedestrians of the next phase (m).

Typical yellow times and all-red times used in practice are in the range of 3–6 and 1–3 seconds, respectively (Austroads, 2016a).

The concept of movement lost time is demonstrated in Fig. 9.5. The intersection lost time (L) is a summation of lost times of critical movements. The total lost time is defined by the number of phases in the cycle and is independent of the cycle time.

9.4.5 Movement saturation flow rate

Saturation flow rate and lost time are measured in the field and the methodology for these is explained in Appendix E of ARR 123. In the case of a new signal

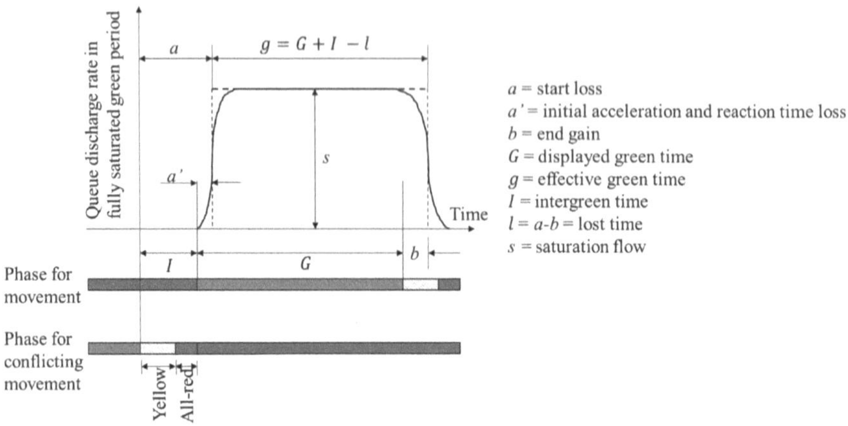

Figure 9.5 Movement lost time.
Adapted from Akcelik, R., 1981. Traffic signals: capacity and timing analysis. Research Report 123, Australian Road Research Board, Melbourne, Australia.

installation or modifications to existing signalized intersection, saturation flow rates need to be estimated or predicted to incorporate the effects of modified conditions at intersections.

In the absence of measured saturation flow, the value can be estimated using base saturation flow rate and the correction factors. The base saturation flow rates specified in ARR 123 by environment class and lane type are based on extensive surveys in Australia (refer to Table 5.1 of ARR 123). The next step is to apply various adjusting factors affecting saturation flow rate and estimate saturation flow rate (s) for a particular movement. The guidelines for adjusting saturation flow rates for factors such as lane width, gradient, traffic composition, turn, lane utilization, etc. are documented in ARR 123 and the equation for estimating saturation flow is as follows:

$$s = \left(f_w f_g / f_c\right) s_b \times N \tag{9.6}$$

where s_b = base saturation flow [though car unit (TCU)/h]; f_w = lane width factor; f_g = gradient factor; f_c = traffic composition factor which is determined by proportions of vehicles types (cars or heavy vehicles) and turning traffic (TCU/vehicle); N = number of lanes; s = estimated saturation flow rate (vehicles/h).

$$f_w = \begin{cases} 1.0 \text{ for } 3.0 \le w < 3.7 \\ 0.55 + 0.14w \text{ for } 2.4 \le w < 3.0 \\ 0.83 + 0.05w \text{ for } 3.7 \le w < 4.6 \end{cases} \tag{9.7}$$

$$f_g = 1 \pm 0.5(G_r/100) \tag{9.8}$$

$$f_c = \frac{\sum e_i q_i}{q} \tag{9.9}$$

where e_i = through car equivalent of vehicle turn type i which is specified in Table 5.2 of ARR 123; q_i = flow in vehicles (regardless of the vehicle types) for turn type i; q = total movement flow (vehicles) = $\sum q_i$; G_r = gradient in percentage ($+G_r$ for downhill gradient and $-G_r$ for uphill gradient).

Note that saturation flow rates for the opposed turn and short lane are a function of signal timings. For designing, an iterative process is adopted where during the first iteration, the short lane is assumed as a full-length lane and the opposed turn is treated as a normal turn and the corresponding base saturation flow rate is assigned to them. In a successive iteration, saturation flow values for the short lane and opposed turn are updated as specified in ARR 123 section 5.

9.4.6 Critical movement identification

For each movement, the following data are required to identify critical movement:

- Starting and terminating phase for each movement;
- Intergreen time (I);
- Minimum displayed green time (G_{min});
- Arrival flow rate (q) and saturation flow rate (s);
- Lost time (l); and
- Practical degree of saturation (x_p).

The minimum displayed green time depends on vehicle or pedestrian requirements. The minimum green time for vehicles should be related to the slowest vehicle likely to use the intersection within 95% probability and is commonly between 6 and 20 seconds.

The minimum displayed green time for pedestrians can be calculated as follows:

$$G_{min} = t_{pw} + W/v_{pw} \quad (9.10)$$

where G_m is displayed green time which is a sum of pedestrian walk time (t_{pw}) and clearance time, that is, W/v_{pw}. W is kerb to kerb crossing distance (m) and v_{pw} (commonly taken as 1.2 m/s) is pedestrian walking speed during clearance. Pedestrian walk times are usually in the range of 5–16 seconds.

To identify critical movement, the required movement time is calculated as follows:

$$t = \max((u \times c) + l, t_m) \quad (9.11)$$

where t_m is the minimum required movement time and can be obtained as follows:

$$t_m = G_{min} + I \quad (9.12)$$

For pedestrian movements, t is equal to t_m. For initial calculation, c in Eq. (9.11) is replaced by 100 and the cycle time is calculated after identifying critical movements. After knowing the required movement time, a critical movement search diagram is prepared, the details for which are presented in section 4 of ARR 123.

In the case of nonoverlap movement, choose the movement with the largest value of t in each phase as the critical movements. For, overlapping movements, keep the movements which have the largest t during each phase and eliminate others. Identify all possible movement combinations that complete one cycle. Sum up the movement times t for each path and find a path with the largest required total time. The movements on that path are critical movements. An example is presented later (Akcelik, 1981).

Fig. 9.6A illustrates a phasing diagram indicating movements receiving the right of way during each phase. Fig. 9.6B represents a critical movement search diagram with all movements as well as respective required green time (in bracket) for each movement, calculated using Eq. (9.11). In the current example, movements 1 and 4 are overlapping movements. Movements 3 and 7 during the corresponding phase required longer green time (t) (Fig. 9.6B) than movements 6 and 5, respectively. Therefore, movements 6 and 5 can be eliminated from the critical path. Now, the possible paths completing a cycle are 1 and 7 or 2, 3, and 7 for node A to A as well as 3 and 4 for node B to B. the longest path can be determined as follows: $T_{1,7} = t_1 + t_7 = 49$, $T_{2,3,7} = t_2 + t_3 + t_7 = 81$, $T_{3,4} = t_3 + t_4 = 97$.

As the path, "$T_{3,4}$" requires the largest time the critical movements are 3 and 4. Following identification of critical movements, intersection parameters such as total lost time for critical movements (L); total flow ratios for the critical movements (Y); and total green time ratios for the critical movements (U) are calculated.

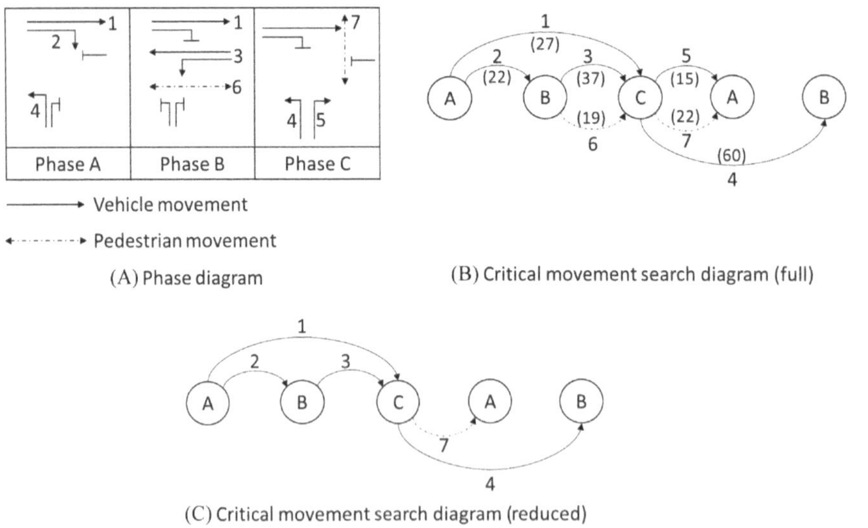

Figure 9.6 Critical movement identification.
Adapted from Akcelik, R., 1981. Traffic signals: capacity and timing analysis. Research Report 123, Australian Road Research Board, Melbourne, Australia.

9.4.7 Calculating signal parameters

9.4.7.1 Cycle lengths

The intersection cycle time is to be defined carefully. Too short cycle length results in inefficiency due to excessive lost time between phases, whereas too long cycle length increases delay and driver frustration. There are two cycle time functions that are used: *practicial cycle time* (c_p) and *optimal cycle time* (c_o).

Practical cycle time (c_p): This is the minimum cycle time for the intersection to operate where the degree of saturation of all movements is below the specified acceptable limit.

$$c_p = \frac{L}{1 - U} \tag{9.13}$$

Optimal cycle time (c_o): This is the optimal cycle time with respect to performance measures (such as minimum delay and minimum number of stops). An approximate optimum cycle time is modeled as:

$$c_o = \frac{(1.4 + k)L + 6}{1 - Y} \tag{9.14}$$

In Eq. (9.14), k refers to stop penalty parameter. Refer to sections 6 and 7 of ARR 123 for the values of the stop penalty parameter. The value of k is based on the performance measure chosen. For instance, $k = 0$ for minimum delay and $k = 0.4$ for minimum fuel consumption. This parameter represents a multiplicity of control/design objectives. In practice, the design objective is chosen based on site conditions during the design period. For example, during the low traffic demand (off-peak) period, aiming at minimum fuel consumption is reasonable, whereas, during the high demand (peak) period, minimum queue length should be aimed for. Hence, depending on the strategy to be adopted, the value of the stop penalty parameter should be chosen.

The cycle times for an intersection is recommended to be between c_p and c_o. Furthermore, the cycle time chosen must be less than the maximum acceptable cycle time c_{max}. In other words, chosen cycle time should satisfy, $c_p \leq c \leq$ minimum(c_{max}, c_o). The r maximum cycle time is recommended as 100–120 seconds for a two-phase intersection and 150–180 seconds for sites with complex phasing systems and high traffic demands (Austroads, 2016a).

9.4.7.2 Green times

For selected cycle time, effective green times (refer to Fig. 9.5) are calculated for critical and noncritical movements and finally phase green times are determined. The following formula is used to calculate critical movement green time:

$$g = \frac{c - L}{U} u \tag{9.15}$$

In absence of overlapping movement, effective green time for noncritical movement sharing the same phase as critical movement is calculated as follows:

$$g = (g_c + l_c) - l \tag{9.16}$$

If there is an overlap movement which is critical, then the green time for noncritical movement is calculated as follows:

$$g = \left(\frac{c^* - L^*}{U^*}\right) u\text{org} = g_m \tag{9.17}$$

where $c^* = g_c + l_c$ is treated as subcycle time and L^*andU^* are the sum of nonoverlap movement lost time and green time ratios, respectively. If overlapping movement is not critical, then noncritical overlap movement green time is calculated as follows:

$$g = \left(\sum g_c + \sum l_c\right) - l \tag{9.18}$$

Finally, a displayed green time (refer to Fig. 9.5) for the phase is calculated as follows:

$$G_d = (g + l) - I \tag{9.19}$$

where $g + l$ is time allocated to a movement that received right of way during that phase only.

Phase change time can be calculated by knowing the green times for all phases. Phase change time is set to zero for the first phase and the same is calculated for successive phases as follows:

$$F_i = F_{i-1} + Y_{i-1} + G_{i-1} \tag{9.20}$$

where F_i = change time for phase i; F_{i-1} = change time for previous phase $i - 1$; Y_{i-1} = intergreen for previous phase $i - 1$; G_{i-1} = green time for previous phase $i - 1$.

Phase change time is of importance while defining offset for coordinated signals which will be discussed in the following section.

For a detailed explanation of formulae mentioned earlier and for calculation of cycle length and green time, refer to sections 6 and 7 of ARR 123. The following section demonstrates the example of designing signal timings for an isolated intersection.

9.5 Working example

The procedure for designing signal timings for an isolated explained in Section 9.4 can be understood in detail with the help of the example presented in this section. Consider the configuration of an isolated intersection shown in Fig. 9.7. The

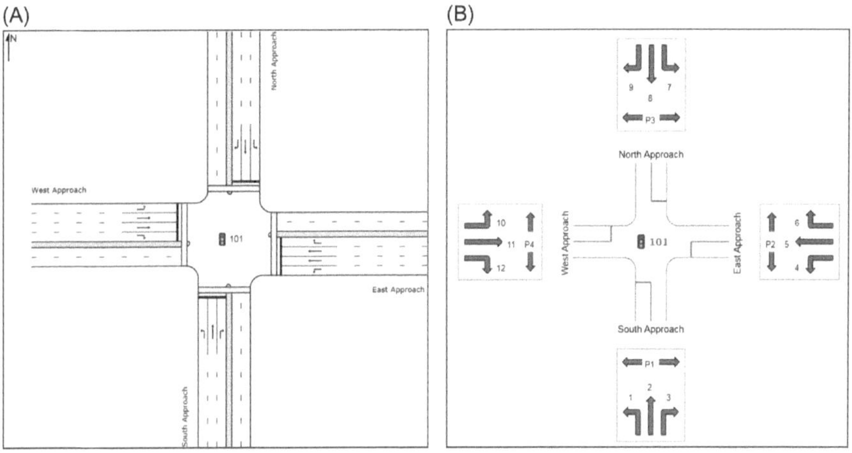

Figure 9.7 Geometry of an intersection and corresponding movements. (A) Intersection configuration. (B) Movement IDS.

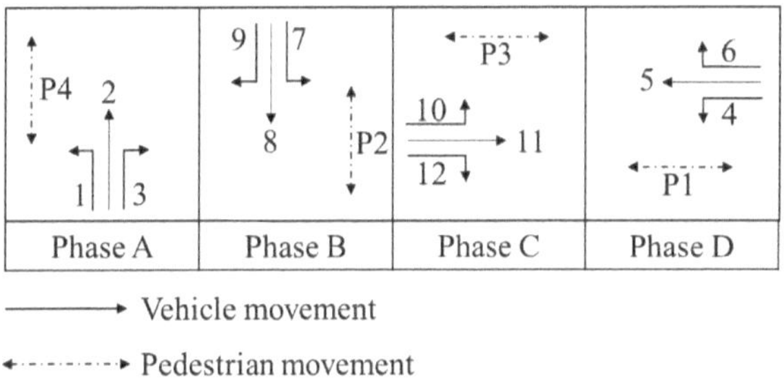

Figure 9.8 Movement IDs and phasing.

corresponding movement ID and phasing are shown in Fig. 9.7B and Fig. 9.8, respectively. Here, P1, P2, P3, and P4 represent pedestrian movements.

9.5.1 Data available

The lane width is 3 m, the median width is 2 m, and the length of each lane is 500 m. The crosswalk widths are as follows:

- East approach: 20 m;
- West approach: 20 m;
- North approach: 17 m;
- South approach: 17 m.

Table 9.2 summarizes the data available for designing of signal timings (cycle time and phase durations) for the intersection shown in Fig. 9.7. Note that there is no demand for heavy vehicles.

Table 9.2 Input data (units are specified in Table 9.1)

Movement	No. of lanes	Start phase	End phase	Arrival flow, q (veh/h)	Sat. flow, s (veh/h)	Intergreen time, I (s)	Min disp. green, G_m (s)	Lost time, l (s)	Min effective green ($g_m = G_{min} + I - l_1$) (s)	Prac. deg. sat., x_p
1	1	A	B	165	1800	5	5	6	4	0.9
2	1	A	B	376	2000	5	10	6	9	0.9
3	1	A	B	137	1800	5	5	6	4	0.9
4	1	D	A	98	1800	5	5	6	4	0.9
5	2	D	A	695	4000	5	10	6	9	0.9
6	1	D	A	169	1800	5	5	6	4	0.9
7	1	B	C	126	1800	5	5	6	4	0.9
8	1	B	C	408	2000	5	10	6	9	0.9
9	1	B	C	154	1800	5	5	6	4	0.9
10	1	C	D	176	1800	5	5	6	4	0.9
11	2	C	D	679	4000	5	10	6	9	0.9
12	1	C	D	239	1800	5	5	6	4	0.9
P1	–	D	A	Pedestrian		5	18	6	17	–
P2	–	B	C	Pedestrian		5	21	6	20	–
P3	–	C	D	Pedestrian		5	18	6	17	–
P4	–	A	B	Pedestrian		5	21	6	20	–

Details of parameters specified in Table 9.2 are as follows:

- Arrival flow is the traffic demand observed during peak hour expressed in vehicles/h.
- Saturation flow rates are estimated using a procedure described in section 5 and Appendix E of ARR 123.
- Intergreen time (I) depicted in Table 9.2 is the sum of yellow time and all-red time, which are 3 and 2 seconds, respectively. Minimum displayed green time depends on vehicle or pedestrian requirements.
- Minimum displayed green time for pedestrians can be calculated using Eq. (9.8). Pedestrian clearance time might overlap with the intergreen time of the terminating phase. In such a case, G_m is reduced by the overlap time, which is assumed as 2 seconds for this example. For movements P1 and P3, $W = 17$ m and for movements P2 and P4, $W = 20$ m. In Table 9.2, $G_{min\text{P1}} = G_{min\text{P3}} = 6 + \left(\frac{17}{1.2}\right) - 2 = 18$ seconds and $G_{min\text{P2}} = G_{min\text{P4}} = 6 + \left(\frac{20}{1.2}\right) - 2 = 21$ seconds.
- Initial loss caused by the driver's reaction and acceleration of vehicles are assumed to be 3 seconds for all movements. Similarly, end gain is assumed as 2 seconds for all movements. Hence, lost time, $l = Y + 3 - 2 = 5 + 3 - 2 = 6$ seconds.
- The practical degree of saturation (x_p) is chosen as 0.9 for all movements.

9.5.2 Step 1: identify critical movements: initial calculation under cycle time of 100 seconds

Two major tools used for identifying the critical diagram are a critical movement search table and critical movement search diagram, which are shown in Table 9.3 and Fig. 9.9, respectively.

Table 9.3 Critical movement search table (units are specified in Table 9.1)

Movement	$y = q/s$	$u = y/x_p$	$100u + l$	$t_m = G_{min} + I$	t^b
1	0.09	0.10	16	10	16
2	0.19	0.21	27	15	27
3	0.08	0.08	14	10	14
4	0.05	0.06	12	10	12
5	0.17	0.19	25	15	25
6	0.09	0.10	16	10	16
7	0.07	0.08	14	10	14
8	0.20	0.23	29	15	29
9	0.09	0.10	16	10	16
10	0.10	0.11	17	10	17
11	0.17	0.19	25	15	25
12	0.13	0.15	21	10	21
P1	–	–	–	23	23
P2	–	–	–	26	26
P3	–	–	–	23	23
P4	–	–	–	26	26

[b]For pedestrian movements, movement time (t) is always equal to the minimum required movement time (t_m).

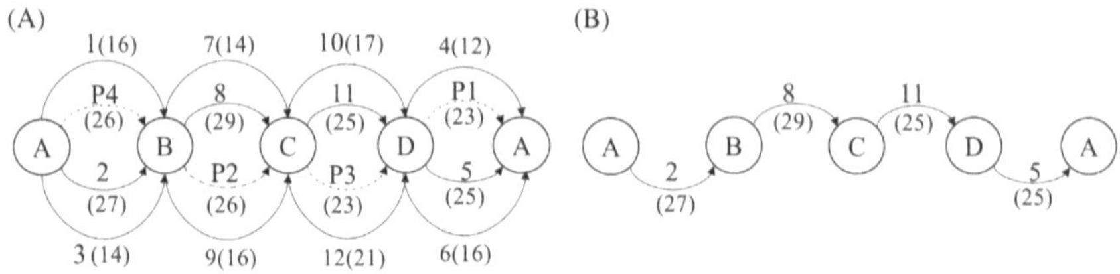

Figure 9.9 Critical movement search diagram. (A) Critical movement search diagram (full). (B) Critical movement search diagram (reduced).

The search diagram is prepared using the required green time (t) for all movements (refer to Fig. 9.9A). As there is no overlap movement, the reduced diagram is obtained by simply keeping the movements receiving the longest green time during each phase and eliminating other movements. Critical movements are the movements remaining in a reduced diagram (refer to Fig. 9.9B). Here, movements 2, 8, 11, and 5 are identified as critical movements during phases A, B, C, and D respectively.

9.5.3 Step 2: calculate intersection parameters

Calculate intersection loss time (L), flow ratio (Y), and green time ratio (U) as the sum of corresponding critical movement parameters.

$$L = l_2 + l_8 + l_{11} + l_5 = 6 + 6 + 6 + 6 = 24$$

$$Y = y_2 + y_8 + y_{11} + y_5 = 0.19 + 0.20 + 0.17 + 0.17 = 0.73$$

$$U = u_2 + u_8 + u_{11} + u_5 = 0.21 + 0.23 + 0.19 + 0.19 = 0.82$$

Refer to Tables 9.2 and 9.3 for critical movement parameters, that is, l, y, and u.

9.5.4 Step 3: calculate cycle time

Calculate practical cycle time using Eq. (9.13).

$$c_p = \frac{L}{1-U} = \frac{24}{1-0.82} = 133.33 \approx 135 \text{ seconds}$$

Calculate optimum cycle time using Eq. (9.14) and stop penalty parameter, $k = 0$ for minimum delay.

$$c_o = \frac{(1.4+k)L+6}{1-Y} = \frac{(1.4+0)24+6}{1-0.73} = 147 \text{ seconds}$$

Maximum cycle time, $c_{max} = 150$ seconds.

We choose cycle time, $c = 140$ seconds which satisfies the condition, $c_p \leq c \leq \text{minimum}(c_{max}, c_o)$

Table 9.4 Check for critical movements (units are specified in Table 9.1)

Movement	$t = uc + l$	$t_m = G_{min} + I$	$t' = max(uc + l, t_m)$
1	20	10	20
2[a]	35	15	35
3	18	10	18
4	14	10	14
5[a]	33	15	33
6	21	10	21
7	17	10	17
8[a]	38	15	38
9	19	10	19
10	21	10	21
11[a]	32	15	32
12	27	10	27
P1	–	23	23
P2	–	26	26
P3	–	23	23
P4	–	26	26

[a]Critical movements identified in Step 1.

9.5.5 Step 4: checking critical movements

Using an estimated cycle time of 140 seconds, check if critical movements identified in Step 2 are still valid. Recalculate required time (t') for each phase as $max(uc + l, t_m)$.

It can be verified from Table 9.4 that during phases A, B, C, and D movements 2, 8, 11, and 5 are still critical.

9.5.6 Step 5: calculate movement and phase green times

9.5.6.1 Movement green times

Firstly, calculate green times for critical movement in each phase using Eq. (9.15).

$$g_2 = \left(\frac{c-L}{U}\right)u = \left(\frac{140-24}{0.82}\right)0.21 = 30 \text{ seconds}$$

$$g_8 = \left(\frac{c-L}{U}\right)u = \left(\frac{140-24}{0.82}\right)0.23 = 32 \text{ seconds}$$

$$g_{11} = \left(\frac{c-L}{U}\right)u = \left(\frac{140-24}{0.82}\right)0.19 = 27 \text{ seconds}$$

$$g_5 = \left(\frac{c-L}{U}\right)u = \left(\frac{140-24}{0.82}\right)0.19 = 27 \text{ seconds}$$

Table 9.5 Summary of movement green time (units are specified in Table 9.1)

Movement	Starting phase	Lost time, l	Effective green, g	Movement time ($Mt = g + l_1$)	Displayed green time ($G = g + l_1 - Y$)
1	A	6	30	36	31
2	A	6	30	36	31
3	A	6	30	36	31
4	D	6	27	33	28
5	D	6	27	33	28
6	D	6	27	33	28
7	B	6	32	38	33
8	B	6	32	38	33
9	B	6	32	38	33
10	C	6	27	33	28
11	C	6	27	33	28
12	C	6	27	33	28
P1	D	6	27	33	28
P2	B	6	32	38	33
P3	C	6	27	33	28
P4	A	6	27	33	28

Table 9.6 Summary of phase green time

Phase	Intergreen time, I	Displayed green ($G = g + l - I$)	Phase change time, F	Green start ($F + I$)	Green end ($F + I + G$)
A	5	31	0	5	36
B	5	33	36	41	74
C	5	28	74	79	107
D	5	28	107	112	140

Check for cycle time, $c = \sum I + \sum G = 20 + 120 = 140$.

$$\text{Check}, g_2 + g_8 + g_{11} + g_5 = 116 = c - L$$

Next, calculate green times for noncritical movements using Eq. (9.16).
For example, $g_1 = (g_2 + l_2) - l_1 = 30 + 6 - 6 = 30$
Movement green times are summarized in Table 9.5.

9.5.6.2 *Phase green times*

Phase change time and displayed green time for each phase are calculated using Eqs. (9.19) and (9.20), respectively. Displayed green time for each phase is displayed green time for nonoverlapping movement receiving green time during that phase, for example, $G_A = G_1 = G_2 = G_3 = G_{P4}$. Phase timings are summarized in Table 9.6.

Table 9.7 Check for movement's degree of saturation

Movement	$x = (c/g)y$	x_p
1	0.43	0.9
2[a]	0.88	0.9
3	0.36	0.9
4	0.28	0.9
5[a]	0.90	0.9
6	0.49	0.9
7	0.31	0.9
8[a]	0.89	0.9
9	0.37	0.9
10	0.51	0.9
11[a]	0.88	0.9
12	0.69	0.9

[a]Critical movements.

9.5.7 *Step 6: movement's degree of saturation*

Calculate movement's degree of saturation (x) using allocated green time and check if $x \leq x_p$ for all movements.

Table 9.7 depicts that for all critical and noncritical movements, the degree of saturation is less than the practical degree of saturation.

9.6 Discussion

This chapter gives a brief overview of the Australian method of fixed-time signal timing design. The two most important points considered while designing this method are:

- The prediction of traffic performance under a given set of conditions including road geometry, arrival flow, signal timings, etc.; and
- Choosing the best possible traffic performance after analyzing changes in the above conditions.

Furthermore, the movement-based method for signal timing design used in Australia and New Zealand is unique. The movement-based approach uses the "movement lost time concept" instead of "phase lost time." This approach helps in the easier understanding of complicated phasing systems or multiphasing systems by facilitating a clear understating of the relationship between the movement and signal phasing characteristics. Though, in practice, there is an emphasis on the use of measured saturation flow, the Australian and New Zealand method also provides for saturation flow estimation that considers the effect of traffic composition, signal timings on saturation flow especially for opposed turn movements, and movements using a short turning lane.

Generally, estimating signal timings that optimize only traffic performance alone may not be sufficient. The traffic signal design often has conflicting objectives, such as safety, operational efficiency, fuel consumption, public transport priority, and constraints regarding pedestrians, and cyclists. The design objectives and constraints vary for each site and the method summarized in this chapter can incorporate various objectives and constraints in signal design.

Bibliography

Akcelik & Associates_Pty.Ltd., 1984. SIDRA Intersection. Retrieved from <http://www.sidrasolutions.com>

Akcelik, R., 1978. X and Y in traffic signal design. In: Paper Presented at the Proc. 9th ARRB Conf, vol. 9, no. 5, pp. 45–60.

Akcelik, R., 1979. Capacity and timing calculation methods for signalised intersections. In: Paper Presented at the International Symposium on Traffic Control Systems, 1979, Berkeley, California, USA.

Akcelik, R., 1981. Traffic signals: capacity and timing analysis. Research Report 123, Australian Road Research Board, Melbourne, Australia.

AS 2578–2009, Traffic signal controllers.

AS 1742.14–2014, Manual of uniform traffic control devices: part 14 – Traffic signals.

AS 2144–2014, Traffic signal lanterns.

AS 2339–2017, Traffic signal posts, mast arms and attachments.

Austroads, 2015a. Guide to traffic management part 2: traffic theory. Second edn, AGTM02-15, Austroads, Sydney, NSW.

Austroads, 2015b. Guide to traffic management: part 13: road environment safety. Second edn, AGTM13-15, Austroads, Sydney, NSW.

Austroads, 2016a. Guide to traffic management part 9: traffic operations. Third edn, AGTM09-16, Austroads, Sydney, NSW.

Austroads, 2016b. Guide to traffic management part 10: traffic control and communication devices. Second edn, AGTM10-16, Austroads, Sydney, NSW.

Austroads, 2017a. Guide to traffic management part 3: traffic studies and analysis. Third edn, AGTM03-17, Austroads, Sydney, NSW.

Austroads, 2017b. Guide to traffic management part 6: intersections, interchanges and crossings. Third edn, AGTM06-17, Austroads, Sydney, NSW.

Lowrie, P., 1990. Scats, Sydney co-ordinated adaptive traffic system: a traffic responsive method of controlling urban traffic. Roads and Traffic Authority NSW, Darlinghurst.

May, A., 2005. Encyclopedia of Melbourne – eMelbourne-traffic lights. Retrieved from <http://www.emelbourne.net.au/biogs/EM01508b.htm>

NSW, R, 2010. Traffic signal operation. Retrieved from <http://www.rms.nsw.gov.au/business-industry/partners-suppliers/documents/technical-manuals/rtatc106b.pdf>

RTA-NSW, 2008. Traffic signal design section-2 warrants. Retrieved from <http://www.rms.nsw.gov.au/documents/business-industry/partners-and-suppliers/guidelines/complementary-traffic-material/tsdsect2v14-i.pdf>

RTA-NSW, 2013. SCATS. Retrieved from <http://www.scats.com.au/how-to-purchase.html>

RTA, N, 2008. Traffic signal design section-12 controller. Retrieved from NSW: <http://www.rms.nsw.gov.au/documents/business-industry/partners-and-suppliers/guidelines/complementary-traffic-material/tsdsect12a-i.pdf>

Japan

10

Hideki Nakamura[1], *Miho Iryo-Asano*[1] *and Takashi Oguchi*[2]
[1]Graduate School of Environmental Studies, Nagoya University, Nagoya, Japan,
[2]Institute of Industrial Science, The University of Tokyo, Tokyo, Japan

10.1 Basic information on traffic signal control in Japan

10.1.1 Historical overview and development of traffic signal control in Japan

It was 1930 when the first color lamp type, automatic traffic signal imported from the United States was introduced in Japan. In this period, domestic manufacturers began to develop domestic traffic lights. Since 1957 the National Police Agency (NPA) has set up plans for the installation of 200 units of traffic lights each year and has advanced signaling equipment development, and in 1965 this number grew to over 8000 nationwide. In 1958, multistage fixed-time control of three stages of morning, daytime and nighttime, and a traffic adaptive signal combined with a vehicle detector were introduced in 1963. Audio signals for the visually impaired were introduced for the first time in Tokyo in 1955 (Japan Society of Traffic Engineers (JSTE), 2006).

The Japan Society of Traffic Engineers (JSTE) was established in 1965, and research and technology development on traffic control by government agencies and traffic signal manufacturers was promoted. In 1966, the area-wide traffic signal control started in the Ginza district of Tokyo. A centralized supervisory control device of automatic responsive signaling equipment was installed in 1967, and centralized control of signal parameters came into operation. Also, "scrambled crossing phase" (see control type 1 of Table 10.1) was first introduced in Kumamoto in 1968, and the movement separation signal control using green arrow light was introduced in Aichi prefecture for the first time in 1972.

In the latter half of the 1960s, the number of traffic fatalities rapidly increased, and in 1970 the number of traffic fatalities who died within 24 hours was recorded at 16,765 persons per year (record high). Under the environment of this period, called the "traffic war," traffic safety facilities including traffic lights were rapidly implemented nationwide, and it contributed greatly to a reduction in the number of traffic fatalities after that.

With the spread of microcomputers in the 1980s, microadaptive functions (terminal adaptive control) and program functions became incorporated into signal controllers. In 1995, the Metropolitan Police Department (MPD) introduced a signal

Global Practices on Road Traffic Signal Control. DOI: https://doi.org/10.1016/B978-0-12-815302-4.00010-8

Table 10.1 Pedestrian–vehicle separation phases

Control types		Phasing plan (ϕi: phase number i)
1. Dedicated pedestrian phase	Dedicated pedestrian phase is provided. During that phase, vehicles should not enter the intersection and pedestrians can cross any crosswalks	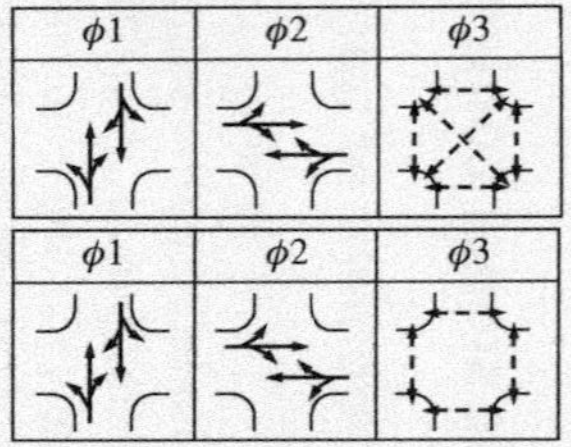
2. Right- and left-turn vehicle separation	Pedestrian phase and conflicting turning vehicle phase are separately provided	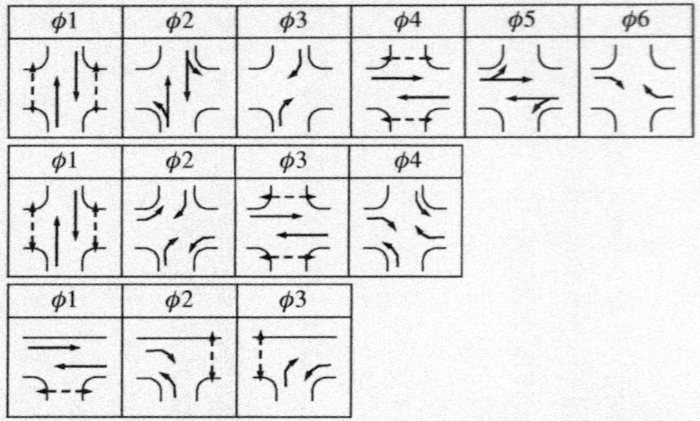 Note: dedicated lanes for every movement are required.

control scheme that automatically generates cycle length and green split from the load factor of each approach at about 300 intersections in Tokyo, and this method was named MODERATO (Management by Origin-DEstination Related Adaptation for Traffic Optimization) in 1998. Also, in 2002, the introduction of a pedestrian–vehicle separation signal control method (see Table 10.1) was started for the purpose of securing pedestrians and cyclists crossing at intersections.

10.1.2 The state-of-the-art

10.1.2.1 General information and guidelines for traffic signal control

Road traffic in Japan is a left-hand side system. All traffic control, including traffic signals, is under the jurisdiction of the police. Although it is said that each prefectural police department has its own guideline, the contents of each guideline are not open to the public. Only one guideline on traffic signal control is open to road traffic engineers and researchers—the "Manual on Traffic Signals" (Japan Society of Traffic Engineers (JSTE), 2006) published by the JSTE. Regarding guidelines for intersection design, "Planning and Design of At-Grade Intersections" (Japan Society of Traffic Engineers (JSTE), 2002) is commonly used by road authorities and engineers. Since considerable contents overlap in the two guidelines as they are closely related, a new guideline combining them, "Basic Manual on Planning, Design and Traffic Signal Control of At-Grade Intersection", was published in 2018 by the JSTE (Japan Society of Traffic Engineers (JSTE), 2018). This chapter is mainly based on the contents of this new guideline. For the planning, design, and operation of roundabouts, the first guideline "Roundabout Manual" (Japan Society of Traffic Engineers (JSTE), 2016) was published in 2016, also by the JSTE.

10.1.2.2 Statistics related to traffic signal control

The total number of signalized intersections in Japan was about 171,000 as of April 2017, which denotes 1.35 signalized intersections per 1000 inhabitants or 454 signalized intersections per 1000 km^2 in Japan. Among this total, traffic responsive and adaptive control using vehicle detectors and controlled by centralized traffic control systems accounts for 43%, isolated actuated control 9%, and fixed-time control 48%, respectively. In addition, there are about 31,000 signalized pedestrian crossings at places other than intersections.

The centralized traffic control (43%, 74,000 intersections) can be categorized into two types: one is the so-called "program formation control" (8%, 13,000) or MODERATO which is an adaptive control and produces a set of signal parameters based on detector information, and the other is the so-called "program selection control" or responsive control (35%, 60,000) which selects a set of predetermined signal parameters.

All control systems in Japan are provided by several Japanese electric companies and their specifications are strictly controlled and standardized by the

NPA. Thus the differences between the products of these companies are very minor and limited.

10.1.2.3 Representative traffic signal heads and arrangements

Signal heads in Japan generally have green, yellow, and red lights arranged laterally from the left (Fig. 10.1A), although the vertical type (Fig. 10.1B) is used in snowy cold regions. The green arrow light indicates that even if a red circle light is displayed at the same time, users can proceed into the direction indicated. A typical

Figure 10.1 Traffic signal heads in Japan. (A) Ordinary horizontal arrangement. (B) Vertical arrangement for snowy and cold regions. (C) Signal head with green arrows. (D) Signal head for pedestrian and bicycle.

usage of this green arrow is for the protected right-turn (PRT) movement. In recent years, arrows for each movement direction are increasingly used for movement separation control at intersections on multilane roads (Fig. 10.1C). The yellow arrow light gives the right of way for trams only. The red arrow light is not provided in Japan. For this reason, signal control based on lane groups by using three color arrow lights only, which is often used in some Western countries, is impossible under the current legislation. In addition, the remaining time of the signal display is not indicated except for the crossing pedestrians (see Fig. 10.8).

Traffic signal heads are generally installed over the roadway on the far side of the intersection to aid visibility. It is also common to additionally implement multiple signal heads to secure visibility. An example of a general type of signal head for pedestrians and cyclists is shown in Fig. 10.1D. The flashing green light is used for the clearance of crossing pedestrians. Under the Road Traffic Law, it is stipulated that crossing pedestrians should not begin crossing when there is a flashing green light, and pedestrians on the crosswalk must promptly complete their crossing or return. Although there is no signal device exclusively for bicycles, when necessary the ordinary signal head is used exclusively for bicycles with the auxiliary signboard "bicycles only." In addition, pushbuttons are widely used where crossing pedestrian demand is low. An acoustic device can be supplementarily used for visually impaired pedestrians.

Since a signal control based on lane groups or the "dual-ring" signal control is not generally used in Japan, channelization of vehicle movements using raised islands and/or two-stage crossing of pedestrians are not yet common. Therefore, intersections are currently designed on the premise of relatively simple phasing plans. Hence, the introduction of a more flexible and enhanced phasing scheme as well as sophisticated intersection design corresponding thereto at the intersection of multilane roads is a particularly significant issue in Japan. The new version of the guideline promotes flexible intersection designs and control schemes.

10.2 Control strategies selection policy

10.2.1 Warrants for the installation of traffic signals

According to the "Notice on Guidelines for Installing Traffic Signals" (2015) by the Traffic Bureau of the NPA, the basic requirement for introducing traffic signal control is to satisfy all of the following conditions: physical spaces available for installing a traffic signal and for road users waiting for a green time, visibility of the signal lights, and the maximum two-way traffic volume of the major road to be 300 veh/h or more. In addition to the requirements above, at least one of the following conditions must also be satisfied in principle.

1. Past history of the number of road accidents

   The case where there have been two or more accidents with casualties in the past year that are considered to have been deterred by the installation of traffic lights at or near the place where the traffic lights are to be installed. And no other measures can substitute for ensuring traffic safety.

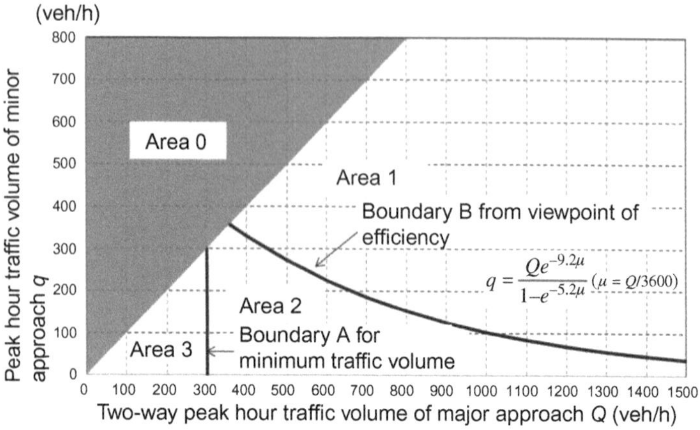

Figure 10.2 Traffic demand requirements for implementing/removing a traffic signal at intersections.
Adapted from National Police Agency of Japan, 2015. Guideline of Traffic Signal Installation. (in Japanese).

2. Road characteristics
   The case where it is necessary to secure traffic safety, especially in the vicinity of elementary and/or junior high schools, kindergartens, nursery schools, child parks, hospitals, nursing homes for the elderly, and so on.
3. Vehicle traffic volume
   At the intersection where a combination of the two-way peak-hour traffic volume of the major road and the peak-hour inflow traffic volume from the minor road is in Area 1 shown in Fig. 10.2.
4. Pedestrian traffic volume
   The case of a heavy pedestrian demand for crossing a roadway but where it is very difficult due to heavy conflicting traffic volume, and no pedestrian bridges in the vicinity.

10.2.2 Selection criteria of actuated or fixed-time control for isolated intersections

Criteria for selecting the traffic signal control method for isolated intersections are as follows.

1. Key intersections of trunk roads with high traffic demands
   In this case the base is fixed-time control. For the case of a multiple phasing plan, adaptive control can be partially applied to some of the phases such as PRT phase, while other phases are kept as fixed time control.
2. Intersections of a major road and a minor road with low traffic demand
   This is based on semiactuated control, which makes the minor road actuated. When necessary to provide the opportunity for pedestrians to cross the major road, a pedestrian pushbutton is provided.
3. Intersections of roads with lower traffic demands
   In general, fixed-time control is sufficient, however when pedestrian and/or vehicle traffic demands are very low, adaptive control with a pedestrian pushbutton is desirable.

10.2.3 Terminal adaptive control (actuated control)

Terminal adaptive control is an actuated control that the terminal controller expands and contracts the green time or such like on the spot, based on various kinds of sensor information connected to it. Full-actuated control is a method where a traffic signal timing is actuated by traffic flows from every entry. This control is suitable for intersections and/or time periods where traffic volume is not high and its variation is irregular, however it is not very commonly used.

On the other hand, semiactuated control by detecting right-turn vehicles and/or pedestrians is widely used in Japan. There are various types of semiactuated control methods that detect specific targets and change the signal timing for them; such as gap actuated for right-turn vehicles, dilemma actuated, bus actuated, high-speed actuated, pedestrian actuated, and elderly actuated, etc. For example, in the case of gap-actuated control, a vehicle detector provided at a position within about 30 m upstream from the stop line detects the approach or presence of a vehicle, thereby prolonging or shortening the green time.

10.2.4 Centralized adaptive control

Centralized adaptive control is a method where the central control unit of the traffic control center determines the signal parameters at a certain time interval (e.g., 1, 2.5, 5 minutes), computes each signal display time, and remotely controls the terminal signal controllers. Conventionally, at traffic control centers in Japan, a program selection control (also referred to as a pattern selection control) has been commonly adopted, which selects a control program designed in advance according to several traffic patterns classified by traffic volume and time occupancy. However the program selection control has the following disadvantages:

- It is very labor-intensive work to design and update signal control parameters;
- The methodology to determine the coefficient for calculating the evaluation index (a weighted sum of traffic volume and time occupancy) is empirical and its grounds are often ambiguous; and
- Cycle length tends to be designed long due to considering some margin, and this is likely to cause unused green times and to increase pedestrian waiting time.

To overcome the weak points mentioned above, a program formation control method MODERATO was developed in the mid-1990s. In this control, signal control parameters such as green split, cycle length, and offset are automatically updated according to traffic conditions.

10.2.4.1 Traffic indicator used for program selection control (traffic responsive control)

Either the weighted sum of traffic volume and time occupancy S_1 shown in Eq. (10.1), or the larger value obtained by normalizing the traffic volume and time occupancy with the reference values S_2 shown in Eq. (10.2), is used for the program selection control.

$$S_1 = \alpha \cdot Q + \beta \cdot O \tag{10.1}$$

$$S_2 = max\left\{\frac{Q}{Q'}, \frac{O}{O'}\right\} \tag{10.2}$$

where Q = measured traffic volume; Q' = reference traffic volume (constant); O = measured time occupancy; O' = reference time occupancy (constant); and α, β = constant.

These indicators consider the characteristic that the time occupancy increases as the queue expands to the vicinity of the vehicle detector position, which is set upstream of the queue length that can be handled in one signal cycle.

10.2.4.2 Load factor used for the program formation control

In the program formation control method MODERATO, a concept of the load factor is applied in order to represent a wide range of traffic conditions including the oversaturation state. The load factor is able to deal not only with the throughput but also with the remaining queues in the oversaturated conditions. The load factor ρ expressed by Eq. (10.3) is the basis of updating cycle length and green split based on the vehicle detector information.

$$\rho = \frac{(Q_{in} + E)/C}{s} \tag{10.3}$$

where Q_{in} is the number of vehicles passed the detector position in the entry section in each cycle, E is the number of remaining vehicles at the end of the green time, C is cycle length, and s is saturation flow rate.

10.2.5 Selection of isolated or coordinated control

It is generally known that when the ratio of the travel time between adjacent intersections to the common cycle length is sufficiently large, the significance of traffic signal coordination is small, and the effect of system control is greater as the traffic volume in the coordinating direction increases. For the coordinated signal control, in sections of the route where the traffic flow is relatively stable and a change in the demand pattern such as uncrowded, normal, and crowded is considerably distinctive, the multistage program selection control is applied. In this case, the terminal adaptive control (e.g., gap-actuated control for right-turn traffic) can also be combined at some intersections. Meanwhile, in the case of such routes as major arterial roads where traffic volume is high, hourly variation is distinctive, and high efficiency is required, the centralized adaptive control that enables advanced traffic signal coordination closely responding to traffic flow variation is applied.

10.3 Signal timing procedure

The procedure for traffic signal control design proposed in the manual is shown in Fig. 10.3.

Step 0: Investigation of the conditions at the intersection

Intersection geometry and traffic condition data are collected as necessary input data. Intersection geometry data include the number of lanes and their configurations, clearance/entering distance (distance from the stop line to the conflict point) for each combination of movements, lengths of turning lanes, and so on. Traffic conditions contain design traffic volume, saturation flow rate, and approaching speed to the intersection.

Step 1: Settings of phasing plan

The phasing plan is determined based on lane configurations and traffic demands. Two-phase control is applied for simple four-leg intersections. When the right-turn traffic demand is high and an exclusive right-turn lane is available, PRT phase can be applied. If the pedestrian demand is high and the impact of the conflict between pedestrians and turning vehicles is expected to be significant, pedestrian–vehicle separation control is considered.

In this step, the phasing plan may not be fixed to one, particularly at multilane intersections where more flexible phasing plans are available. Therefore, multiple phasing plan options can be given.

Step 2: Intergreen time and lost time calculation

Yellow and all-red intervals are calculated using the clearance/entering distance and the approaching speed. Lost time is also calculated based on these.

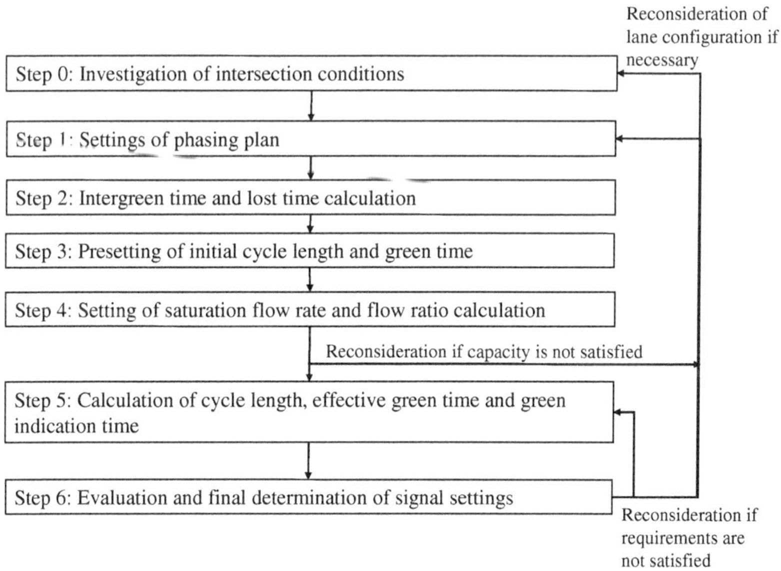

Figure 10.3 Flowchart of signal control design.

Step 3: Presetting of initial cycle length and green time

For the calculation of intersection capacity, the number of right turners passing at the end of the permissive green phase is necessary. In order to obtain this, initial cycle length is arbitrarily determined. Accordingly, the remaining time, calculated by subtracting the lost time from the initial cycle length, is assigned to each phase as the initial green time in proportion to the design traffic volume. This information is also utilized for calculating the saturation flow rate of a shared through/turn lane.

Step 4: Settings of saturation flow rate and flow ratio calculation

After settings of saturation flow rate, the intersection flow ratio is calculated based on the intersection geometry, the design traffic volume, and the phasing plan.

Step 5: Calculation of cycle length, effective green time, and green indication time

Cycle length is determined using total lost time and intersection flow ratio. Then, the effective green time and green indication time for each phase are calculated.

Step 6: Evaluation and final determination of signal settings

When the pedestrian green is given at the same time as vehicle green, the vehicle green time is examined from the viewpoint of minimum pedestrian green time. If this requirement is not satisfied, it goes back to Step 5 and the effective green time and the cycle length are reconsidered so that the effective green time guarantees the pedestrian green time, keeping the green split. The volume to capacity ratios of all movements are also examined. Finally, among the multiple options of phasing plan, the most appropriate phasing plan is selected considering overall safety and efficiency.

This chapter only focuses on the settings of traffic signal timing, while phasing plans and signal timing settings actually depend on intersection geometry, such as lane configuration and length of turning lanes. Therefore, the manual recommends reconsidering the geometry conditions in case the capacity cannot meet the demand only by the improvement of signal control design.

10.4 Signal phasing

At typical four-leg or three-leg intersections, the phasing plan is determined mainly based on how to give the right of way to pedestrians and right-turning vehicles. When the flow ratio of an intersection is small and the demands of pedestrians and right turners are low, two-phase control can be applied as shown in Fig. 10.4. The

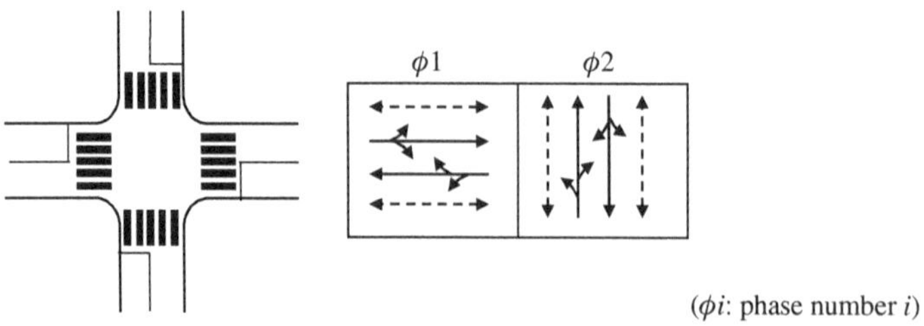

Figure 10.4 Two-phase signal control.

arrows drawn by the solid lines and dashed lines are the vehicle and pedestrian movements, respectively, which have the right of way in each phase.

In the simple two-phase control, turning vehicles and opposing through traffic or pedestrians have conflicts in the same phase. When the demand becomes larger, more phases are required to have enough capacity for these conflicting traffics. Protected (or permissive-protected) right-turn phase, lagging phase, and pedestrian–vehicle separation phase are commonly used as typical multiphase plans in Japan.

PRT phase is often given after permissive right-turn phase or after protected through and left-turn phase. The upper left of Fig. 10.5 shows the typical phasing plans of permissive-protected phase applied to west and east approaches. Providing the PRT phase before the permissive phase is not recommended because of the possible collisions with opposing through traffic. If protected phase for right-turn vehicles as in the upper right of Fig. 10.5, the right-turning vehicles have full priority in $\phi1$ but they need to yield to opposing through traffic and pedestrians in $\phi2$. In this case, drivers of right-turning vehicles approaching the intersection in the beginning of $\phi2$ may think they are still protected from all other movements. In this way, generally speaking in Japan, switching from a high-priority state to a low-priority state is avoided when the right of way of certain movement is consecutively given. This is because the road users may not properly recognize the priority change from higher to lower and enter the intersection by misunderstanding their priority state. When only the protected phase is applied without the permissive phase, the protected phase is usually inserted after the through and left-turn phase as in the center of Fig. 10.5.

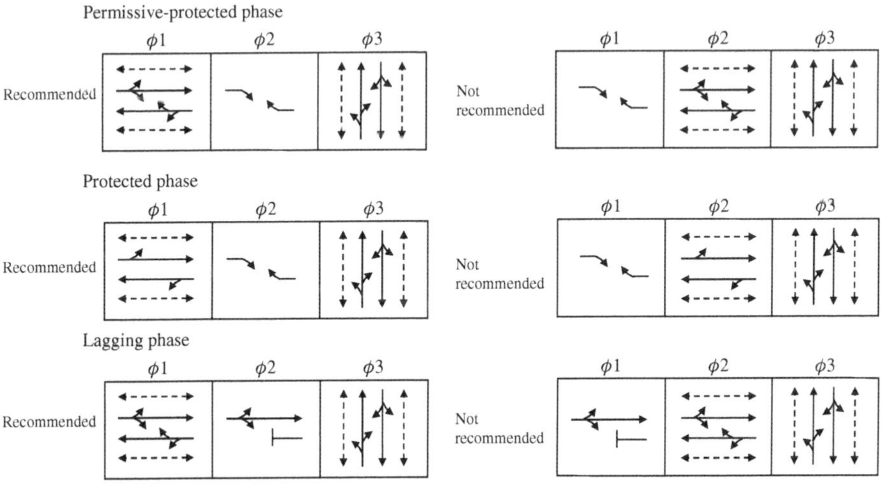

(ϕi: phase number i)

Figure 10.5 Phasing plans for right-turn vehicles (permissive-protected, protected, and lagging).

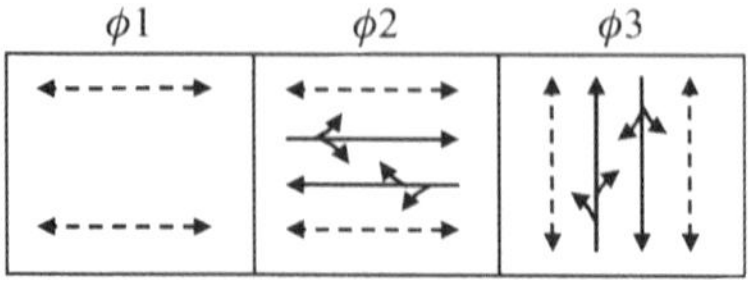

(ϕi: phase number i)

Figure 10.6 Leading pedestrian interval (LPI).

Lagging phase is used when traffic lights for different directions from one approach cannot be separately provided. In this case, the green time of an approach with higher demand is extended while the green of the opposing approach is terminated earlier, as in the bottom of Fig. 10.5. For the same reason as the permissive-protected phase, the "leading phase" of one approach to the opposite approach is not recommended.

As a pedestrian-oriented control type, the pedestrian–vehicle separation phase is has become popular recently. Examples of pedestrian–vehicle separation phases are shown in Table 10.1. They are installed when capacity or safety problems are expected because of conflicts between pedestrians and turning vehicles. In addition to the phase types listed in Table 10.1, leading pedestrian interval (LPI) control, which starts with pedestrian green a few seconds earlier than the conflicting vehicle phase, is also starting to be installed at a few intersections (Fig. 10.6).

10.5 Parameter determination

10.5.1 Saturation flow rate

In the past, saturation flow rate was derived using a formula multiplying the base saturation flow rate by adjustment factors considering geometry and traffic characteristics. However, the results often deviate from the actual saturation flow rates. The new manual (Japan Society of Traffic Engineers (JSTE), 2018) recommends using the values observed on site. If the observation is difficult because the intersection is newly implemented, observation data at other sites with similar traffic and geometry conditions may be carried out. In case that similar intersections are not available, the calculation formula with the adjustment factors can be utilized.

The formula is given by multiplication of the base saturation flow rate and adjustment factor ratios. The basic saturation flow rate is given as 2000 pcu/h for through traffic and 1800 pcu/h for turning traffic. Adjustment factors include lane width, gradient, heavy vehicle percentage, impact of turning vehicles, and the presence of a bus stop near the intersection.

Among these, the impact of turning vehicles is further classified in the case of dedicated lanes with protected phases and that of shared lanes or dedicated lanes without protected phases. In the case of shared lanes or permitted phases, the saturation flow rate is adjusted considering the impact of opposing through vehicles or pedestrians which may block the movement of turning vehicles considering the degree of saturation of the conflicting movements. This analysis needs cycle length and green time information to obtain the degree of saturation. Therefore, the preset cycle length and green time given in Step 3 of Section 10.3 are utilized for the calculation. After obtaining the signal parameters, saturation flow rate needs to be evaluated again in Step 6 to confirm whether the calculated signal parameter can satisfy the capacity requirement.

Theoretically speaking, the conditions when the vehicles are blocked by other movements do not achieve the saturation flow rate. However, in the Japanese manual, such an approximation method is applied to roughly calculate the necessary green time for each phase.

10.5.2 Intergreen times (yellow and all-red)

In principle, vehicles approaching the intersection have to stop at the stop line when they face a yellow indication. They can pass through the stop line only when they cannot safely stop. The yellow interval is determined based on this principle. Suppose that vehicles approaching the subject intersection run at velocity V. The minimum stopping distance D_s for approaching vehicles to stop is expressed by Eq. (10.4) using maximum deceleration d (m/s^2):

$$D_s = V\tau + \frac{V^2}{2d} \tag{10.4}$$

where τ is the reaction time of drivers. This means that if the distance from vehicles to the stop line is more than D_s, they can safely stop. Meanwhile, vehicles whose distance is less than D_s should be able to pass through the stop line before the red indication starts. Therefore, the yellow interval Y should be more than or equal to D_s/V. If the yellow interval is much larger than D_s/V, vehicles at D_s' ($VY > D_s' > D_s$) will have two options: safely stop at the stop line or safely pass through the stop line before red starts. As providing two options will confuse driver decision making, usually Y is determined so that $Y = D_s/V$. Considering Eq. (10.4), the yellow interval can be derived by Eq. (10.5), where $d = 3.0$ (m/s^2) and $\tau = 0.7$ (s) are utilized for calculation in the Japanese manual.

$$Y = \frac{D_s}{V} = \tau + \frac{V}{2d} \tag{10.5}$$

The all-red time should be designed so that vehicles entering at the end of yellow time can pass through the conflict point with the path of road users entering at the onset of the consecutive green phase. Here, clearance time t_c and entering time t_e are defined as the travel times from the stop line to the conflict point for the vehicles in the previous phase and in the next phase, respectively. The difference $\Delta T = t_c - t_e$ is the all-red time, which is the required time gap between the end of the yellow time and the onset of the consecutive green time. In practice, all-red time is sometimes determined assuming $t_e = 0$ considering the safety margins.

10.5.3 Lost time (start-up and clearance lost times)

The lost time at each intergreen is estimated depending on the conditions of phases as below.

1. The case that yellow and all-red intervals are inserted between phases

If the yellow time is longer than or equal to 3 seconds, the lost time at intergreen is determined as 1 seconds less than summation of yellow and all-red intervals. Thus, the effective green time is 1 seconds less than the green indication time. If the yellow time is less than 3 seconds, the lost time is equivalent to the summation of yellow and all-red time.

2. The case that the green indication is followed by a green arrow indication (permissive-protected phasing)

This is the additional treatment given in the new manual (Japan Society of Traffic Engineers (JSTE), 2018). The order of the indications is "green circle (permissive phase)→yellow→green arrow + red circle (protected phase)→yellow→red." There are no "red-only" indications between the permissive and protected phases. In this case, some of the vehicles which have the right of way at the protected phase enter the intersection before the onset of the protected phase. Therefore, the lost time L_j (second) of phase j is given as follows.

$$L_j = Y_j - 1 - \frac{D}{v} \quad \text{if } Y_j \geq 3 \tag{10.6}$$

$$L_j = Y_j - \frac{D}{v} \quad \text{if } Y_j < 3 \tag{10.7}$$

where Y_j (s) is the yellow time of phase j, D (m) is the distance between the stop line and the yield line in the intersection as in Fig. 10.7, and v (m/s) is the entering speed.

3. The case that the green phase of the approach with lower flow ratio terminates before the other approach in the lagging phasing plan

The yellow time between the two phases is effectively used by the approach with the higher flow ratio. Therefore, this interval is not considered as the lost time.

10.5.4 Cycle length

To obtain the cycle length, the flow ratio needs to be calculated. The lane flow ratio λ_{ij} of lane j at phase i is given as

$$\lambda_{ij} = \frac{q_{ij}}{s_{ij}} \tag{10.8}$$

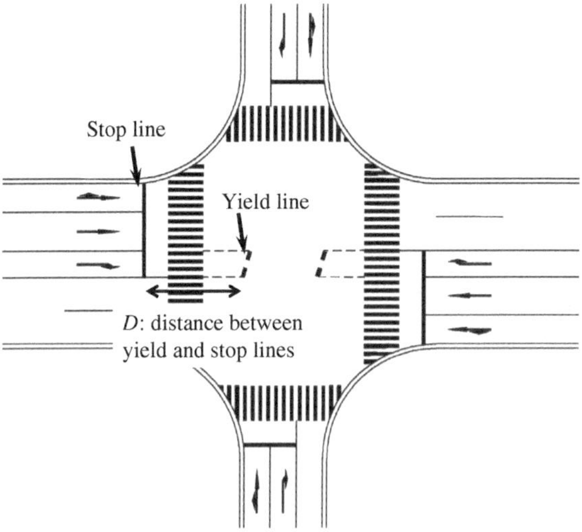

Figure 10.7 Stop line and yield line for right-turning vehicles.

where q_{ij} and s_{ij} are the design traffic volume and saturation flow rate of lane j at phase i, respectively. The phase flow ratio of phase i, λ_i, is defined as the maximum value of the lane flow ratio. The movement which has the maximum lane flow ratio in a phase is the critical movement of the phase.

$$\lambda_i = \max(\lambda_{i1}, \ldots, \lambda_{ij}, \ldots, \lambda_{in}) \tag{10.9}$$

where n is the number of lanes which have the right of way during phase i. The intersection flow ratio λ is given as the summation of the phase flow ratios.

$$\lambda = \sum_i \lambda_i \tag{10.10}$$

Theoretically, minimum cycle length $C_{\min}$ under uniform arrivals at an isolated intersection is given as Eq. (10.11):

$$C \geq \frac{L}{1 - \lambda} = C_{\min} \tag{10.11}$$

where L is the intersection lost time (summation of lost time for each phase). As a practical cycle length, following cycle length is utilized considering the randomness of arrival traffic. α is a parameter and the value of 0.9 is proposed in the manual.

$$C \geq \frac{L}{1 - \lambda/\alpha} = C'_{\min} \tag{10.12}$$

Too long a cycle length increases not only the delay but also the possibility of illegal user behavior due to irritation. The maximum cycle length is recommended as 120 seconds. In the case of exceptional conditions, at most 160 seconds can be allowed.

However, in practice, cycle lengths much longer than 120 seconds are commonly used, mainly at intersections on urban arterials and suburban highways. The size of these intersections is usually large due having many number lanes but channelization with raised splitter or corner islands is not commonly used. Thus, crosswalks at those intersections are apt to be long and the necessary pedestrian time is likely to cause long cycle lengths. Two-stage pedestrian crossing at signalized intersections is uncommon in Japan.

10.5.5 Green times for vehicular traffic (minimum, maximum)

Effective green time is determined so that the degrees of saturation of each phase are equivalent. For that, the difference between the cycle length and the lost time is allocated in proportion to phase flow ratios. When vehicle green time and pedestrian green time are given simultaneously, the summation of the minimum green time for pedestrians and the "pedestrian-red/vehicle-green time" becomes the minimum vehicle green time.

The "pedestrian-red/vehicle-green time" (PR) is the time for turning vehicles to pass through the crosswalk when the turning vehicles and pedestrians are given the right of way in the same phase and the pedestrian demand is high. PR interval is given dependent on the turning vehicle demand.

10.5.6 Pedestrian times (minimum green, clearance time)

The minimum pedestrian time is explained as the summation of the discharge time and the clearance time. The discharge time is the necessary time that all pedestrians waiting at the onset of pedestrian green phase in a queue can start crossing. The clearance time is the time for pedestrians to complete the crossing. Therefore, the minimum pedestrian time t_p is given as Eq. (10.13).

$$t_p = \frac{L_p}{V_p} + \frac{p}{s_p \times W} \tag{10.13}$$

where L_p is the crosswalk length, p is the number of pedestrians in the queue at the onset of green, s_p is the saturation flow rate of pedestrians per unit width, W is the effective crosswalk width, and V_p is the crossing speed.

The indications of pedestrian phase in Japan consist of steady green and flashing green. The meaning of steady green is that pedestrians can start crossing. The flashing green means that pedestrians should not start crossing and those who are on the crosswalk have to either complete the crossing or return to the original side immediately. Considering this definition, the minimum flashing green interval is given as

Figure 10.8 Example of countdown pedestrian signal head (International Association of Traffic and Safety Sciences (IATSS), 2007).

the time pedestrians can cross half of the crosswalk length. The minimum steady green interval is the remaining time of t_p by subtracting the minimum flashing green interval. It is also noted that the steady green interval should be at least 5 seconds.

Although the Japanese manual recommends the settings mentioned above, in practice the minimum steady green interval is set as the same value as t_p in Eq. (10.13) or even longer. The flashing green time is given according to the settings above, in addition to t_p. Therefore the total time required for pedestrians including flashing green tends to be much longer than the recommended value.

The pedestrian flashing green time is not sufficient for the pedestrian crossing time. Thus, pedestrians who start crossing at the end of pedestrian green time cannot complete crossing before the onset of red. To improve pedestrian safety, countdown pedestrian signals are widely utilized. There are several types of countdown signals, with the most common type displaying the remaining time by the number of dots displayed inside the traffic light, as shown in Fig. 10.8 (International Association of Traffic and Safety Sciences (IATSS), 2007). At the beginning of pedestrian green, all the dotted lights turn on as in the left-hand side of Fig. 10.8. The dots disappear one by one as the time proceeds, and all the dots disappear when the pedestrian flashing green starts. Similarly, the countdown red dots are displayed during the red time.

10.5.7 Capacity calculation for determining green splits

To determine green splits, the remaining time subtracting the lost time from the cycle length is assigned to each phase in proportion to their flow ratios. The capacities of protected phases are calculated by multiplying the green split and the saturation flow rate. Meanwhile in the permissive phases, vehicle can pass through the intersection by searching for gaps in conflicting traffic. The capacity of the permissive phase can be given based on the gap acceptance theory. In addition to this, there are additional vehicles which can be stored between the stop line and the yield line during the green phase. These vehicles are able to pass the intersection at the intergreen, even if there are no available gaps within the green time. Therefore, this amount of vehicles is added for the calculation of capacity.

Since the vehicles stored before the yield line appear in each cycle, this number of vehicles per unit time is dependent on the cycle length. Thus, there is a need to preset the initial cycle length for checking the capacity of the permissive phase. Furthermore, the capacity of the permissive phase should be evaluated again after the calculation of cycle length.

10.5.8 Red times (minimum, maximum)

The red time of each movement or approach is given as the results of the above calculation procedure. There are no configurations of minimum or maximum red time.

10.6 Special considerations

Japan faces frequent natural disasters, such as earthquakes, storms, and floods. In the case of disasters, it is essential to control the traffic for smooth evacuation and for emergency vehicle access into densely inhabited districts. For that, some of the prefectures in Japan consider disaster-responsive traffic signal control systems. When a disaster, such as a major earthquake, occurs, signal phasing plans at the intersections with the disaster-responsive system switch to emergency modes from ordinary settings so that the traffic to central downtown is banned or limited.

10.7 Example

The signal timing plan of the example intersection is designed based on the procedure of the Japanese manual (Japan Society of Traffic Engineers (JSTE), 2006), as well as its current application in practice. Although the manual provides the recommended procedure, it is often simplified in practice. Therefore, the typical procedure used in practice is applied in this section.

The geometry and saturation flow settings are given as in Fig. 10.9. In addition to the common settings, standard crosswalk width (4 m) and typical settings of vehicle approaching speed are applied as in this figure.

Step 0: Investigation of the conditions at the intersection

Intersection geometry, traffic conditions, and other necessary data are as given in Fig. 10.9.

Step 1: Settings of phasing plan

This is a multilane intersection and thus the phasing plan can be flexible. However, since the four-phase plan with permissive-PRT phase is widely used in Japan, it is also applied to this example, as in Fig. 10.10.

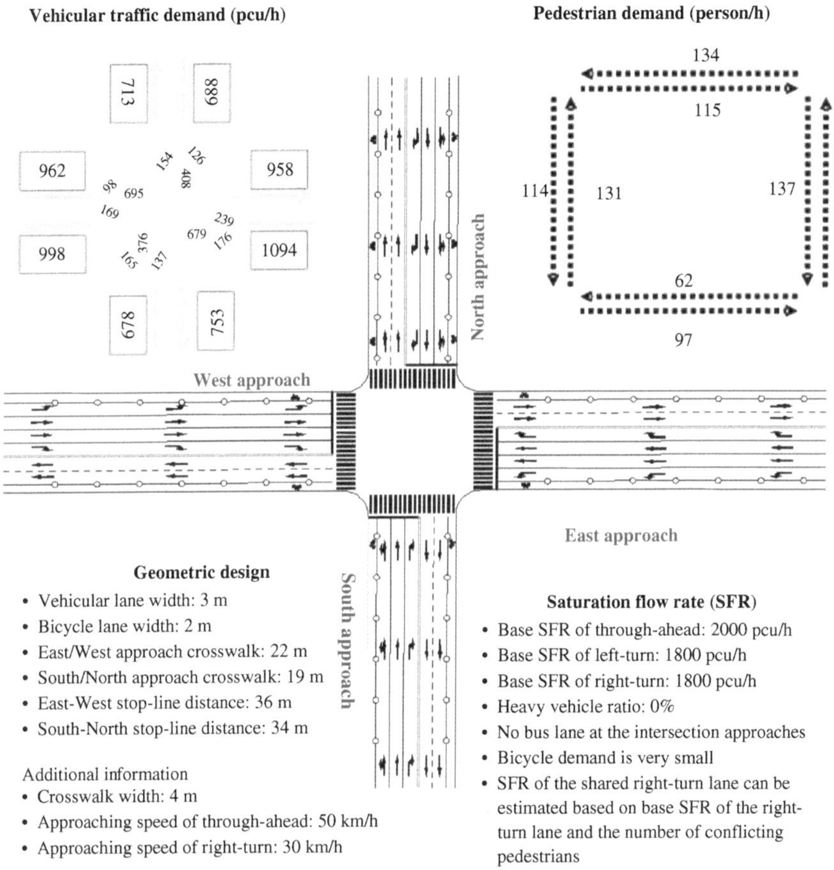

Figure 10.9 Example intersection and hypothesized condition.

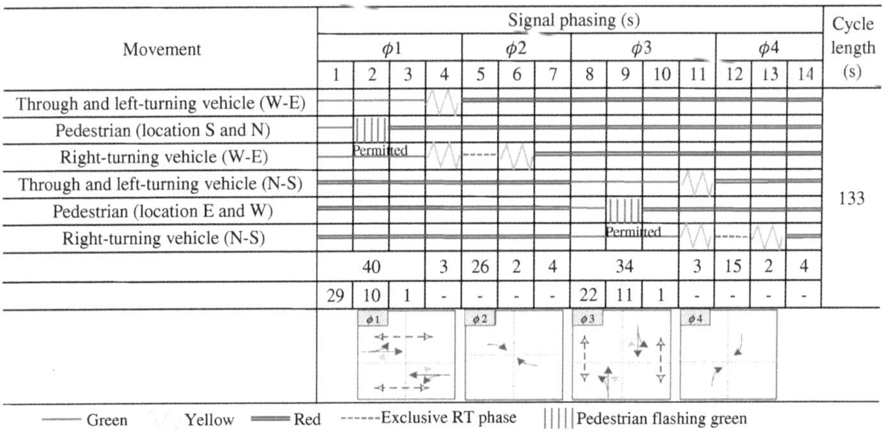

Figure 10.10 Four-phase signal timing plan based on the procedure of the current Japanese manual.

Step 2: Intergreen time and lost time calculation

The yellow interval is calculated using Eq. (10.5). Yellow time for through-ahead (phases 1 and 3 in Fig. 10.10) and for right-turn (phases 2 and 4) are determined as 3 and 2 seconds, respectively. The all-red interval is determined by the clearing distance over the approaching speed. Although precise measurement is recommended, the distance between stop lines is often utilized as the clearing distance in practice. By using this, the all-red times for both approaches are given as 4 seconds.

The lost time between phases is calculated following Section 10.5.3. The intergreen between phases 2 and 3 as well as phases 4 and 1 contains yellow and all-red intervals. Thus, total lost time of the intersection becomes 16 seconds $(=(2+4)\times 2+(3-1)\times 2)$.

Step 3: Presetting of initial cycle length and green time

The initial cycle length is arbitrarily determined. In this example, a cycle length of 90 seconds is given considering the intersection type (multilane intersection). By subtracting the all-red and yellow times from the cycle length, the total initial effective green is obtained as 74 seconds $(=90-(2+4+(3-1))\times 2)$. In this step, the split is roughly assigned in proportion to the demands of each lane. Accordingly, the green times of phases 1–4 are preset as 26, 17, 20, and 11 seconds, respectively.

Step 4: Settings of saturation flow rate and flow ratio calculation

The manual recommends measuring the actual saturation flow rate of the site or that of other intersections with similar geometry and traffic conditions. Meanwhile, there is another method applying adjustment factors to the base saturation flow rate. The latter method is utilized in this example. Left-turn movements are affected by the existence of pedestrians in the given phasing plan. The saturation flow rates of dedicated left-turn lanes in east and west approaches are calculated as 1645 and 1570 pcu/h, respectively. Those of shared left-turn and through lanes in south and north approaches are obtained as 1698 and 1757 pcu/h, respectively.

Flow ratios are calculated by Eq. (10.8). Table 10.2 is the list of the flow ratios at each lane. Therefore, flow ratios of phases 1–4 are 0.17, 0.11, 0.14, and 0.06, respectively. Finally, the intersection flow ratio becomes 0.49 as the summation of the phase flow ratios.

Table 10.2 Flow ratios of each lane at the example intersection

<table>
<tr><th>Approach</th><th>Left-turn</th><th>Through-ahead</th><th>Right-turn</th></tr>
<tr><td>East</td><td>0.11</td><td>0.17</td><td>0.11</td></tr>
<tr><td>West</td><td>0.06</td><td>0.17</td><td>0.07</td></tr>
<tr><td>South</td><td colspan="2">0.14</td><td>0.05</td></tr>
<tr><td>North</td><td colspan="2">0.14</td><td>0.06</td></tr>
</table>

Note: The numbers in the bold fonts represent the maximum values of lane flow ratios within a phase that corresponds to the phase flow ratios.

Step 5: Calculation of cycle length, effective green time, and green indication time

The minimum cycle length is calculated as 36 seconds by Eq. (10.12) and the green times are assigned in proportion to the phase flow ratio. Then, the green times of phases 1–4 are calculated as 7.0, 4.5, 5.9, and 2.6 seconds, respectively.

Regarding the minimum pedestrian green time, the second term of Eq. (10.13) is often assumed to be 0 if pedestrian demand is not so high. Therefore, that for the south/north approach crosswalk is 19 seconds and that for the east/west approach crosswalk is 22 seconds. This time is practically given as pedestrian steady green (PG). In addition, the flashing green (PFG) times of 10 and 11 seconds for the south/north and east/west approach crosswalks, respectively, are given as the time that pedestrians can cross half of the crosswalk. Therefore, the minimum pedestrian green times of $\phi 1$ and $\phi 3$ are $19 + 10 = 29$ seconds and $22 + 11 = 33$ seconds, respectively. Furthermore, pedestrian-red (PR) times of 1–5 seconds are given between the end of pedestrian flashing green and vehicle yellow. Since the demands of left turners and pedestrians are not so high, a minimum additional time (1 seconds) is applied in this case. Overall, minimum vehicle green times of $\phi 1$ and $\phi 3$ are $29 + 1 = 30$ and $33 + 1 = 34$ seconds due to the restrictions of pedestrian crossing.

Obviously, the green times based on the minimum cycle length do not satisfy the minimum effective vehicle green times above. Therefore, the cycle length is enlarged so that all phases satisfy the minimum pedestrian green. Since the minimum vehicle green time of $\phi 3$ is the maximum among all minimum vehicle green times, $\phi 3$ is considered as the critical phase in determining the cycle length. The green time of $\phi 3$ is set as the minimum vehicle green indication time (34 seconds). Then, green times of other phases are extended proportionally. Therefore, the green times of $\phi 1$, $\phi 2$, and $\phi 4$ are 40 seconds ($=7.0/5.9 \times 34$), 26 seconds ($=4.5/5.9 \times 34$), and 15 seconds ($=2.6/5.9 \times 34$), respectively. As a result, the cycle length becomes 133 seconds.

Step 6: Evaluation and final determination of signal settings

The saturation flow rate and capacity are calculated again by using the traffic signal parameters determined in Step 5. The degree of saturation of each phase is checked to see whether it is less than 1.0. In this example, all phases satisfy this requirement and thus the signal parameters are determined as summarized in Fig. 10.10.

Acknowledgment

The authors are grateful to Dr.-Eng. Xin Zhang of Nagoya University for her contributions to the example calculations.

Bibliography

International Association of Traffic and Safety Sciences (IATSS), 2007. H856 Project Report "Research into practical optimized signal control with an emphasis on actual conditions on pedestrian crossings", (in Japanese).

Japan Society of Traffic Engineers (JSTE), 2002. Planning and Design of At-Grade Intersections. JSTE (in Japanese).

Japan Society of Traffic Engineers (JSTE), 2006. Manual on Traffic Signal Control. JSTE, revised edition, (in Japanese).

Japan Society of Traffic Engineers (JSTE), 2016. Roundabout Manual. JSTE (in Japanese).

Japan Society of Traffic Engineers (JSTE), 2018. Basic Manual on Planning, Design and Traffic Signal Control of At-Grade Intersection. JSTE, (in Japanese).

National Police Agency (NPA) of Japan, 2015. Guideline of Traffic Signal Installation. National Police Agency of Japan (in Japanese).

China

11

Keshuang Tang[1], Keping Li[1], Meng Li[2] and Dongbo Liu[3]
[1]College of Transportation Engineering, Tongji University, Shanghai, P.R. China, [2]Department of Civil Engineering, Tsinghua University, Beijing, P.R. China, [3]Traffic Management Research Institute, Ministry of Public Security, Wuxi, P.R. China

11.1 History

In the 1920s, a manually controlled traffic light with green and red indicators was first implemented in Shanghai, marking the start of road traffic signal control history on the Chinese mainland. In the 1960s, the Chinese mainland started installing traffic lights with green, yellow, and red indicators in accordance with "*Road Traffic Signs and Traffic Signal Agreements*" of the United Nations (United Nations, 1968). Traffic lights were mostly operated with a manual control mode before the 1970s, as shown in Fig. 11.1A.

Since the 1970s, a few scholars in the Chinese mainland, for example, Prof. Peikun Yang from Tongji University, began to study signal control theory and methods adopted in the United Kingdom, the United States, and Australia. A few original theories and methods, such as the delay estimation method based on traffic conflict points and the bicycle flow model were proposed and validated in the 1970s and the 1980s. After more than 10 years' accumulation in research achievements and practical experiences, basic principles, procedures, and methods of signal control were established, and elaborated on in one of the first textbooks about transportation engineering in the Chinese mainland, published in 1987, that is, *Traffic Engineering* (Zhou, 1987).

In 1973, the first fixed-time signal controller (Fig. 11.1B) and the first actuated signal controller for isolated intersections were developed by the Research Institute of Highway of Ministry of Transport, and tested at several intersections in Beijing (Research Institute of Highway of Ministry of Transport, 2016). This initiated the research and development history of traffic signal controllers and systems in the Chinese mainland. In 1979, a national workshop on urban road traffic management organized by the Ministry of Public Security was held in Guangzhou, at which the necessity of research and development of traffic signal controllers and systems was intensively discussed. Since then, many local authorities, research institutes, and industrial enterprises have begun importing foreign signal controllers and systems or developing Chinese ones.

In the 1980s and 1990s, the Chinese mainland adopted a policy of "incorporating independent development and abroad import" for the signal control industry.

Global Practices on Road Traffic Signal Control. DOI: https://doi.org/10.1016/B978-0-12-815302-4.00011-X

(A) (B)

Figure 11.1 Traffic lights and signal controllers in the Chinese mainland in the 1970s. (A) Manually controlled traffic lights. (B) The first fixed-time signal controller.

Supported by a "*7th Five-Year Plan*" National Key Scientific and Technological Project "*Urban Traffic Control System*," a research team led by Tongji University developed the first real-time adaptive signal controller in the Chinese mainland in 1990. Meanwhile, Shanghai introduced the Australian SCATS (Sydney Coordinated Adaptive Traffic System) in 1985, Beijing introduced the British SCOOT (Split Cycle Offset Optimizing Technique) in 1986, and Shenzhen introduced the Japanese KATNET in 1987. Afterward, the Spanish ITACA (Intelligent Adaptive Traffic Control Agent) and the German ACTRA (Advanced Control & Traffic Response Algorithm) were introduced to China in 1995 and 2005, respectively.

In addition, a notable event occurred in 1985, in that Traffic Management Research Institute (TMRI) of the Ministry of Public Security was established in the city of Wuxi. This is a public welfare research institute that is responsible for the enactment and revision of national and industry standards in the field of road traffic signal control, as well as the testing of signal control products. Under the leadership of TMRI, the establishment of national and industry standards on road traffic signal control was brought onto the agenda.

From the late 1990s to the early 2000s, research institutes such as Tongji University, Jilin University, and the Institute of Automation of the Chinese Academy of Sciences developed a number of new traffic signal controllers and systems, while accounting for the characteristics of mixed traffic flow prevailing in the Chinese mainland. Most of the above controllers and systems are used for research purposes only, and their practical applications are fairly limited. Meanwhile, major domestic manufacturers have developed their own signal controllers and systems, such as HiCon (Qingdao HISENSE), MiTCO (Shanghai BAOKANG), NATS (Nanjing LES), SCRCS (Wuxi HUATONG), MATHS (Lianyungang JARI), SMOOTH (Shenzhen GREENWAVE), DW-UTC (Jiangsu DAWAY), Intelliffic (Zhejiang SUPCON), and KELI-UTC (Anhui KELI).

In the late 2000s, many big Intelligent Transportation Systems (ITS) companies, such as Beijing E.HUALU, Hangzhou ENJOYS, and Shanghai SEISYS entered the signal control industry, motivated by the rapid development of ITS technologies and the growth of the market. These companies focus more on traffic control and management platforms than traditional signal controllers and systems. Meanwhile, many domestic manufacturers have released new signal controllers, systems, and relevant products to accommodate the demand of integrated traffic control and management platforms as well as emerging traffic detection technologies such as geomagnetism, microwave radar, and video.

In the 2010s, HIKVISION, the largest video surveillance products supplier in China, started research, development, and servicing of signal control based on video surveillance data. Several large Internet companies in China, such as BAIDU and DIDI, proposed a concept of "Internet + Signal Control" aiming to enhance signal control by the use of Internet crowdsourcing data, and they became important players in the signal control industry. To some extent, the use of video surveillance data and internet crowdsourcing data has promoted technological innovations and the advancement of signal control evaluation and optimization.

11.2 State of the art

11.2.1 Quantity of traffic signals

According to an incomplete statistic, in the Chinese mainland, more than 600 cities have established traffic command and dispatch systems integrating traffic violation enforcement, traffic flow monitoring, and signal control, which is operated and administered by traffic police. More than 400 cities have established traffic control centers by networking traffic signals in urban areas. The total number of traffic signals in the Chinese mainland reached 85,500 at the end of 2016. It is estimated that the actual number of traffic signals exceeded 100,000 at the end of 2018. The networked traffic signals in urban areas of the four largest cities, that is, Shanghai, Beijing, Shenzhen, and Guangzhou, are approximately 4000, 3000, 2500, and 1300, respectively. The regional networking of traffic signals has been realized for urban areas in most prefecture-level cities and relatively developed country-level cities (TMRI, 2015; Seven Transportation, 2017).

11.2.2 Modes of signal control

In the Chinese mainland, traffic signals with an isolated time-dependent fixed-time control account for about 71% of the total, those of time-dependent fixed-time arterial coordination control account for about 24%, and those of actuated and adaptive signal control account for about 5%. In 2015, there were about 4200 arterials in total operated with time-dependent fixed-time arterial coordination control in the Chinese mainland (TMRI, 2015). The extremely low percentage of actuated and adaptive signal control is partly due to the fact that many traffic signal controllers and systems with functions of actuated and adaptive control, for example, SCATS and SCOOT, are often operated in a fixed-time control mode, due to poor maintenance and the high fault rate of traffic detectors, as well as the use of green and red countdown signals for vehicular traffic. In addition, most local authorities and traffic engineers in the Chinese mainland agree that the fixed-time arterial coordination control is an effective and economic alternative for the improvement of traffic congestion, in the context of the current demand for traffic management and available technologies. Thus, fixed-time arterial coordination control has been actively promoted in the last 10 years.

11.2.3 Signal controllers market

At present, there are more than 100 signal controller manufacturers in the Chinese mainland, of which around 20 are comparatively large in terms of market share. The predominant domestic manufacturers include Qingdao HISENSE, Nanjing LES, Shanghai BAOKANG, Wuxi HUATONG, Lianyungang JARI, Chongqing ULIT, Hangzhou ENJOYOR, Shenzhen GREENWAVE, Jiangsu DAWAY, Zhejiang SUPCON, SIEMENS China, and Anhui KELI, which collectively control more than 60% of the entire market share. Regarding signal controller types, the time-dependent fixed-time signal controllers have the largest market share at about 60%, followed by the centralized coordinated signal controllers with about 32%, and the actuated or adaptive signal controllers with about 8% (TMRI, 2015; Seven Transportation, 2017).

Fig. 11.2 presents the sales trend for the domestic signal controller market from 2010 to 2016, in which an average annual growth rate of over 30% can be seen. With regard to the newer signal controllers, the proportion of centralized coordinated signal controllers was highest, followed by the time-dependent fixed-time signal controllers and the actuated and adaptive signal controllers. Taking 2016 as an example, a total of approximately 27,518 sets were sold, of which domestic products accounted for nearly 85%. More specifically, the centralized coordinated controllers totaled about 51%, and the time-dependent fixed-time signal controllers about 35%, while the actuated signal controllers were about 14% (Seven Transportation, 2017).

The main market for foreign controllers is concentrated in a few large cities, where foreign signal control systems, such as SCOOT and SCATS, were introduced

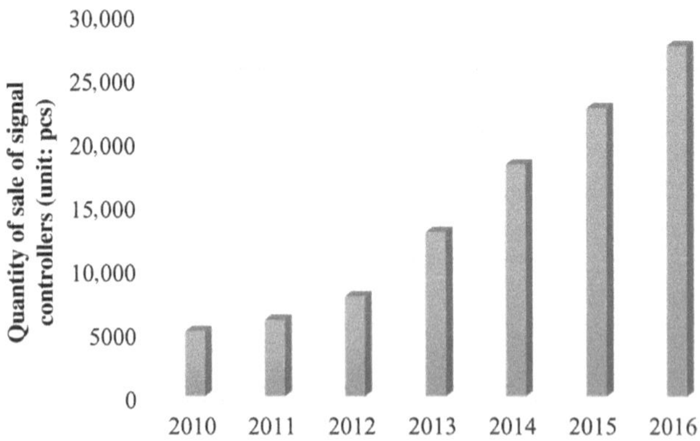

Figure 11.2 Quantity of new signal controllers sold between 2010 and 2016.

earlier, for example, in the 1980s and 1990s. For example, approximately 6500 SCATS controllers have been implemented in more than 16 cities in the Chinese mainland, including Shanghai, Tianjin, Guangzhou, and Hangzhou. Owing to continuous improvements in product quality, as well as relatively low purchase and maintenance costs, domestic controllers have occupied dominant positions in the signal controller market, and the market share of foreign controllers has been gradually decreasing in last 10 years.

11.2.4 Design, implementation, operation and maintenance roles

In Chinese cities, road traffic signal control is generally operated and managed by the traffic police department under the administration of the Bureau of Public Security, while road transport infrastructure and traffic facilities are usually in the charge of the Transport Commission or the Transport Bureau. Usually, signal control manufacturers or other service providers are responsible for the initial design and implementation of signal control schemes, in addition to providing signal controllers, systems, and related devices. Maintenance service providers, which are often the signal control manufacturers or the service providers responsible for initial design and implementation, monitor and maintain signal control schemes regularly according to the needs of the traffic police. Due to limited budgets, professional consulting companies and research institutes are rarely involved in the design, implementation, operation, and maintenance of signal control schemes, unlike many other countries such as the USA and Germany. In this context, as an authorized research institute directly administered by the Ministry of Public Security, TMRI has been playing a critical role in the establishment of technical standards for all the above stages since its foundation.

11.2.5 Relevant laws, regulations, and technical standards

Currently, the main traffic laws and regulations in the Chinese mainland are "*Road Traffic Safety Law of the People's Republic of China*" and "*Regulation on the Implementation of the Road Traffic Safety Law of the People's Republic of China*" (the Ministry of Public Security, 2011a,b,c). These two documents have legally defined the restrictions, limitations, restraints, and indicative functions of traffic signals, as well as relevant traffic rules. Under the traffic laws and regulations, technical standards give instructive recommendations or mandatory requirements as regard to the design, implementation, operation, and maintenance of road traffic signal control. By 2017, more than 20 national and industry standards had been issued and published, and a relatively complete framework of technical standards had been well established. Major existing standards on signal controllers and systems, control devices, traffic detectors, performance evaluation, communication protocols, etc., are summarized in Fig. 11.3.

As shown in Fig. 11.3, the existing standards are more oriented toward signal control products as well as their installation and communication. A national or industry standard specifically addressing traffic signal timing remains absent thus

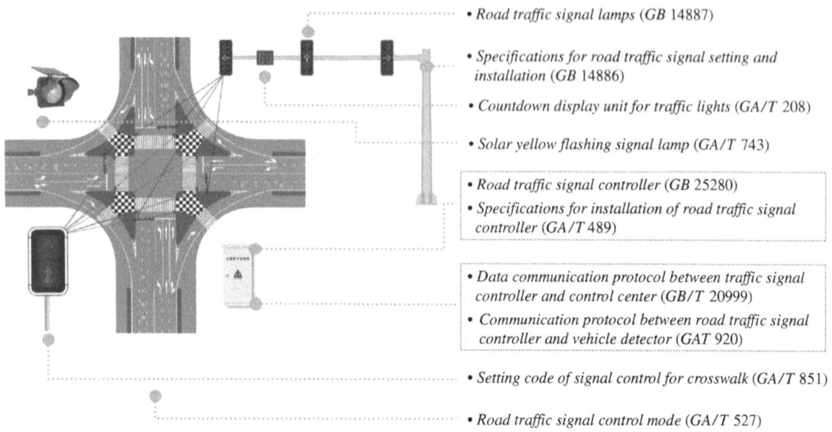

Figure 11.3 Major existing national and industry standards regarding road traffic signal control.

far in the Chinese mainland, though many of the items related to signal timing have been included in various existing standards. As a result, most signal timing practitioners have been referring to textbooks or foreign guidelines and manuals, for example, Traffic Signal Timing Manual (FHWA, 2008), Highway Capacity Manual 2010 (TRB, 2010), RilSA (FGSV, 2015), and Manual on Traffic Signal Control (JSTE, 2006), based on their own empirical experiences.

In light of this, a research team led by Tongji University issued the first province-level standard for traffic signal timing in the Chinese mainland in 2015, that is, "*Design Specification for Signal Control of Urban Intersections in Anhui Province*" (Department of Public Security of Anhui Province, 2015). Based on successful experiences in developed countries and unique situations in the Chinese mainland, this standard systematically summarizes signal control strategies, signal timing procedure, methods, and timings for urban intersections. It reflects the overall level of signal control practice in the Chinese mainland to some extent. However, as a local standard, it has few mandatory clauses and no binding force on areas outside Anhui Province, and its nationwide impact on signal timing practice is fairly constrained. Therefore, TMRI and Tongji University are now collaborating on a technical guideline for traffic signal timing design, which aims to provide more authoritative requirements and instructions for signal timing practitioners in the Chinese mainland.

11.2.6 Signal displays

Technical criteria for lens color, indication, arrangement, and placement of signal displays is specified in the national standards "*Road Traffic Signal Lamps*" (the Ministry of Public Security, 2011a,b,c) and "*Specification for Setting and Installation of Road Traffic Signals*" (the Ministry of Public Security, 2016).

However, a lot of improper use of signal displays and inconformity with the standards still exists, particularly in developing regions, due to ineffective implementation and supervision of national standards. In view of this fact, the Ministry of Public Security launched a campaign called "*Promoting Intellectualization of Urban Road Traffic Signal Timing and Standardization of Traffic Signs and Markings*" at the start of 2016, to inspect and eliminate those improper uses of signal displays and improve signal timings. A nationwide investigation by TMRI identified more than 71,000 problematic cases regarding traffic signals by the end of 2016 (Gu et al., 2016).

11.2.6.1 Lens color and indication

In the national standards, various types of lens color and indication of signal heads are regulated for motor vehicles, nonmotor vehicles, and pedestrians, respectively, as illustrated in Table 11.1. Basically, the flashing green light is recommended for the end of the green phase for pedestrians, but is not explicitly prohibited for motor vehicles and nonmotor vehicles. The green or red countdown is to inform the end

Table 11.1 Lens color and indication of signal displays regulated in national standards

Signals	Lens color	Indication
(a) Circular signal for motor vehicles		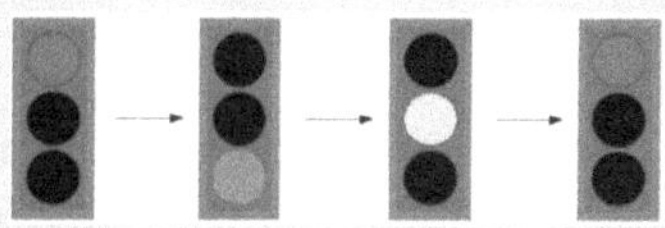
(b) Arrow signal for left-turning motor vehicles		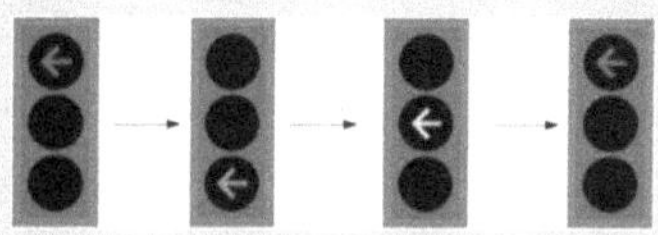
(c) Arrow signal for right-turning motor vehicles		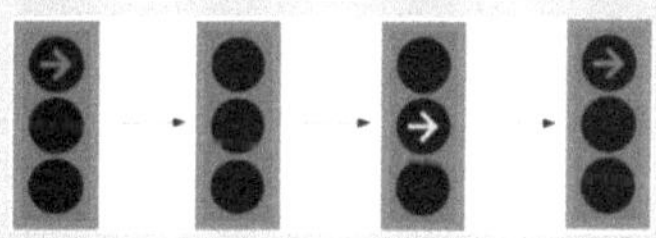
(d) U-turn signal for motor vehicles		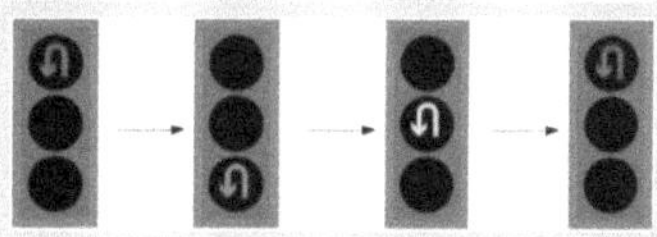
(e) Graphic signal for nonmotor vehicles		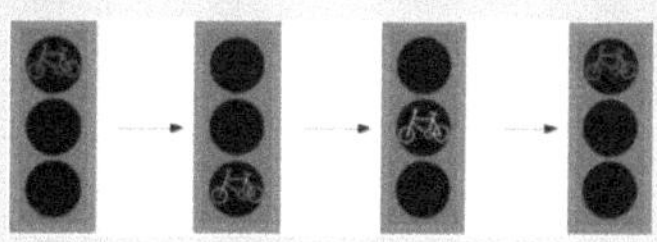
(f) Graphic signal for pedestrians		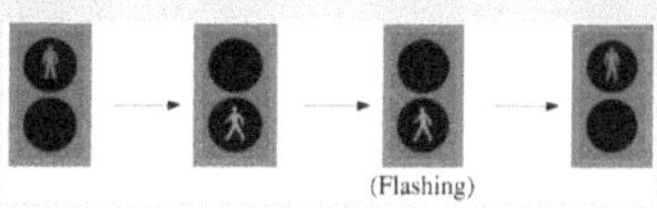

of a green or red phase for drivers, riders, and pedestrians. An arrow signal for right-turning motor vehicles is not permitted to display green, but yellow and red only. In addition, exclusive signals for motor vehicles and pedestrians are mandatory, while exclusive signals for nonmotor vehicles are optional.

In addition, a red/yellow indication of 1–2 seconds, and not greater than 3 seconds in any case, is recommended to indicate the start of green for motor and nonmotor vehicles in the local standard of Anhui Province. It is also stated that flashing green is permitted for termination of the green phase for motor vehicles.

However, in practice, the signal indication sequence for motor vehicles often appears as green→green countdown→green countdown with flashing green→yellow→red→red countdown→green, or green→green countdown→yellow→red→red countdown→green, in addition to the pattern presented in Table 11.1 (Tang et al., 2016a,b). The green and red signal countdowns are commonly displayed on separate LED screens or directly on signal heads, as exhibited in Fig. 11.4. The length of the flashing green usually falls between 3 and 6 seconds dependent on the speed limit (Tang et al., 2016a,b). Flashing green prevails in many cities, though it is not recommended in the national standards. In addition, a flashing red of 3 seconds or a red/yellow of 2–3 seconds is applied at the end of the red phase and to alert the start of the green phase in very few cities. Notably, the national standards do not explicitly prohibit simultaneous release of nonmotor vehicles or pedestrians with the conflicting left-turning or right-turning motor vehicles given green arrow indicators. Therefore, in practice, improper uses of green arrow indicators can be seen in some cities, and they have caused many potential safety hazards for conflicting nonmotor vehicles and pedestrians.

The nonmotor vehicles generally follow the indications for motor vehicles in the same direction in the case that there are no exclusive signals for nonmotor vehicles. Quite a few cities, such as Shanghai, Beijing, Shenzhen, and Hangzhou, install exclusive signals for nonmotor vehicles when the demand of nonmotor vehicles is high. Meanwhile, independent signal timings for nonmotor vehicles are often adopted to more efficiently and safely accommodate the nonmotor traffic demand, by using early-onset or early ending of green signals for nonmotor vehicles

Figure 11.4 Green and red signal countdown displays implemented in the Chinese mainland.

(Wang et al., 2014). In addition, green and red signal countdowns directly displayed on signal heads are used in some cases, as illustrated in Fig. 11.4.

In practice, the signal indication sequence for pedestrians is generally set as green→green countdown→green countdown with flashing green→red→red countdown→green. Green and red signal countdowns directly displayed on signal heads are used in most cities, as illustrated in Fig. 11.4. The duration of the flashing green signal for the clearance of pedestrians is determined by the width of the crosswalk and the crossing speed of pedestrians. However, it is not clearly stated in the national standards that pedestrians are not allowed to enter the crosswalk after the onset of flashing green. Consequently, the role of the flashing green signal as the clearance of pedestrians has not well followed in practice, leading to many late exits and flashing-green-running of pedestrians at the change of signal phases. Thus, a few cities such as Yinchuan have attempted to implement a flashing red signal for the clearance of pedestrians.

11.2.6.2 Arrangement and placement

The arrangement and placement of signal displays should consider intersection geometry, signal phasing, as well as visibility, according to the above national standards, and proper patterns for various situations are illustrated in Fig. 11.5. Basically, a three-section circular display only is recommended for small intersections with a two-phase control, as shown in Fig. 11.5A. They can also be clustered with another three-section arrow display for left-turning or right-turning traffic at

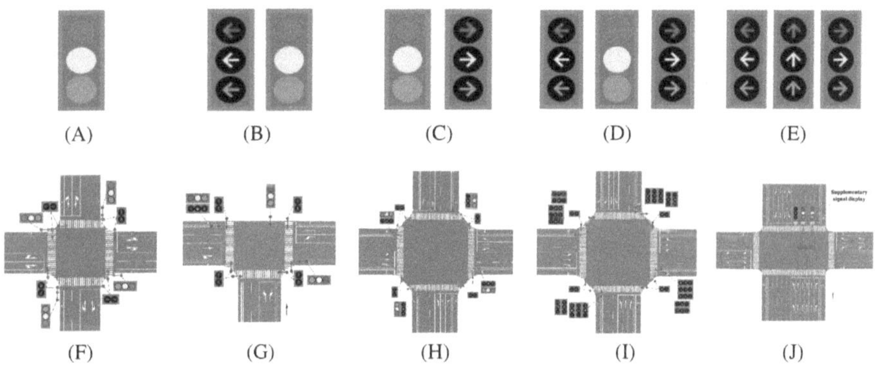

Figure 11.5 Recommended arrangement and placement of signal displays in national standards. (A) One three-section circular display only; (B) a three-section circular display clustered with a three-section left-turn arrow display; (C) a three-section circular display clustered with a three-section right-turn arrow display; (D) a three-section circular display clustered with a three-section left-turn and right-turn arrow display; (E) a cluster of 3 3-section arrow displays for all directions; (F) placement at small intersections with a two-phase control; (G) placement at T-type intersections with a three-phase control; (H) placement at large intersections with a four-phase control; (I) placement at large intersections with a multiple-phase control; (J) placement of supplementary signal displays.

(A) (B) (C)

Figure 11.6 Typical examples of improper arrangement and placement of signal displays in practice. (A) Improper arrangement of arrow indications; (B) improper cluster of signal displays; (C) improper cluster and placement of signal displays.

large intersections with a three-phase or four-phase control, as shown in Fig. 11.5B–D. A cluster of three 3-section arrow displays for all directions is usually used at large intersections with multiple-phase control, as shown in Fig. 11.5E. Note that both vertical and horizontal arrangements of signal displays are permissive in the national standards. Regarding the placement of signal displays, the approach-based and far-side pattern is generally recommended, as shown in Fig. 11.5F–I. Supplementary signal displays can be installed if one cluster of signal displays cannot properly cover all the approach lanes, due to the existence of a wide approach or obstacle blockage, as shown in Fig. 11.5J. In addition, another group of near-side signal displays can be added if the distance between the approach stop-line and the signal displays installed at the exit is greater than 50 m.

In practice, the arrangement and placement of signal displays generally complies with the requirements of national standards in most cities. However, some exceptions exist in small cities, particularly in developing regions. Typical improper arrangements of signal displays include three arrow indications for different directions on a single signal display (e.g., Fig. 11.6A), and four-section, five-section, or six-section vertical clusters of signal displays with circular and arrow indications (e.g., Fig. 11.6B and C). There are two aspects to the main problems for the placement of signal displays. One is that signal displays are placed and mounted improperly and cause poor visibility for road users, due to insufficient height of the traffic signal pole (e.g., Fig. 11.6C), far distance between the signal displays and the stop-line, or blockage by roadside obstacles. The other is that an unnecessarily large number of signal displays are installed at one intersection, which sometimes leads to confusion of road users.

11.2.7 Software tools

Software tools for signal timing design and optimization are not recommended in the above standards. In practice, a few signal control manufacturers and service providers have developed simplified software to aid their engineers in the design and optimization of fixed-time signal control schemes. Those software are not commercial, but for their own use only. As for the commercial software, Synchro has been widely utilized for fixed-time signal timing design and optimization at isolated intersections by a large number of practitioners. A few skillful practitioners are able to adopt CROSSIG for the design and optimization of fixed-time control schemes at coordinated intersections. In addition, microscopic traffic simulation tools are often used for the evaluation of signal timing plans. VISSIM has the largest market share in the Chinese mainland, followed by TransModeler, AIMSUN, Paramics, and CORSIM.

11.2.8 Maintenance

It is recommended in the national standards that the operational performance of signal control schemes be monitored and optimized regularly. However, it is not specified how often such processes should be conducted. Therefore, in practice, signal control maintenance is usually done based on the maintenance contract agreement, the additional needs of traffic police, and the available budget. Significant differences in the maintenance cycle exist across regions and cities. Generally, maintenance is more frequent in developed regions and cities, for example, on average every 6–12 months, while the average maintenance cycle is usually greater than 12 months in most developing regions and cities. Due to the limited budget, major maintenance works are related to equipment failure, and there are less signal retiming works.

11.3 Selection of signal control strategies

11.3.1 Signal warrants

Whether to install a traffic light at an intersection is largely dependent upon the following three types of conditions, according to the national standard "*Specification for Setting and Installation of Road Traffic Signals.*"

1. Road level. Basically, the intersection of a major or minor arterial and another major or minor arterial, or the intersection of a first-class highway and another first-class highway, should have a traffic light installed;
2. Traffic demand. A traffic light should be installed at an intersection if traffic demand during a successive 8-hour time period or traffic demand during the peak hours reaches the levels required by the standard;
3. Safety requirement. A traffic light is essential at an intersection if more than five traffic accidents or one fatal traffic accident on average have occurred in a successive 3-year time period.

In addition, a traffic light is also required if both traffic demand and the number of traffic accidents at an intersection reaches 80% of the required quantities in points (2) and (3) above. Traffic lights should also be installed at some special intersections even if none of the above conditions could be satisfied, for example, intersections located in a coordinated arterial, intersections with unconventional geometry, intersections with more than four legs, and intersections with a special need for the protection of pedestrians and nonmotor vehicles.

In practice, the installation of traffic lights at reconstructed intersections basically follows the above technical criteria. However, the installation of traffic lights at new intersections is more often decided by the intersection size or available budget, as accurate traffic flow and accident information is usually difficult to obtain. Therefore, traffic lights are unnecessarily installed at some new intersections, leading to increased delays for road users. In contrary, a number of intersections with an essential need for traffic lights have been operated with a nonsignalized mode due to limited budget, resulting in traffic safety hazards.

11.3.2 Fixed-time control versus actuated or adaptive control

The applicable conditions of various control strategies are described in the industry standard "*Road Traffic Signal Control Mode*" (the Ministry of Public Security, 2015). It is recommended that a fixed-time and isolated control is utilized under the following conditions.

1. Traffic demands are relatively stable at each approach and direction during the concerned time periods;
2. The subject intersection is far from its surrounding intersections, or serves relatively minor traffic function in the road network;
3. The subject intersection is operated by a signal controller without an online connection, or running in an offline mode.

In contrast, an isolated actuated or adaptive signal control should be considered under the following conditions: the intersection is undersaturated and comparably independent in terms of traffic flow correlations with its surrounding intersections; traffic demands are more fluctuated and unbalanced during the concerned time periods; its signal controller has the function of actuated or adaptive control; and essential traffic detectors are available.

In practice, however, regardless of whether it is a new or reconstructed intersection, fixed-timed signal control rather than actuated or adaptive control has been selected in most cases, as indicated by the statistics presented in Section 11.2.2. In the context of fixed-time signal control, traffic flow dynamics at key intersections are usually handled by utilizing more time-of-day (TOD) signal timing plans or multiple-phase signal phasing plans. The underlying reasons for the number of fixed-time signal controls are threefold: firstly, technical requirements and general costs for the design, implementation, operation, and maintenance of the actuated or adaptive signal control are relatively high. There is often a shortage of available budgets and experienced technicians for the operation and maintenance of the actuated or adaptive signal control systems; secondly, frequent road construction and maintenance works due to rapid economic development in the Chinese mainland have resulted in a fairly high damage rate to traffic detectors implemented for actuated or adaptive signal control in many cities. For example, it is roughly estimated that more than 50% of the loop detectors at the SCATS controlled intersections in

Shanghai have been incapable of reporting reliable detection data; thirdly, limited coverage of traffic detectors as well as unsatisfactory quality of detection data have often been a significant obstacle to the implementation of actuated or adaptive signal control.

However, economic development in the Chinese mainland is shifting from an extensive economy to an intensive economy, leading to rising demands for high efficiency and intelligentization of road traffic signal control. Thus, government, industry, and academia have invested a lot of resources and made tremendous efforts into the research and development of new signal control systems and technologies, in order to accommodate various demands. Successful attempts have included data-driven signal control based on Big Data and Artificial Intelligence, and emerging traffic detection methods such as geomagnetic, video, microwave radar, and connected vehicles (Li et al., 2017). Some of these technologies have been successfully applied in a few cities, for example, Shenzhen, Guangzhou, Hangzhou, Nanjing, and Jinan. It is widely expected that those advanced signal control technologies will be gradually extended throughout the Chinese mainland in the near future, though the fixed-time signal control is still dominant at present.

11.3.3 Isolated control versus coordinated control

It is recommended in the standard that a fixed-time coordination control be used under the following conditions.

1. The coordinated traffic flows should accommodate the main traffic loads of the entire road network;
2. Traffic demands of different paths and streams within the coordinated area are relatively stable during the concerned time periods;
3. Traffic flow patterns and characteristics at the coordinated intersections are similar and strongly correlated with each other, and the spacing between the coordinated intersections is ideally less than 800 m;
4. There is little roadside disturbance and less traffic flow breakdowns between the coordinated intersections;
5. Signal controllers at the coordinated intersections are capable of automatic clock synchronization.

Likewise, the actuated or adaptive coordination signal control could be considered under the following conditions: the coordinated approaches are undersatured or have no queue spillbacks; traffic demands of different paths and streams within the coordinated area are more fluctuated and unbalanced; reliable communication between the signal controllers and the traffic control center can be realized; and essential traffic detectors are available.

In practice, the fixed-time arterial coordination control is most widely implemented at both new and reconstructed intersections in the Chinese mainland, followed by actuated arterial coordination control. However, adaptive arterial coordination control as well as actuated or adaptive network coordination control are very rare, for the reasons explained in Section 11.3.2. The fixed-time arterial coordination control is often implemented at the urban commuting corridors with tidal traffic flows. Three to four TOD plans including the morning peak, the afternoon peak, and 1−2 off-peaks are usually used based on the traffic flow characteristics. In addition, one-way progression is mostly applied for the peaks, with two-way progression for the off-peaks. Actuated arterial coordination control is often implemented at the major arterials with low traffic demand at the crossing roads. It is normally realized in a way that time splits are dynamically adjusted based on the detected traffic demands at the crossing roads, while leaving the common cycle length, signal phasing, and offsets unchanged.

In recent years, due to increased attention given to traffic congestion and vehicle exhaust pollution, fixed-time arterial coordination control has been greatly promoted in the Chinese mainland, as an effective means of improving traffic efficiency. In particular, many cities such as Shenzhen, Shanghai, Wuhan, Jinan, and Taiyuan have arranged specific appropriation for the improvement of current traffic signal timings, after the campaign "*Promoting intelligentization of Urban Road Traffic Signal Timing and Standardization of Traffic Signs and Markings*," launched by the Ministry of Public Security at the start of 2016. In those cities, many of the efforts focused on the implementation of fixed-time arterial coordination control, and achieved quite satisfactory results.

11.4 Signal timing procedure

As previously described, a national standard for traffic signal timing design is still currently needed in the Chinese mainland. However, many items such as basic principles signal phasing, signal timing parameters, capacity estimation, and performance evaluation have been addressed in the existing standards. In addition, a local standard of Anhui Province, called the "*Code for Signal Control of Intersections on Urban Roads*," comprehensively describes how to design traffic signal timings, based on existing standards, research findings, and practical experiences.

The flow chart for fixed-time signal timing design at isolated intersections is presented in Fig. 11.7, as recommended by the local standard in Anhui Province. Major steps include the partition of TOD, determination of design traffic volumes, design of signal phasing schemes, and calculation of signal timing parameters. Generally speaking, its basic philosophy is similar to those recommended by FHWA and JSTE. The overall procedure is commonly followed by traffic police and practitioners in the determination of signal timings, while those specific calculation steps are often slightly modified or simplified based on actual situations. For example, essential empirical observations or exact computations for saturation flow rate, intergreen time, lost time, and cycle length, are skipped in many cases. Each of the major steps is presented in the following sections and a working example is given in Section 11.11.

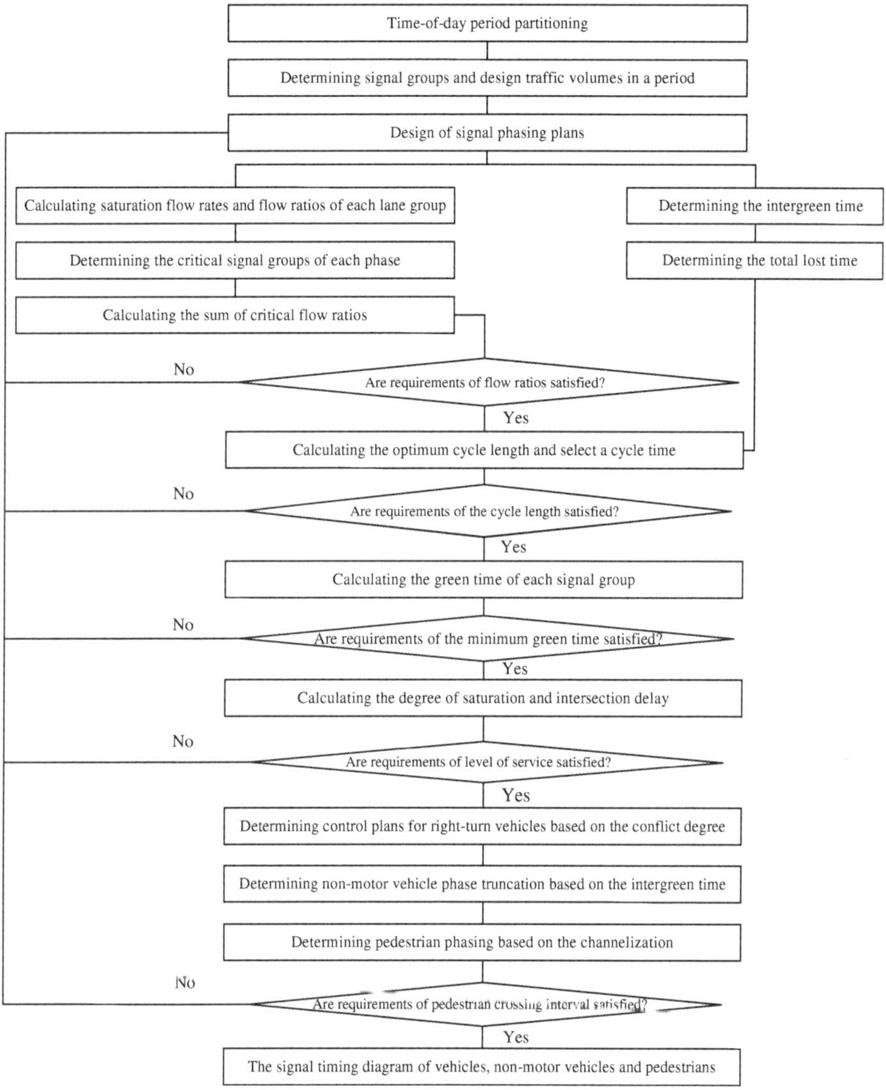

Figure 11.7 Flow chart of fixed-time and isolated signal timing design.

11.5 Partition of time-of-day

TOD should be partitioned into several characteristic time periods to facilitate time-dependent signal control schemes, based on traffic flow patterns at the subject intersection. It is required in an industry standard that at least three types of time periods, that is, peak, off-peak, and night time, need to be allowed for. It is recommended in the local standard in Anhui Province that at least four time periods, that is, AM peak, PM peak, off-peak, and night time, should be considered. In addition, a traffic volume observation of 16-hour is required for the partition of TOD at the reconstructed intersections.

In practice, for new intersections, TOD is basically partitioned into four time periods, that is, AM peak, PM peak, off-peak and night time, due to the lack of traffic volume observations. For the reconstructed intersections, 1 or 2 hours of traffic counting for each of the representative time periods as listed above is normally conducted. In some cases, TOD is partitioned based on the onsite traffic police's daily experiences. The number of TOD plans is largely dependent upon the available budget and human resource for signal timing design and maintenance, and varies significantly from city to city. Usually, 6–10 TOD plans are used in urban areas, while 4–6 TOD plans are more often seen in suburban areas. More TOD plans might be implemented for a few intersections with important traffic functions and significant traffic demand fluctuations.

11.6 Determination of design traffic volumes

The design traffic volume needs to be determined for each TOD plan. This should be the hourly traffic volume that is converted from the peak 15-min flow rate, based on the recommended method in the highway capacity mannual (HCM) 2010. In a case where the peak 15-minute flow rate is not used, peak hour factors of 0.75 for the major approaches and 0.80 for the minor approaches are recommended in the national standard. The design traffic volume of motor vehicles is measured by the passenger car unit (pcu). Seven vehicle types, that is, passenger car, motor cycle, van (>7 seats), minivan, heavy truck, bus, and tram, are considered in the determination of design traffic volume. The equivalent factor for each vehicle type is presented in Table 11.2.

In addition, nonmotor vehicle and pedestrian volumes should be observed for each direction, and special attention should be given to left-turn nonmotor vehicles.

11.7 Design of signal phasing plans

The Chinese mainland operates a right-hand traffic rule. Phasing plans are generally determined based on intersection geometry, conflicting traffic volumes, and safety requirements. Typical phasing plans for a four-leg intersection recommended in the

Table 11.2 Equivalent factor of different types of vehicles (pcu)

Passenger car	Motorcycle	Van (>7 seats)	Minivan	Heavy truck	Bus	Tram
1.0	0.5	1.2	2.0	3.0	3.0	3.5–6.0

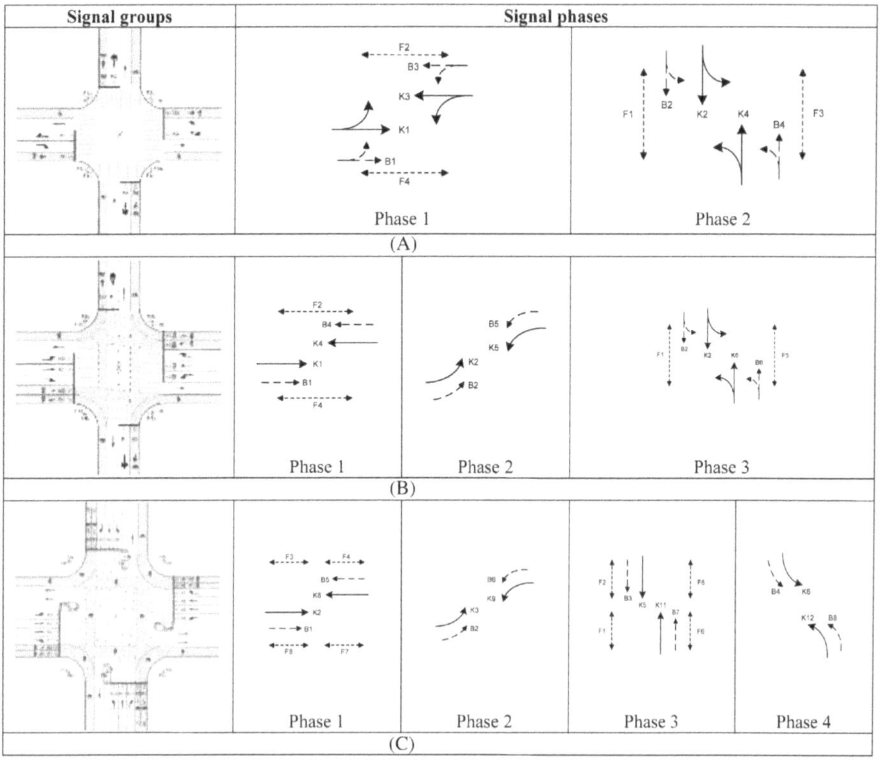

Figure 11.8 Typical phasing plans recommended in the national and local standards. (A) Two-phase plan; (B) three-phase plan; (C) four-phase plan.

national and local standards are presented in Fig. 11.8. An even phase structure with a lagging left-turn is common, regardless of it being part of a two-phase or four-phase plan. Nonmotor vehicles are often released, along with the motor vehicles in the same direction, and pedestrians are released together with the through traffic in the same direction. Two-stage crossing of pedestrians is also recommended for the channelized intersections with median islands or corner islands of sufficient refuge space. More specific treatments of left-turn motorized traffic, right-turn motorized traffic, and nonmotor vehicles are explained in the following subsections.

In practice, typical two-phase or three-phase control is commonly applied at most small and medium intersections, while the typical four-phase control is applied at most large intersections. Special four-phase control, such as approach-based release and the overlapping phase, can be seen at a few coordinated intersections with unbalanced traffic demands. Multiple-phase control with more than four phases is rarely used, except at the large intersections with unconventional geometry or with more than four legs. The exclusive pedestrian phase is sometimes applied at the intersections with an extremely high volume of pedestrians.

11.7.1 Treatment of left-turn motorized traffic

It is stated in the national standard that the protected left-turn phase for motor vehicles can be applied if any of the following conditions is satisfied.

1. There is more than one exclusive left-turn lane, or more than one opposite through lane;
2. The left-turn volume is high, or the conflicting traffic volumes in the opposite through direction are high;
3. There is more than one stream conflicting with the left-turn movements;
4. The speed of the opposite flow is relatively high;
5. Protected left-turn is required as the left-turn lane is placed on the right or a shared U-turn and left-turn lane exists.

In practice, the lagging left-turn is usually adopted for the protected left-turn phase in the case of isolated and fixed-time control, while the leading left-turn is also allowable in the case of coordinated control or actuated/adaptive control. In addition, the leading left-turn is sometimes applied at approaches where the left-turn queue spillbacks easily occur due to insufficient queue storage space.

11.7.2 Treatment of right-turn motorized traffic

Three types of right-turn phases, including the permissive right-turn phase, the prohibited-permissive right-turn phase, and the protected right-turn phase, are suggested in the local standard of Anhui Province. It is stated in the standard that the permissive right-turn phase, that is, right-turn on red, is applicable when the conflicting traffic volumes between the right-turn motorized traffic and crossing pedestrians and nonmotor vehicles are low. However, it has been most widely implemented in practice in the Chinese mainland, regardless of the conflicting traffic volumes.

The prohibited–permissive right-turn phase is applicable when the right-turn traffic volume is high, and the conflicting pedestrians are greater than 20 persons/cycle as well as the conflicting nonmotor vehicles are greater than 10 vehicles/cycle. The prohibited duration can usually be set at between 5 and 20 seconds, dependent on the volumes of pedestrians, nonmotor vehicles, and right-turn vehicles, queue storage space for right-turn traffic, and pedestrian green time. Right-turn arrow indicators, that is, red and yellow arrows, should be used when applying the prohibited–permissive right-turn phase, and an example with a clustered signal display of circular and arrow indications is illustrated in Fig. 11.9A.

The protected right-turn phase is applicable when the right-turn traffic demand and the left-turn traffic demand at the crossing road are balanced, and meanwhile the conflicts among right-turn movements, pedestrians, and nonmotor vehicles are trivial. Right-turn arrow indications, that is, red, green, and yellow arrows, should be used when applying the protected right-turn phase, and an example is illustrated in Fig. 11.9B. A typical four-phase plan with protected right-turn traffic is presented in Fig. 11.10.

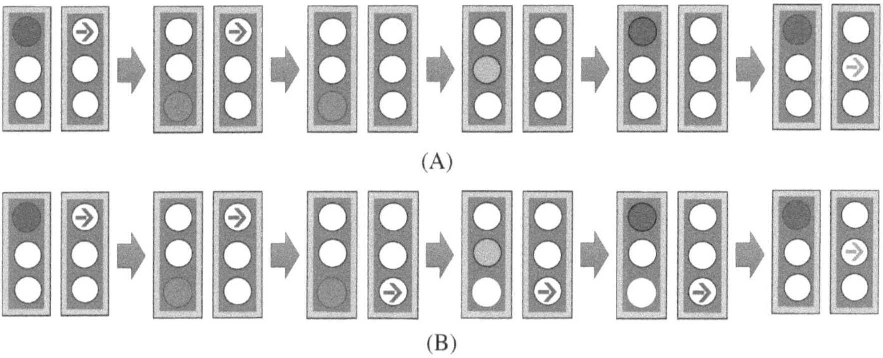

Figure 11.9 Signal indications for prohibited–permissive right-turn and protected right-turn. (A) Signal indication for the prohibited-permissive right-turn. (B) Signal indications for the protected right-turn.

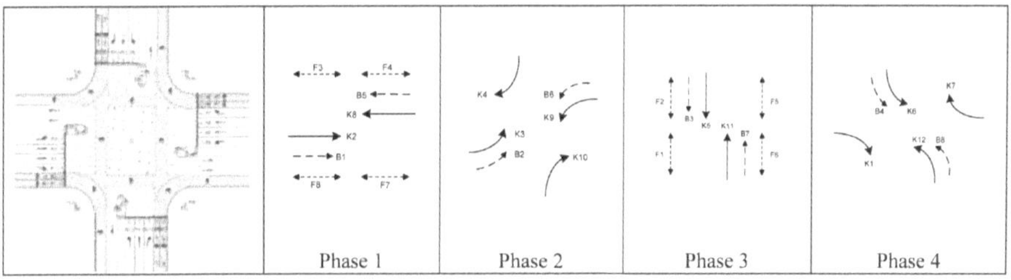

Figure 11.10 A typical four-phase plan with protected right-turns.

11.7.3 Treatment of nonmotor vehicles

For two-phase intersections with exclusive signals or exclusive phases, they should be set up for left-turn nonmotor vehicles if the average flow rate during the red interval is relatively high, for example, greater than 15 passenger cars per cycle. Alternatively, other methods of street crossing organization for nonmotor vehicles should be adopted.

For four-phase intersections, nonmotor vehicle exclusive signals should be set up when the green interval needed by the nonmotor vehicles is greater than for the motor vehicles in the parallel direction by over 3 seconds. An "early break" mode should also be adopted by the nonmotor signals.

For three-phase intersections, signals of the minor traffic direction can refer to the setting conditions of left-turn nonmotor vehicle exclusive signals at two-phase intersections, while signals of the major road can refer to the setting conditions of nonmotor vehicle exclusive signals at four-phase intersections.

For intersections without a nonmotor vehicle exclusive signal, the nonmotor vehicles go through the intersections according to the motor vehicle signals, and the green interval of the signal timing scheme should meet the requirements of safe clearance of nonmotor vehicles.

At intersections without a protected left-turn phase for the nonmotor vehicles or inhibiting left-turn motor vehicles, two-step crossing can be adopted for the left-turn nonmotor vehicles, when the flow is not high and the waiting area of left-turn nonmotor vehicles is adequate. That is, the left-turn nonmotor vehicles go through the intersection according to the through phase of motor vehicles and wait at the waiting area. Appropriate signs and marks must be adopted for auxiliary management.

11.8 Calculation of signal timing parameters

Signal timing parameters for fixed-time and isolated control mainly include inter-green times, cycle length, green splits, etc. To calculate the signal timing parameters, saturation flow rate, capacity, flow ratio, minimum green time, and maximum red time should also be determined.

11.8.1 Saturation flow rate

It is recommended in the national standard that empirical data should be used to determine the saturation flow rate. If empirical data are unavailable, the saturation flow rate can be determined based on the ideal saturation flow rate shown in Table 11.3, and adjustment factors dependent upon the local conditions, including lane width, approach grade, heavy vehicle ratio, turning radius, turning ratio, and interference of pedestrians and nonmotor vehicles (Ministry of Housing and Urban-Rural Development, 2011).

11.8.2 Capacity

According to the above standards, the capacity of a signalized intersection can by calculated as follows. It is basically similar to the method recommended in the HCM 2010 (TRB, 2010), and followed by signal timing practitioners.

$$c = \sum_i c_i = \sum_i \sum_j S_j^i \times \frac{g_i}{C} \tag{11.1}$$

Table 11.3 Ideal saturation flow rates recommended in the national standard (pcu/lane/h)

	Minimum	Mean	Maximum
Through lanes	1550	1650	1750
Left-turn lanes	1450	1550	1650
Right-turn lanes	1350	1450	1550

$$g_i = G \times \frac{y_i}{Y} \tag{11.2}$$

$$G = C - L \tag{11.3}$$

$$Y = \sum_i y_i = \sum_i \frac{q_k^i}{S_k^i} \tag{11.4}$$

where c is the intersection capacity of a particular cycle (pcu/h); c_i is the capacity of phase i (pcu/h); S_j^i is the saturation flow rate of lane group i (pcu/h) released in phase i; g_i is the green split of phase i (s); C is the cycle length (s); G is the total green time of a cycle (s); g_i is the flow ratio of phase i; Y is the flow ratio of a cycle; L is the total lost time of a cycle (s); q_k^i is the design traffic volume of the critical lane group k released in phase i (pcu/h); S_k^i is the saturation flow rate of the critical lane group k released in phase i (pcu/h).

11.8.3 Lost time

For the planning stage, the lost time for a particular phase is set as the intergreen time of the phase, that is, the yellow time plus the all-red time, according to the above standards. The total lost time for a particular cycle is thus the sum of intergreen times of all phases, as shown in Eq. (11.5). In practice, 3 seconds of yellow time and 1−2 seconds of all-red time are often assumed when determining the lost time of a phase in the planning stage. For the operation stage, it is recommended in the standards that empirical data should be used for the calculation of lost time. However, the sum of intergreen time is still predominantly applied in practice.

$$L = \sum_i L_i = \sum_i I_i \tag{11.5}$$

where L_i is the lost time of phase i (s); I_i is the intergreen time of phase i (s).

11.8.4 Intergreen time

Two methods types for the calculation of intergreen times are recommended in the local standard of Anhui Province. One is a general method based on the method recommended in the Traffic Signal Timing Manual (FHWA, 2008). The other is a sophisticated method based on the method recommended in the RiLSA (FGSV, 2015).

In the general method, the yellow time is determined according to the vehicle's approach speed, the driver's perception-and-reaction time, and the maximum deceleration rate, which is similar to the FHWA method. To better account for traffic situations in the Chinese mainland and for simplicity in practice, the following recommendations are also made in the standard: the yellow time for motor vehicles can be set as 3 seconds if the speed limit of the intersection is not greater than 50 km/h, or the speed limit is beyond 50 km/h but a flash green is used before the yellow indicator; with every incremental increase of 10 km/h in the speed limit, the yellow time for motor vehicles can be extended by 1 second for the intersections with a speed limit beyond 50 km/h and without a flash green. In addition, a yellow time of 2 seconds is recommended for nonmotor vehicles. The all-red time is determined according to the vehicle length, the clearance distance and speed, and the entering distance and speed. To facilitate the application of this proposed method, a working table is also provided in the standard that recommends the all-red times against the difference of the clearance distance and the entering distance as well as the types of entering and clearing vehicles (Table 11.4).

The sophisticated method that generally follows the German method is recommended for use at the following types of intersections: intersections with a high nonmotor traffic volume and demanding special protection for safety reasons; medium or large intersections with unconventional geometry; intersections existing with at least one combination of the clearing and entering flows with a difference between the clearing distance and the entering distance of more than 40 m.

In practice, none of the above methods has been taken seriously by traffic police or signal timing practitioners, that is, thoughtful considerations of local conditions and exact computations are often omitted, although the basic principles are widely followed. Most cities implement a yellow time of 3 seconds regardless of the speed limit and the use of a flash green signal, followed by an all-red time of 1 or 2 seconds dependent upon the intersection size, for motor vehicles. Very few cities use a yellow time only, that is, an all-red time of 0 second. In addition, intergreen times for motor vehicles and nonmotor vehicles released in the same phase are generally identical, leading to many late exits of nonmotor vehicles and traffic conflicts at the intersections with a high nonmotor traffic volume.

Table 11.4 A working table for the determination of all-red times provided in the local standard of Anhui Province

All-red time (s)	**Difference between the clearance distance and the entering distance (m)**			
	<10	**10–20**	**20–30**	**30–40**
Motor vehicles clearing—motor vehicles entering	1	2	3	3
Motor vehicles clearing—nonmotor vehicles entering	1	1	2	3
Nonmotor vehicles clearing—motor vehicles entering	1	3	5	7
Nonmotor vehicles clearing—nonmotor vehicles entering	1	2	4	6

11.8.5 Cycle length

It is recommended in the local standard of Anhui Province that the optimum cycle length of delay minimization for the fixed-time and isolated control can be calculated by the Webster formula shown in Eq. (11.6).

$$C_O = \frac{1.5L + 5}{1 - Y} \tag{11.6}$$

where C_O is the optimum cycle length (s); L and Y are as defined previously.

It is recommended in the national standard that a feasible range of cycle length is 30–150 seconds for the undersaturated intersections, while the maximum cycle length should not exceed 180 seconds for the saturated intersections. Countermeasures, such as coordinated control, two-stage crossing of pedestrians, and improvement of intersection geometry may be considered if the above conditions cannot be fulfilled. It is recommended in the local standard of Anhui Province that the minimum cycle length should not be less than 40 seconds, and the maximum cycle length should not exceed 150 seconds in any case.

In practice, as discussed earlier, exact computations are skipped and the determination of cycle lengths is totally reliant on the experiences of traffic police or practitioners in some cases, particularly when technicians or the available budget for maintenance are inadequate. If exact computations are conducted, a cycle length slightly greater than the calculated optimum cycle length is often applied for the isolated intersections, in order to reserve capacity. In general, cycle lengths often fall between 90 and 180 seconds in the daytime in megacities, for example, Beijing, Shanghai, Shenzhen, and Guangzhou. At a few intersections with high traffic demands or multiple-phase control, cycle lengths may even exceed 200 seconds. The cycle length usually varies from 90 to 150 seconds during the daytime at urban intersections in medium-sized and large cities. Meanwhile, most urban intersections operate with short cycle lengths in the range of 60–120 seconds in small cities.

11.8.6 Minimum green time

It is specified in the national standard that the minimum green time for the through motorized traffic shall not be less than 15 seconds at the major arterials, and 8 seconds at the minor arterials. It is further recommended in the local standard of Anhui Province that the minimum green time for other motorized traffic should not be lower than 10 seconds. The minimum green time for nonmotor vehicles should be greater than 5 seconds.

The minimum green time for pedestrians, that is, the duration of pedestrian green (not including flash green), should be greater than 5 seconds, and the clearance time for pedestrians, that is, the duration of pedestrian flash green, is calculated as below that according to the local standard of Anhui Province. The minimum green time for a particular phase should not be shorter than the sum of green time and clearance time of pedestrians released in this phase.

$$t_{FG} = \frac{L_0}{V_p} \tag{11.7}$$

where t_{FG} is the clearance time for pedestrians (seconds); L_0 is the crossing distance of one time (m); V_p is the crossing speed of pedestrians (m/s), and 1.0–1.2 m/s is recommended.

In practice, the requirements regarding the minimum green times for motor vehicles and nonmotor vehicles are commonly satisfied. However, two major problems exist for the minimum green time and clearance time for pedestrians in many cities. One problem is insufficient green time for pedestrians to safely clear the intersection, due to large intersection size and inflexible signal phasing plans as presented in Section 11.7. For example, the crossing distances at many large intersections are longer than 40 m, where there are no refuge islands and the two-stage crossing of pedestrians is not possible. The other problem is unreasonable allocation of time durations of pedestrian green and pedestrian flash green, due to misunderstandings of traffic police or practitioners on the concept of pedestrian clearance time. More specifically, the duration of pedestrian green time can be set unnecessarily long, while the duration of pedestrian flash green is too short for safe clearance, though the total green time of the phase is long enough. These two problems have resulted in many late exits of pedestrians and traffic conflicts, mostly at large intersections.

11.8.7 Maximum red time

Factors such as acceptable waiting time and available spaces for pedestrians and nonmotor vehicles, queue storage space for motor vehicles, and travel time of public transportation significantly affect the determination of maximum red times. It is stated in the national standard that the maximum red time for motor vehicles should not be greater than 120 seconds in undersaturated conditions, and 150 seconds in saturated intersections. Meanwhile, the maximum red times for nonmotor vehicles and pedestrians should not exceed 90 seconds. It is further recommended in the local standard of Anhui Province that the maximum red time for pedestrians should be less than 60 seconds in small or medium-sized cities.

In practice, extraordinarily long waiting times, for example, greater than 90 seconds, often exist for pedestrians and nonmotor vehicles at busy intersections during peak time periods, since traffic operators always attempt to allocate more green splits to motorized traffic in order to alleviate traffic congestion. As a result, traffic violations of bicycle riders and pedestrians, for example, red-light-running, occur frequently. Those traffic violations reduce the capacity of the intersection and further deteriorate traffic congestion, in addition to having negative impacts on safety. Such a vicious circle often happens in many large cities.

11.9 Special considerations

11.9.1 Traffic operations with inside waiting areas

As shown in Fig. 11.11A, the left-turn waiting area inside the intersection often exists at signalized intersections in the Chinese mainland, which is similar to Japan (it is a right-turn waiting area in the Japanese case) (Tang et al., 2013). Some traffic police and practitioners in the Chinese mainland support the use of waiting areas, as being particularly helpful for the prevention of queue spillbacks, and sometimes also improving discharge flow rates. Therefore, it has been implemented in many cities for the past 20 years, in order to provide additional space for queuing vehicles at large intersections or intersections with insufficient queue storage space in the approach lanes. Intersections with left-turn waiting areas are commonly operated by a typical four-phase plan, as presented in Fig. 11.8C. At an intersection approach with a left-turn waiting area, the left-turning vehicles are allowed to enter the intersection and wait at the designated area (i.e., the left-turn waiting area) for green onset when the adjacent through traffic is discharged, even if the left-turn signal indication is red. Meanwhile, it is not allowed when the conflicting traffic movements at other approaches are released.

In recent years, a number of cities, including Shanghai, Beijing, Chengdu, Yantai, and Luzhou, have started implementing through waiting areas together with a left-turn waiting area, as shown in Fig. 11.11B. Similarly, under a typical four-phase plan, the through traffic is permitted to enter the intersection and wait at the designated area (i.e., the through waiting area) for green onset, when the left-turn traffic movements at the crossing road are released, even if the signal indication is red. In this case, an LED screen is commonly mounted, together with the signal displays to alert drivers to the proper time to enter the intersection, as illustrated in Fig. 11.11B.

Though such left-turn and through waiting areas have been implemented for quite a long time in the Chinese mainland, their actual performance, as well as their advantages and disadvantages, are still being intensively argued over in academia and industry. A number of researchers and practitioners believe that they might

(A) (B)

Figure 11.11 Examples of left-turn and through waiting areas inside the intersection. (A) Left-turn waiting area. (B) Left-turn and through waiting areas.

impose negative impacts on safety performance, especially during the signal change and clearance intervals, increase vehicle fuel consumption due to there being more frequent stop-and-go incidents, and also greatly constrain the flexibility of signal phasing for coordinated control (Ma et al., 2017).

11.9.2 Signalized roundabout

A few large roundabouts located in urban areas are operated by traffic signals in some cities in the Chinese mainland, for example, Shanghai. Fig. 11.12 shows an example of Wujiaochang roundabout with five legs and three circulating lanes. It is widely accepted by most traffic police and practitioners in the Chinese mainland that, given a certain proportion of undisciplined road users, signal control has considerable potential to prevent self-locking of the roundabout, as well as improving safety and capacity, compared with nonsignalized control (Ma et al., 2013). Thus, unlike many other countries such as Germany, Japan, and the USA, quite a few large roundabouts are operated by traffic signals.

11.10 Performance evaluation

In the above national and local standards, the level of service (LOS) of signalized intersections is measured by control delay during the 15-minute peak time period, which is basically estimated based on the methods recommended in the HCM 2010 (TRB, 2010). The classification criteria of LOS for motor vehicles, pedestrians, and public transit are listed in Table 11.5. In addition to control delay, other recommended performance measures in the standards include queue length, number of stops, maximal flow ratio, throughput, and degree of saturation, etc.

In practice, in the planning stage, control delay of motor vehicles is usually considered when determining signal timings, while those of pedestrians and nonmotor vehicles are omitted in many cases. Control delay and degree of saturation are most often used for performance evaluation of undersaturated intersections, while queue length and capacity or throughput are used for saturated intersections.

Figure 11.12 A large roundabout in Shanghai (Wujiaochang Roundabout) operated by traffic signals.

Table 11.5 LOS classification criteria for various road users at signalized intersections

	Motor vehicles (s/vehicle)	Pedestrians (s/person)	Public transit (s/vehicle)
A	0–10	0–15	0–10
B	10–20	15–20	10–20
C	20–35	20–30	20–30
D	35–55	30–45	30–45
E	55–80	45–60	45–60
F	≥80	≥60	≥60

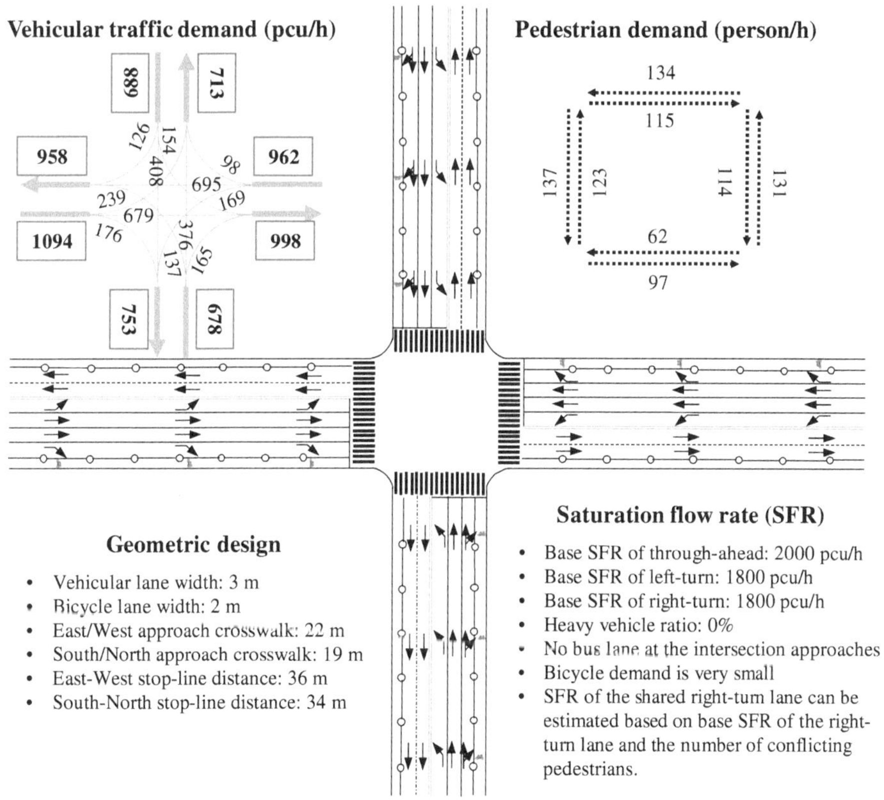

Figure 11.13 A typical four-leg intersection and common settings for the working example.

11.11 Working example

A typical four-leg intersection, as presented in Fig. 11.13, is considered in this working example of signal timing design, while following the procedure and methods introduced above.

11.11.1 The first trial calculation

1. Signal phasing

   A typical four-phase plan with protected left-turn, as presented in Fig. 11.8C, is used for this example, according to traffic demands and intersection geometry.
2. Through and left-turn vehicles

   The yellow time is 3 seconds, accounting for the speed limit. The all-red time is 1 seconds based on the critical distance, that is, the length difference for vehicles from the stop line to leaving the conflict point and vehicles in conflicting movements from the stop line to just accessing the conflict point. Thus, the intergreen time I is 4 seconds. In addition, the flashing green time included in the green interval is 3 seconds.

   Saturation flow rates are adjusted based on lane widths and the effects of pedestrians. The critical flow ratios are calculated as follows:

- Phase 1 (WBT/EBT): $y_1 = max\left\{\frac{q_5}{S_5}, \frac{q_1}{S_1}\right\} = max\left\{\frac{695}{2000}, \frac{679}{2000}\right\} = 0.17$
- Phase 2 (WBL/EBL): $y_2 = max\left\{\frac{q_6}{S_6}, \frac{q_2}{S_2}\right\} = max\left\{\frac{169}{1800}, \frac{239}{1800}\right\} = 0.13$
- Phase 3 (NBT/SBT): $y_3 = max\left\{\frac{q_7}{S_7}, \frac{q_3}{S_3}\right\} = max\left\{\frac{271}{2000}, \frac{106}{1711}, \frac{267}{2000}, \frac{141}{1518}\right\} = 0.14$
- Phase 4 (NBL/SBL): $y_4 = max\left\{\frac{q_8}{S_8}, \frac{q_4}{S_4}\right\} = max\left\{\frac{137}{1800}, \frac{154}{1800}\right\} = 0.09$

   Furthermore, the sum of critical flow ratios is calculated as
- $Y = 0.13 + 0.17 + 0.09 + 0.14 = 0.53$

   The total lost time per cycle is calculated as
- $L = 4 \times 4 = 16$ seconds

   Thus, the optimum cycle time is calculated as
- $C_0 = \frac{1.5L + 5}{1 - Y} = \frac{1.5 \times 16 + 5}{1 - 0.53} = 62$ seconds

   The total green time is then
- $G = C_0 - L = 62 - 16 = 46$ seconds

   Finally, the phase durations are known as follows:
- Phase 1 (WBT/EBT): $g_1 = G \times \frac{y_1}{Y} = 46 \times \frac{0.17}{0.53} = 15$ seconds
- Phase 2 (WBL/EBL): $g_2 = G \times \frac{y_2}{\mathrm{Y}} = 46 \times \frac{0.13}{0.53} = 11$ seconds
- Phase 3 (NBT/SBT): $g_3 = G \times \frac{y_3}{\mathrm{Y}} = 46 \times \frac{0.14}{0.53} = 12$ seconds
- Phase 4 (NBL/SBL): $g_4 = G \times \frac{y_4}{Y} = 46 \times \frac{0.09}{0.53} = 8$ seconds < 10 seconds

   The computing result of green time for phase 4 north-bound left-turn/south-bound left-turn (NBL/SBL) is less than the minimum green time, notably 10 seconds for the minor arterial, so the green time should be adjusted.

3. Right-turn vehicles

   Since the conflicting pedestrian volume is small (<3 persons/cycle), right-turn vehicles in dedicated lanes are free of signal control and other right-turn vehicles in shared lanes move with through vehicles.

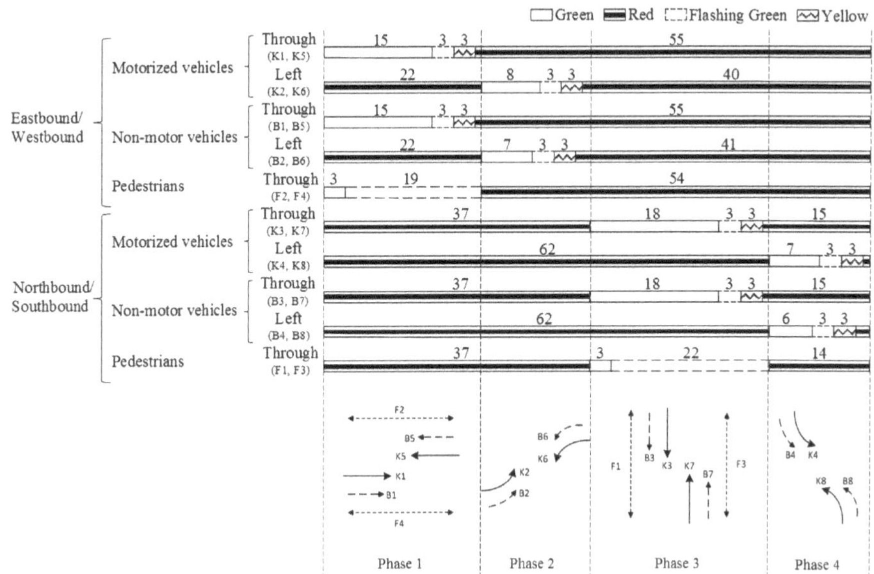

Figure 11.14 Final signal timing plan for the working example.

4. Nonmotor vehicles

Bicycles are released with motorized traffic. The green time for nonmotor vehicles is early cut according to the intergreen time needed by the bicycles and motorized vehicles. The signal timing plan for nonmotor vehicles is shown in Fig. 11.14.

5. Pedestrians

The green time for pedestrians is

- $g_{f2a} = g_{f2b} = g_{f4a} = g_{f4b} = g_1 + I = 15 + 4 = 19$ seconds
- $g_{f1a} = g_{f1b} = g_{f3a} = g_{f3b} = g_3 + I = 12 + 4 = 16$ seconds

The minimum green time for pedestrians to cross the street is calculated as

- E–W/W–E: $g_{min1} = 7 + \frac{L_f}{V_f} - I = 7 + \frac{5 \times 3 + 2 \times 2}{1} - 4 = 22$ seconds $>$ 19 seconds
- S–N/N–S: $g_{min2} = 7 + \frac{L_f}{V_f} - I = 7 + \frac{6 \times 3 + 2 \times 2}{1} - 4 = 25$ seconds $>$ 16 seconds

where V_f denotes the speed of pedestrians and L_f denotes the pavement length (m).

Computing results of the green interval for pedestrians are shorter than the necessary time to cross the street, therefore the green time has to be adjusted.

11.11.2 The second trial calculation

1. Through and left-turn vehicles

For vehicular safety reasons, the green time of phase 4 (NBL/SBL) is extended to 10 seconds.

To ensure safety for pedestrians in the east–west/west–east direction, the green time of phase 1 west-bound through/east-bound through (WBT/EBT) is extended as

- $g_1' = g_1 + (g_{min1} - g_{f2a}) = 15 + (22 - 19) = 18$ seconds

  Similarly, for pedestrians in the north–south/south–north direction, the green time of phase 3 north-bound through/south-bound through (NBT/SBT) is extended:
- $g_3' = g_3 + (g_{min2} - g_{f1a}) = 12 + (25 - 16) = 21$ seconds

  Hence, the new cycle length is
- $C = g_1' + g_2 + g_3' + g_4' + L = 18 + 11 + 21 + 10 + 16 = 76$ seconds

2. Pedestrians

Green time for pedestrians is
- $g_{f2a} = g_{f2b} = g_{f4a} = g_{f4b} = g_1' + I = 18 + 4 = 22$ seconds
- $g_{f1a} = g_{f1b} = g_{f3a} = g_{f3b} = g_3' + I = 21 + 4 = 25$ seconds

  The maximal waiting time for pedestrians is
- $r_{f2a} = r_{f2b} = r_{f4a} = r_{f4b} = C - g_{f2a} = 76 - 22 = 54$ seconds < 90 seconds
- $r_{f1a} = r_{f1b} = r_{f3a} = r_{f3b} = C - g_{f1a} = 76 - 25 = 51$ seconds < 90 seconds

  Thus, the maximal waiting time is satisfied for the requirement of the maximal red time. The flashing green time for pedestrians uses the clearing time:
- $fg_{f2a} = fg_{f2b} = fg_{f4a} = g_{f4b} = \frac{L_f}{V_f} = \frac{5 \times 3 + 2 \times 2}{1} = 19$ seconds
- $fg_{f1a} = fg_{f1b} = fg_{f3a} = fg_{f3b} = \frac{L_f}{V_f} = \frac{6 \times 3 + 2 \times 2}{1} = 22$ seconds

where fg denotes the flashing green time.

Hence, the final scheme is presented in Fig. 11.14.

Acknowledgments

The authors express their gratitude to *Seven Transportation Website* for their support of part of the statistics data in this chapter, and are also thankful to Ms. Jiarong Yao and Ms. Yanning Wei for their support in translating and editing the chapter.

Bibliography

Department of Public Security of Anhui Province of the People's Republic of China, 2015. Code for Signal Control of Intersections on Urban Roads, China. Department of Public Security of Anhui Province of the People's Republic of China (in Chinese).

Federal Highway Administration of United States, 2008. Traffic Signal Timing Manual. Federal Highway Administration of United States, USA.

FGSV, 2015. Richtlinien für Lichtsignalanlagen - RiLSA (German Standard for Traffic Signals). FGSV-Verlag, Cologne, Germany, Edition 2015. (in German).

Gu, J., Lu, J., Wang, Y., 2016. Thinking and advice on promoting intelligence of traffic signal timing and standardization of traffic signs and markings on urban roads. Road Traffic Manage. 11, 22–24 (in Chinese).

Japan Society of Traffic Engineers (JSTE), 2006. Manual on Traffic Signal Control. JSTE, Japan (in Japanese).

Li, F., Tang, K., Yao, J., Li, K., 2017. Real-time queue length estimation for signalized intersections using vehicle trajectory data. Transp. Res. Rec. (2623), 49–59.

Ma, W., Liu, Y., Head, L., Yang, X., 2013. Integrated optimization of lane markings and timings for signalized roundabouts. Transp. Res. C 36, 307–323.
Ma, W., Liu, Y., Zhao, J., Wu, N., 2017. Increasing the capacity of signalized intersections with left-turn waiting areas. Transp. Res. A 105, 181–196.
Research Institute of Highway of Ministry of Transport, 2016. Opening the Book to Recall Distant Memories: Collection for the 60th Anniversary of Research Institute of Highway Ministry of Transport. China Communications Press (in Chinese).
Seven Transportation, 2017. <http://www.7its.com/>. (in Chinese).
Tang, K., Ono, T., Tanaka, S., Kuwahara, M., 2013. A new loss time estimation method for right-turn traffic in Japan considering signal phasing and sneakers. J. Adv. Transp. 47 (8), 704–719.
Tang, K., Zhu, S., Xu, Y., Wang, F., 2016a. Modelling drivers' dynamic decision-making behavior during the phase transition period: an analytical approach based on hidden markov model theory. IEEE Trans. Intelligent Transp. Syst. 17 (1), 206–214.
Tang, K., Dong, K., Chung, E., 2016b. Queue discharge patterns at signalized intersections with flashing green signals and long cycle length. J. Adv. Transp. 50 (8), 2100–2115.
The Ministry of Housing and Urban–Rural Development of the People's Republic of China, 2011. Code for Planning of Intersections on Urban Roads. China Planning Press (in Chinese).
The Ministry of Public Security of the People's Republic of China, 2011a. Road Traffic Safety Law of the People's Republic of China. The Ministry of Public Security of the People's Republic of China, China (in Chinese).
The Ministry of Public Security of the People's Republic of China, 2011b. Regulation on the Implementation of the Road Traffic Safety Law of thePeople's Republic of China. The Ministry of Public Security of the People's Republic of China, China (in Chinese).
The Ministry of Public Security of the People's Republic of China, 2011c. Road Traffic Signal Lamps. The Ministry of Public Security of the People's Republic of China, China (in Chinese).
The Ministry of Public Security of the People's Republic of China, 2016. Specifications for Road Traffic Signal Setting and Installation. The Ministry of Public Security of the People's Republic of China, China (in Chinese).
The Ministry of Public Security of the People's Republic of China, 2015. Road Traffic Signal Control Mode. The Ministry of Public Security of the People's Republic of China, China (in Chinese).
Traffic Management Research Institute of the Ministry of Public Security, 2015. Report on Construction and Application of National Public Security Traffic Control System. Traffic Management Research Institute of the Ministry of Public Security, China (in Chinese).
Transportation Research Board (TRB), 2010. Highway Capacity Manual 2010. National Research Council, Washington, DC, USA.
United Nations, 1968. Convention on Road Signs and Signals. United Nations, Vienna, Austria.
Wang, F., Tang, K., Mei, Y., Kong, T., Li, K., 2014. A multi-objective group based signal control optimization model for mixed traffic flows at isolated intersections. In: Proceedings of the 93rd Transportation Research Board Annual Meeting, Washington, D.C., USA.
Zhou, S., 1987. Traffic Engineering. Tongji University Press (in Chinese).

India

12

Shriniwas Arkatkar[1], *Sudeshna Mitra*[2] *and Tom Mathew*[3]
[1]Department of Civil Engineering, Sardar Vallabhbhai National Institute of Technology, Surat, India, [2]Civil Engineering Department, Indian Institute of Technology, Kharagpur, India, [3]Department of Civil Engineering, Indian Institue of Technology Bombay, Mumbai, India

12.1 Background

The traffic on roads of developing countries like India is highly heterogeneous (mixed) in nature. Vehicles with diverse static (length and width) and dynamic (acceleration, deceleration, and speed) properties characterize this mixture of traffic. Heterogeneity exists not only in terms of vehicle characteristics, but also in the form of diversity of road users (vehicle users and pedestrians). Another peculiarity associated with heterogeneous traffic is the absence of proper lane markings and lane discipline (which reflects the noncompliance of drivers), consequently making traffic movement a rather haphazard and complex phenomenon. Unprecedented vehicular growth (average growth rate of 8%) and rapid rate of urbanization in many Indian cities have resulted in considerable traffic and travel growth on the roads of metropolitan cities and have subsequently resulted in vehicular delays, long queues, and traffic congestion. Growing concerns over negative impacts of traffic congestion and vehicular delays on society and community as a whole warrant the development of sustainable methodologies to regulate traffic on Indian roads. Typically, in Indian conditions, traffic demand is regulated by providing different traffic control measures and strategies ranging from stop-sign control to grade separators. Depending upon the type of control employed at the intersection, it can be termed as uncontrolled, stop controlled, roundabout, signalized, and grade-separated intersection. Intersections, specifically signalized intersections, have become vulnerable with respect to nonlane-based mixed traffic operations with complex vehicular maneuvers and movements in and out of intersections, leading to excessive delays, conflicts, and accidents.

12.1.1 Signalized intersection

Signalization is an intermediate solution, which works on the principles of allocating right-of-way to each or combinations of nonconflicting movements in a cyclic manner. The prime objective of traffic signal control is to minimize vehicular

Global Practices on Road Traffic Signal Control. DOI: https://doi.org/10.1016/B978-0-12-815302-4.00012-1

Figure 12.1 Nonlane-based heterogeneous traffic at signalized intersections in India.

conflict and delay, thereby providing an equitable balance between safe and efficient movement of both vehicular and pedestrian traffic through the intersection. Pretimed or fixed-time, semiactuated, and fully actuated are the different signalization techniques which are generally adopted. In pretimed control cycle length, phase plan and phase times are predetermined and fixed, whereas, actuated signals can change phases before they reach their time limits if the demand is low. They can even skip phases if there is no demand for that phase. In addition to the differences in the controller type, signalized intersections could either be operated in an isolated or coordinated manner. Signal indications include green, amber, and red time for individual approaches once or more than once in each cycle.

In India, urban arterial roads are frequently characterized by fixed-time isolated signalized intersections. For Bangalore city, there are about 400 traffic signals for around 10 million people, which translates to about one traffic signal for every 25,000 people (Report on Best Practices for Traffic Signal Operations in India, 2016). The trends in number of signals per thousand people are similar across many Indian cities. Intersections in most Indian cities are fixed-time isolated signalized intersections, unlike in many other developing and developed countries. However, corridors in certain cities are provided with coordinated signal plans. Furthermore, traffic police also actuate signal timing based on their experience. Fig. 12.1 represents a typical signalized intersection in Indian condition.

12.1.2 State of the art on traffic signal control

A document entitled Guidelines on Design and Installation of Road Traffic Signals (IRC 93-1985), published by the Indian Roads Congress is followed as the manual for the design and installation of road traffic signals in India. IRC 93-1985 provides guidelines and warrants for signalized intersection and design procedures. Another important document referred to is the recently published Indian Highway Capacity Manual (Indian Highway Capacity Manual INDO-HCM, 2018). Details regarding saturation flow and its measurement along with delay and level of service at signalized intersections are well reported in INDO-HCM. The IRC 93-1985 contains guidelines for the installation and design specifications of traffic signal heads under Indian conditions. However, the code practices for traffic signal heads is

Figure 12.2 Typical arrangement of traffic signal heads under Indian conditions.

under revision presently. Nevertheless, Fig. 12.2 shows the typical arrangements of traffic signal heads currently adopted in Indian conditions.

12.2 Control strategies

12.2.1 Warrants for installation of traffic signals

The justification of the need for traffic signal control at any intersection must be evaluated in relation with the warrants laid down by authorities. The Indian Roads Congress, which acts as the governing and regulatory body in India, has developed warrants for signalization, which are published in IRC 93-1985. Of the warrants listed below, at least one of the specified warrants or a combination of more than one (in which case at least 80% of the specified thresholds of the concerned warrants is reached) need to be satisfied for the application of signal control at any intersection.

1. Minimum vehicular volume;
2. Interruption to continuous traffic;
3. Minimum pedestrian volume;
4. Accident experience;
5. Combination of warrants.

12.2.1.1 Warrant 1: minimum vehicular volume

The traffic volume on the major street and the highest-volume minor street for each of any 8 h of an average day should be equal to the values as indicated in Table 12.1.

12.2.1.2 Warrant 2: interruption to continuous traffic

The traffic volume on the major street and the highest-volume minor street for each of any 8 h of an average day should be equal to the values as indicated in Table 12.2.

Table 12.1 Minimum vehicular volume warrant

Number of lanes on each approach		Vehicles per hour on major street (both approaches)	Vehicles per hour on minor street (one-direction only)
Major street	Minor street		
1	1	650	200
2 or more	1	800	200
2 or more	2 or more	800	250
1	2 or more	650	250

Source: IRC 93-1985.

Table 12.2 Interruption of continuous traffic

Number of lanes on each approach		Vehicles per hour on major street (both approaches)	Vehicles per hour on minor street (one-direction only)
Major street	Minor street		
1	1	1000	100
2 or more	1	1200	100
2 or more	2 or more	1200	150
1	2 or more	1000	150

Source: IRC 93-1985.

12.2.1.3 Warrant 3: minimum pedestrian volume

For each of any hour of any 8 h of an average day, the following traffic volume must exist:

1. On the major street, 600 or more vehicles per hour enter the intersection (both approaches); or where there is a raised median island 1–2 m or more in width, 1000 or more vehicles per hour (both directions) enter the intersection.
2. During the same 8 h as in (1) above there are 150 or more pedestrians per hour on the highest volume cross-walk crossing the major street.

12.2.1.4 Warrant 4: accident experience

The accident experience warrant is satisfied when:

1. Adequate trial of less-restrictive remedies with satisfactory observance and enforcement have failed to reduce the accident frequency.
2. Five or more reported accidents, of types susceptible to correction by traffic signal control, have occurred within a period of 12 months, each accident involving personal injury or property damage to an apparent extent of Rs. 2000 or more.
3. The signal installation will not seriously disrupt traffic flow.

12.2.2 *Vehicle-actuated control and signal coordination*

The implementation of vehicle-actuated control in nonlane-based heterogeneous traffic composed of passenger cars, trucks, two- and three-wheelers, and nonmotorized vehicles is more challenging than their implementation in homogeneous traffic conditions. One issue is the subpar performance of predictive techniques using upstream detectors in determining the turning ratios accurately. The varying static and dynamic characteristics result in higher lane-change behavior as the vehicles approach the intersection. The higher number of nonconventional vehicle types also results in the clustering of vehicle at the approaches, which makes the accurate estimation of demand difficult. The conventional loops fail to count and classify vehicles with the required degree of accuracy in such conditions. One innovation in this regard has been the use of multiple loops (Ali et al., 2012), which have shown considerable improvements in the estimation of data. A notable method adopted and successfully implemented in the field is the use of stop-line detectors and detectors at the downstream of the stop-line to determine the turning ratios (Ravikumar and Mathew, 2011).

The issues related to heterogeneous traffic conditions discussed under the section "Vehicle-actuated control" are also influential in the coordination of traffic signals. Studies on platoon ratio and platoon dispersal elaborate the adverse effects of vehicles traveling at different speeds on possible signal coordination (Arasan and Kashani, 2003; Saha et al., 2017a,b). Highly favorable platoon ratios are generally observed when the proportion of passenger cars in the stream is more than 40%, which is not the case in most of situations.

Although there have been notable studies in vehicle actuation and coordination of traffic signals, field implementations of the same have not been as widely carried out as has been in homogeneous traffic owing to the problems discussed before. The guidelines for traffic control in India (IRC Code:093) were last revised in 1985 and contain only general guidelines of signal coordination with the definition of basic objectives and explanations on the major types of coordinated signal systems, and no guidelines on actuated controls. Furthermore, IRC 93-1985, does not explicitly provide specific guidelines concerning the nature of traffic control, that is, isolated or coordinated to be adopted at a particular location.

12.3 Signal timing procedure

The prime objective of phase design is to separate the conflicting movements in an intersection into various phases, so that movements in a phase will have no conflicts. If all the movements are to be separated with no conflicts, many phases are required. In such a situation, the objective is to design phases with minimum conflicts or with less severe conflicts. The signal design procedure involves six major steps. They include: (1) phase design, (2) determination of amber time and clearance time or the "intergreen" time, (3) determination of cycle length, (4) apportioning of green time, (5) pedestrian crossing requirements, and (6) performance evaluation of the design obtained in the previous steps. Signalized intersections are designed based on the principles laid down in IRC 93-1985. In addition, Webster's method for deciding optimum signal length is also adopted in the design of signalized intersection in India.

12.3.1 Steps for traffic signal design as per IRC guidelines

Source: Khanna et al. (2014).

1. The pedestrian green times required for major and minor roads are calculated based on a walking speed of 1.2 m/s and initial walk time of 7.0 s. These are minimum green times required for the vehicular traffic on minor and major roads, respectively.
2. The green time required for the vehicular traffic on the major road is increased in proportion to the traffic on the two approach roads.
3. The cycle time is calculated after allowing an amber time of 2 s each.
4. The minimum green time required for clearing vehicles arriving during a cycle is determined for each lane of the approach road assuming that the first vehicle will take 6 s and the subsequent vehicles or the passenger car unit (PCU) of the queue will be cleared at a rate of 2 s/vehicle. The minimum green time required for the vehicular traffic on any of the approaches is limited to 16 s.
5. The optimum cycle time is calculated using Webster's formula. The saturation flow values may be considered from the section later. The lost time is calculated from the amber time, intergreen time, and the initial delay of 4 s for the first vehicle on each leg.
6. The cycle time and the phases may be revised keeping in view the green time required for clearing the vehicles and the optimum cycle length.

12.3.2 Input information for signal timing

The following primary and secondary surveys required for signal warrant analysis shall be conducted for justifying a traffic signal:

1. Primary: Speed-delay survey for two-wheelers, cars, and buses with at least three trips in each peak hour (morning, afternoon, and evening). Classified traffic volume count survey including pedestrian counts.
2. Secondary: Accident analysis.

The following primary and secondary surveys shall be conducted for implementing a traffic signal including installation of signal and timing at the junction:

1. A 24-h turning movement count which includes all entering traffic for at least 4 typical days (two typical weekdays, Saturday and Sunday);
2. Determine a.m. and p.m. peak hour traffic volumes;
3. Pedestrian volumes;
4. Percentage of trucks and buses;
5. If a 24-h count is not possible, factors of 1.20 and 1.35 can be applied on 16- or 12-h counts, respectively, to obtain the 24-h counts. If historic data are available, these factors should be revised per local data;

6. Condition diagram, which includes roadway geometrics, parking, driveways, sidewalks, signing pavement markings, development of intersection quadrants, and any other features pertinent to the study;
7. A conflict analysis;
8. Approach speed limits and/or approach speeds;
9. Analysis of the existing progression in a coordinated system;
10. Capacity analysis of the intersection for current and future years using growth expectations of the location;
11. Other data, which are desirable for a more precise understanding of the operation of the intersection.

A comprehensive explanation of the above method of signal design is provided in a section 12.5.

12.4 Signal phasing

12.4.1 Phase structure

The driving conditions in India are set to the keep-left rule, and hence, left-turn movements for most of the intersections are permissive. Therefore, right-turning movements are conflicting movements, which need to be considered. Although two-phase and four-phase plans are the most common phasing techniques adopted, the two-phase system is not prevalent in India, as it requires lane discipline and higher compliance from drivers.

12.4.2 Four-phase signals

There are at least three possible phasing options. For example, Fig. 12.3 (Option 1) shows the simplest and trivial phase plan. Here, flow from each approach is put into a single phase avoiding all conflicts. This type of phase plan is ideally suited to urban areas where the turning movements are comparable, with through movements and when through traffic and turning traffic need to share the same lane. This phase plan could be very inefficient when turning movements are relatively low. Fig. 12.3 (Option 2) shows a second possible phase plan option where opposing through traffic is put into the same phase. The nonconflicting right turn flows 7 and 8 are grouped into a third phase. Similarly, flows 5 and 6 are grouped into the fourth phase. This type of phasing is very efficient when the intersection geometry permits having at least one lane for each movement and the through traffic volume is significantly high. Fig. 12.3 (Option 3) shows yet another phase plan. However, this is rarely used in practice. There are five phase signals, six phase signals, etc. They are normally provided if the intersection control is adaptive, that is, the signal phases and timing adapt to the real-time traffic conditions. The most commonly adopted phase plan in India is Option 1. The reason for this is the fact that well-defined lanes are absent in Indian traffic conditions. In a typical intersection, the second option provides the best results since the major flows are grouped together. The third option is also possible.

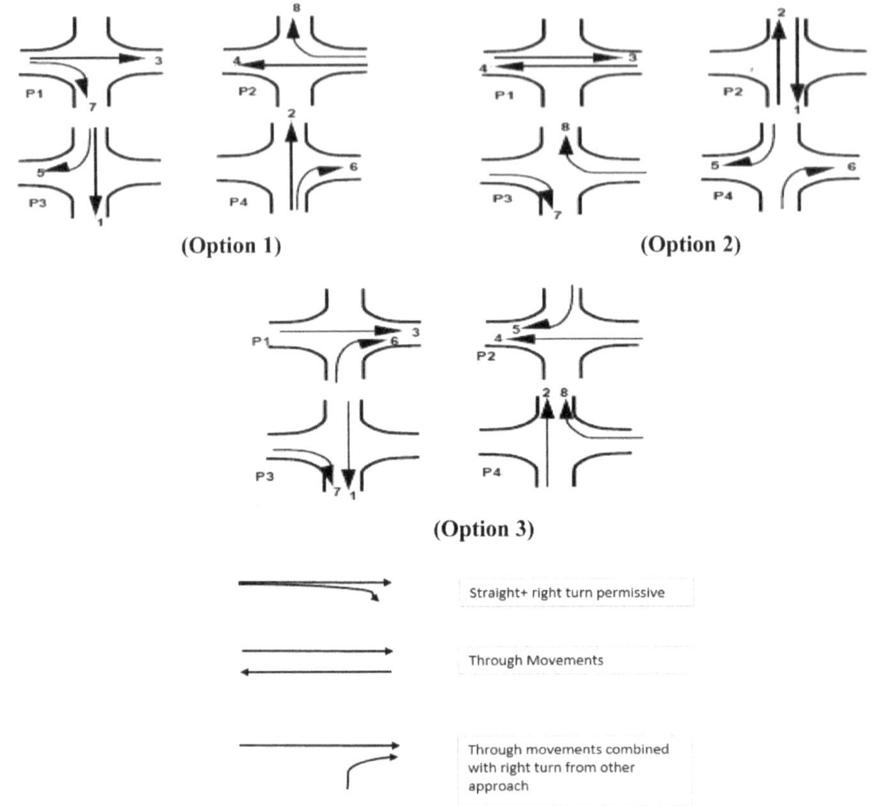

Figure 12.3 Four-phase plan options available.
Source: Chapter 34: Design of traffic signal, Tom Mathew, NPTEL lecture notes.

12.5 Parameter determination

12.5.1 Saturation flow

The signal cycle and the corresponding green times are decided by many factors, of which saturation flow is of prime importance. The saturation flow is the hourly rate at which vehicles can pass a stop line during the green phase. This maximum flow rate is present only a short duration after the start of the green phase and is assumed to continue until the queued vehicles have discharged. At the beginning of the green phase the drivers take time to react and accelerate. That transitional phase is not effectively used to serve the traffic. Similarly, at the end of the green phase the traffic flow can still continue during the first seconds of the yellow phase, giving an extension of the effective time that a flow has access to the intersection. The saturation flow occurs between these two time periods. The saturation flow in vehicles per hour can be represented as shown in Eq. (12.1).

$$SF = \frac{3600}{h_s} \tag{12.1}$$

where SF is the saturation flow (vehicles/h); h_s is the saturation headway (s).

As per Highway Capacity Manual (HCM 2010), the first four vehicles are ignored from saturation flow considering the start-up lost time, and the saturation flow is measured until the last vehicle joining the queue during the red time is cleared. The HCM 2010 has defined a base saturation flow, which should be adjusted to the prevailing conditions using several multiplicative factors.

In the Indian context, as mentioned earlier, the presence of vehicles with varying static and dynamic properties exponentially increases the interactions. Estimation of saturation becomes highly critical in a mixed traffic condition, as the stabilization of flow during effective green period is not similar to that under homogeneous conditions.

What is seen in Fig. 12.4 is unique to heterogeneous traffic and the primary reason for this is that the smaller vehicles, especially two-wheelers, come to the front of queues at signals. At the start of green, there is an initial surge of vehicles because of the smaller vehicles coming to the front of the queue and this is reflected in the high discharge seen in the first 5 s in Fig. 12.4. Thereafter, the vehicle discharge decreases and the same is reflected in Fig. 12.4. In homogeneous traffic conditions on the other hand, the discharge is low at the beginning owing to acceleration delay. This increases and reaches saturation flow and the flow remains the same until the queue is dissipated. This difference in discharge pattern adds complexity to the measurement of saturation flow under heterogeneous traffic conditions when compared to homogeneous traffic conditions. To achieve homogeneity, the vehicles are converted to PCUs based on the impact they have on the traffic in comparison to the impact a passenger car has. Using this concept, the PCU

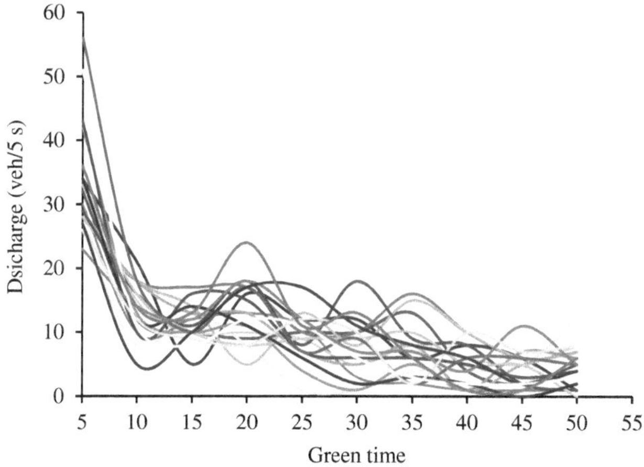

Figure 12.4 Discharge pattern as the sum of all vehicles for several cycles. *Note*: Vehicle discharge is measured every 5 s.

values are obtained by minimizing the variation in saturation flow of a given approach among several cycles. The PCU values to be used for converting the volumes of all vehicle types into an equivalent volume of passenger cars are presented in Table 12.3.

Saturation flow is defined as "The steady state discharge rate of queued vehicles from an approach at a signalized intersection with continuous green and an infinite queue. For the purpose of measurement, it is taken as the maximum departure rate of queued vehicles from an approach during the green interval measured at the stop line under prevailing conditions. It is expressed in passenger cars per hour." In other words, saturation flow is a flow, which may be obtained when there is a continuous queue of vehicles, and they are given 100% green time. It is expressed in vehicles per hour in green time. When the green period commences, vehicles take some time to start and accelerate to the normal running speed, but after a few seconds, the queue discharges at a constant rate, which is known as the saturation flow.

In the Indian context, due to the presence of heterogeneity, the calculation of saturation flow becomes highly crucial. Moreover, a change in the composition of vehicles significantly affects saturation flow values. In addition, the start time of vehicles to accelerate is also a crucial criterion for calculation of saturation flow compared to lost time as considered to developed countries.

The unit base saturation flow rate is given in Eq. (12.2).

$$USF_0 = \begin{cases} 630; \text{if} w < 7.0 \\ 1140 - 60w; \text{if} 7.0 \leq w \leq 10.5 \\ 500; w > 10.5 \end{cases} \quad (12.2)$$

Table 12.3 Passenger car unit (PCU) values for signalized intersections

Vehicle type	PCU values
Motorized two-wheelers	0.38
Motorized three-wheelers	0.45
Passenger car	1.00
Light commercial vehicle (LCV)	1.02
Heavy commercial vehicle (HCV)[a]	1.52[a]
Bus[a]	1.56[a]
Bicycle	0.33
Cycle rickshaw	1.84
Hand/animal-drawn cart	4.00

[a]PCU value is valid when the proportion of this vehicle type is not more than 15%.

Source: Indian Highway Capacity Manual (INDO-HCM), 2018. CSIR-Central Road Research Institute (CRRI), New Delhi.

where USF_0 = unit base saturation flow rate (PCU/h/m); w = effective width of approach (m).

The prevailing saturation flow of the intersection approach for the movement group under consideration is then obtained as:

$$SF = w \times USF_0 \times f_{bb} \times f_{br} \times f_{is} \tag{12.3}$$

where SF = prevailing saturation flow rate (PCU/h); w = effective width of the approach in meter used by the movement group; USF_0 = unit base saturation flow rate; f_{bb} = adjustment factors for bus blockage due to kerbside bus stop; f_{br} = adjustment factors for blockage of through vehicles by standing right-turning vehicles waiting for their turn; f_{is} = adjustment factor for the initial surge of vehicles due to approach flare and anticipation.

Instead of estimating the saturation flow using this model, for an existing intersection, analysts have the option of measuring the saturation flow in the field. The measured saturation flow in vehicles/h can be converted into PCU/h using the PCU values in Table 12.3. The details of these adjustment factors are provided later.

12.5.1.1 Adjustment for bus blockage

The adjustment factor for bus blockage accounts for the reduction in saturation flow due to the presence of bus stops within a 75-m vicinity of the intersection (Indian Highway Capacity Manual INDO-HCM, 2018). This adjustment factor is computed as

$$f_{bb} = \frac{w - 3\left(t_b n_B / 3600\right)}{w} \tag{12.4}$$

where f_{bb} = adjustment factor for bus blockage; w = approach width (m); t_b = average blockage time (s) during green; n_B = number of buses stopping in an hour (buses/h). If there are more than 200 buses/h, then n_B should be limited to a practical value of 200.

The adjustment factor for bus blockage is adapted from HCM (2010) with appropriate modifications. The average width of the road section blocked by the buses is taken as 3 m and the value of the average blockage time t_b during green is taken as 18 s. This value is arrived at based on reasonably large data of bus dwell times collected at various locations having different land uses in urban areas. This factor is be used only if the stops made by the buses at the intersections hinder the discharge during the green interval of the signal, otherwise the factor is taken as 1. In addition, in the case where exclusive bus bays are present, the factor is taken as 1.

12.5.1.2 Adjustment for blockage by standing right-turn vehicles

This adjustment to base saturation flow is applied when the through movement cannot use the full width of approach due to the presence of standing vehicles along the median lane waiting for their exclusive green phase. In the absence of field-observed data, the default value for the width of approach that is occupied by the right-turning vehicles (w_r) may be taken as 2.5 m. This adjustment factor is given in Eq. (12.5).

$$f_{br} = \frac{w - w_r}{w} \tag{12.5}$$

where f_{br} = adjustment factor for standing right-turn vehicles; w = approach width of the through movement (m); w_r = width of approach along the median occupied by the standing vehicles waiting for the exclusive right-turn phase.

This factor is to be applied only if exclusive lanes are not present for the exclusive right-turn phase. $f_{br} = 1$ if exclusive right-turn lanes are present. This factor is applicable to approaches with a width 7 m or more.

12.5.1.3 Adjustment factor for the initial surge of vehicles

This adjustment factor is applied when the flare effect or anticipation effect is observed at the intersection. Due to the presence of these effects, there will be a higher discharge of vehicles during the first few seconds of the green interval. The adjustment factor for only the effect that is present is applied. The default values of adjustment factors for these situations are given in Table 12.4. The adjustment factor depends on the green interval (G) of the phase and the surge ratio (S_R). Surge ratio (S_R) is the ratio of the initial higher discharge rate due to either the flare effect or anticipation effect or both to the uniform saturation flow. For practical purposes, the initial higher discharge rate or the surge flow is worked out based on the discharge rate observed during the first 5 s of green initiation. The adjustment factors for the initial surge of vehicles (f_{is}) is summarized in Table 12.4.

Table 12.4 Adjustment factor for initial surge of vehicles (f_{is})

Green time of phase (s)	**Only anticipation effect**	**Only approach flare effect**		**Anticipation and approach flare effect**	
		Low (S_R = 1.15)	**High (S_R = 1.35)**	**Low (S_R = 1.15)**	**High (S_R = 1.35)**
<15	1.133	1.020	1.047	1.153	1.180
15–30	1.067	1.010	1.023	1.077	1.090
30–45	1.044	1.007	1.016	1.051	1.060
45–60	1.033	1.005	1.012	1.038	1.045
60–75	1.027	1.004	1.009	1.031	1.036
75–90	1.022	1.003	1.008	1.026	1.030
90–120	1.017	1.002	1.006	1.019	1.023
>120	1.000	1.000	1.000	1.000	1.000

f_{is} = 1 when no surge flow is observed on the approach.
Source: Indian Highway Capacity Manual (INDO-HCM), 2018. CSIR-Central Road Research Institute (CRRI), New Delhi.

The factors applicable for the estimation of adjusted saturation flow are summarized in Table 12.5.

Table 12.6 summarizes some specific contributions of a few researchers in the direction of saturation flow values under Indian heterogeneous conditions.

12.5.2 Capacity

The capacity (c) of an approach of a signalized intersection can be expressed as shown in Eq. (12.6).

$$c_A = SF_A\left(\frac{g_A}{C}\right) \tag{12.6}$$

where c_A = capacity of approach A (PCU/h); SF_A = prevailing (after adjustments) saturation flow of the approach (PCU/h); g_A = effective green time for approach A (s); C = cycle time (s).

12.5.3 Cycle length

Cycle time is the time taken by a signal to complete one full cycle of iterations, that is, one complete rotation through all signal indications. Webster's method is

Table 12.5 Saturation flow adjustment factors

Sl. no.	Factor	Formula	Requirement	Other situations
1	Bus blockage	$f_{bb} = \frac{w - 3(t_b n_B/3600)}{w}$	$f_{bb} \leq 1$ $n_B \leq 200$ $t_b = 18$	$f_{bb} = 1$ in any of the following situations: 1. Presence of bus bays 2. Buses do not hinder the discharging of vehicles
2	Blockage by standing right-turning vehicles	$f_{br} = \frac{w - w_r}{w}$	$f_{br} \leq 1$ $w \geq 7$	$f_{br} = 1$ in any of the following situations: 1. Exclusive right-turning lane is present 2. No right-turning vehicles
3	Initial surge	Refer to Table 12.4	$f_{is} \geq 1$	$f_{is} = 1$ if initial surge is not present

Table 12.6 Saturation flow values reported by researchers for Indian conditions

Methodology	Saturation flow values	Authors
–	1890 PCU/h	Indian Road Congress (1994)
Simulation	2196 PCU/h	Arasan and Vedagiri (2006b)
Optimization	1920 PCU/h	Radhakrishnan and Mathew (2011)
HCM model	5514–12,505 vehicles/h	Anusha et al. (2013)[a]
Multilinear regression	2265 PCU/h	Patel et al. (2015)
–	1869–2030 PCU/h/lane	Chand et al. (2016)
Multilinear regression	1896–2310 PCU/h/lane	Kulakarni et al. (2017)
Multilinear and Kriging technique	1800–2100 PCU/h/lane	Biswas et al. (2018)

[a]Not converted to PCU/h.

adopted for calculating the optimum cycle length. The optimum cycle length is given by the equation

$$C_o = \frac{1.5L + 5}{1 - Y} \tag{12.7}$$

where L = total lost time per cycle (s); Y = maximum ratios of flow to saturation flow.

The minimum cycle length preferably recommended is 120 s, which is the maximum delay acceptable to pedestrians and drivers (Indian Road Congress 1994, SP-41).

12.5.4 Green time

The effective green time is the actual time available for the vehicles to cross the intersection. It is the sum of actual green time (G_i) plus the yellow time, minus the applicable lost times. This lost time is the sum of start-up lost time (l_1) and clearance lost time (l_2) denoted as t_L. Thus, effective green time can be written as shown in Eq. (12.8).

$$g_i = G_i + Y_i - t_L \tag{12.8}$$

Minimum green time is considered based on the pedestrian walking speed and initial walk time. An initial walk time of 7 s is considered. For calculating pedestrian walk interval, pedestrian speed of 1.2 m/s is considered.

Minimum green time = pedestrian walk interval + initial walk time

12.5.5 Red time

This is the red indication for a particular movement or set of movements and is denoted by R_i. It is the actual duration the red light of a traffic signal is turned on, to the onset of green. Presently, no well postulated guideline is available to determine red time at signalized intersections in Indian conditions. Usual practice, of subtracting green time and amber time from total cycle time, is adopted to account for the red time for any given phase.

12.5.6 Intergreen time

Gazis et al. proposed the theoretical basis for the calculation of intergreen time in 1960, based on the hypothesis that a driver approaching a signalized intersection during the yellow interval will either have to stop at the stop line or proceed to cross the intersection. It should not be so small that the drivers are forced to respond hastily, thus, undertaking an abrupt stopping maneuver or cause red light running. Good design of intergreen time ensures elimination of the dilemma zone and provides enough time for vehicles near the stop line to clear the intersection safely and for vehicles that are far away from the intersection to come to a safe stop. Thus, based on these criteria, stopping distance to the stop line and crossing distance of the intersection are equated to obtain the yellow time. Additionally, for intersections catering to a heavy traffic flow, it is generally recommended to provide an all-red time, which allows additional time for motorists already in the intersection to clear the intersection on the red indication before conflicting traffic movements are released. Following this, Gazis et al. (1960) proposed a relation by which intergreen time can be obtained by equating the stopping distance required by the vehicle to the clearing distance allowed to the vehicle during yellow interval. Based on this, the relation is given as seen in Eq. (12.9).

$$I = t + \frac{v}{2a} + \frac{W + L}{v} \tag{12.9}$$

where I = intergreen time (s); v = clearance speed (m/s); W = intersection width (m); L = length of the vehicle (m); t = reaction time of the driver (s); a = deceleration rate of the driver (m/s^2).

While intergreen time calculation is critical for the safety of signalized intersections, Indian Road Congress (1985) does not provide any recommendation on calculation of the same. As a result, any one of the two distinct approaches, that is, (1) the German methodology suggested by Retzko and Boltze (1987) or (2) the methodology originally suggested by Gazis et al. (1960) and adopted by the Institute of Transportation Engineers (ITE), United States, is commonly used to estimate intergreen intervals. Arasan et al. (2006a,b) compared these two established methods and a probabilistic method of intergreen time design for heterogeneous traffic condition and arrived at the conclusion that these three methods provide very different estimates of intergreen time. Hence, the authors suggested that the maximum and minimum values from these three methods should be considered and further field trial or simulation methods should be used to decide the suitable value of intergreen time to ensure both capacity and safety (Arasan et al., 2006a,b).

A suitable method combining the ITE method and German method has also been proposed for Indian conditions by Kedia and Mitra, based on a study conducted in Kolkata, India (working paper based on the thesis by Kedia, 2014). Following the German method of intergreen time design, three important components are used to calculate the intergreen time. These are crossing time, clearing time, and entering time. Crossing time (t_a) is defined as the time taken by the last vehicle in the current stream to pass the stop line after green has been taken away. Clearance time (t_c) is defined as the time taken by the last vehicle of the ending phase, to clear off the conflict point at the intersection, and entering time (t_e) is defined as the time taken by the first vehicle of the next phase to reach the conflict point after green has been given to that phase. Following these terminologies, the intergreen time is thus given as seen in Eq. (12.10).

$$T = t_a + t_c - t_e \tag{12.10}$$

where T = intergreen time (s); t_a = crossing time (s); t_c = clearance time (s); t_e = entering time (s).

While the German guideline provides empirically derived values for crossing time, clearance time, and entering time for typical German scenarios, the same are derived for the heterogeneous traffic in Indian conditions and are given as follows.

12.5.6.1 Crossing time

Crossing time (t_a) is calculated based on the assumption that a vehicle traveling at a speed u_0 can come to a complete stop if its distance from the stop line is farther than the stopping sight distance when it decelerates with rate d.

Consider a vehicle traveling at a speed u_0, the stopping sight distance required for the vehicle is given in Eq. (12.11).

$$S.S.D. = (u_0 \times \tau) + \frac{u_0^2}{2d} \tag{12.11}$$

where $S.S.D.$ = stopping sight distance (m); u_0 = clearance speed (m/s); τ = reaction time (s); d = deceleration rate (m/s^2).

Now, in a critical setting, there may be a vehicle approaching the intersection, which will not want to stop at the stop line and it will be the last vehicle in the current stream to pass the stop line after green has been taken away. Hence, the crossing time by the vehicle may be obtained by equating it to $S.S.D.$, if it travels at uniform speed. Thus, if t_a is the time traveled by the last vehicle in uniform speed, then (see Eqs. (12.12) and (12.13)):

$$u_0 t_a = (u_0 \times \tau) + \frac{u_0^2}{2d} \quad (12.12)$$

$$t_a = \frac{S.S.D.}{u_0} \quad (12.13)$$

where all the parameters are as defined previously.

For this purpose, the reaction time is assumed to be 1 s and the deceleration rates for various vehicle categories are given in Table 12.7.

Tables 12.8 and 12.9 give the stopping sight distance and the corresponding crossing time values, respectively, obtained using the aforementioned formulation for different vehicle categories.

12.5.6.2 Clearance time and entering time

While calculation of the crossing time followed a theoretical approach, an empirical approach has been followed for the purpose of calculating clearance as well as

Table 12.7 Deceleration rates for different vehicle categories[a]

Vehicle type	Deceleration (m/s^2)
Passenger cars	2.529
Auto rickshaws	1.777
Two-wheelers	2.317
Buses	2.228

[a]*Source*: Undergoing Research in IIT Bombay on "Calibration and Validation of a Mixed Traffic Simulator – SiMTraM and Implementation of Vehicle Type Dependency in SUMO."

Table 12.8 Stopping sight distance for different vehicle categories

Stopping sight distance (*S.S.D.*) (m)				
Vehicle category / Speed category (km/h)	Passenger car/taxi/van	Trucks	Two-wheelers	Three-wheelers
5	–	–	1.8	1.9
10	4.3	4.5	4.4	4.9
15	7.6	8.1	7.9	9.1
20	11.7	12.5	12.2	14.2
25	16.5	17.8	17.4	20.5
30	22.1	23.9	23.3	27.9
35	28.4	30.9	30.1	36.3
40	35.5	38.8	37.8	45.8
45	43.4	47.6	46.2	–
50	52.0	–	55.5	–

Table 12.9 Crossing times for different vehicle categories

Crossing time (s)				
Vehicle category / **Speed category (km/h)**	**Passenger car/taxi/van**	**Trucks**	**Two-wheelers**	**Three-wheelers**
5	–	–	1.3	1.4
10	1.5	1.6	1.6	1.8
15	1.8	1.9	1.9	2.2
20	2.1	2.3	2.2	2.6
25	2.4	2.6	2.5	3.0
30	2.7	2.9	2.8	3.3
35	2.9	3.2	3.1	3.7
40	3.2	3.5	3.4	4.1
45	3.5	3.8	3.7	–
50	3.7	–	4.0	–

entering times for intersections of varying sizes as well as different operating speeds. Based on the survey conducted at the signalized intersections in Kolkata the values for t_c and t_e for each phase were obtained. As clearance time is defined as the time taken by the last vehicle of the ending phase, to clear off the conflict point at the intersection, it depends on the vehicle category. Using the survey data, average clearance time (μ_{tc}) has been calculated for all categories of vehicles at each intersection. Also, [average − standard deviation ($\mu_{tc} - \sigma_{tc}$)] and [average + standard deviation ($\mu_{tc} + \sigma_{tc}$)] for clearance time have been calculated. These values have been plotted against vehicular speed and different clearing distances (dc) to directly represent the clearance time for any given intersection for all vehicle categories (Fig. 12.5).

Furthermore, to obtain a suitable entering time, an empirical approach has been undertaken, wherein entering times have been plotted against entering distances for the desired level of safety. This has been done due to the fact that the entering maneuver is influenced by several factors, such as pedestrian interference, drivers' behavior, and the degree of heterogeneity in vehicular composition. Hence, entering speed fluctuates, making the empirical approach more suitable for this purpose. Finally, for safety, 40th and 50th percentile entering times were recommended for all intersections and for all vehicle categories. This is because a higher value will result in a less conservative design by lowering the intergreen time, whilst too low a value will make the design inefficient by resulting in a high value of intergreen time, which will have an adverse effect on the capacity. Thus, a 50th percentile value is suitable here as it accounts for the median scenario. A 40th percentile value provides a slightly conservative estimate, thus has also been considered as a plausible case. These entering time values have been plotted against the entering distance. Hence, knowing the entering distance for any given intersection and the desired level of safety, the entering times can be computed. These graphs are shown in Fig. 12.6.

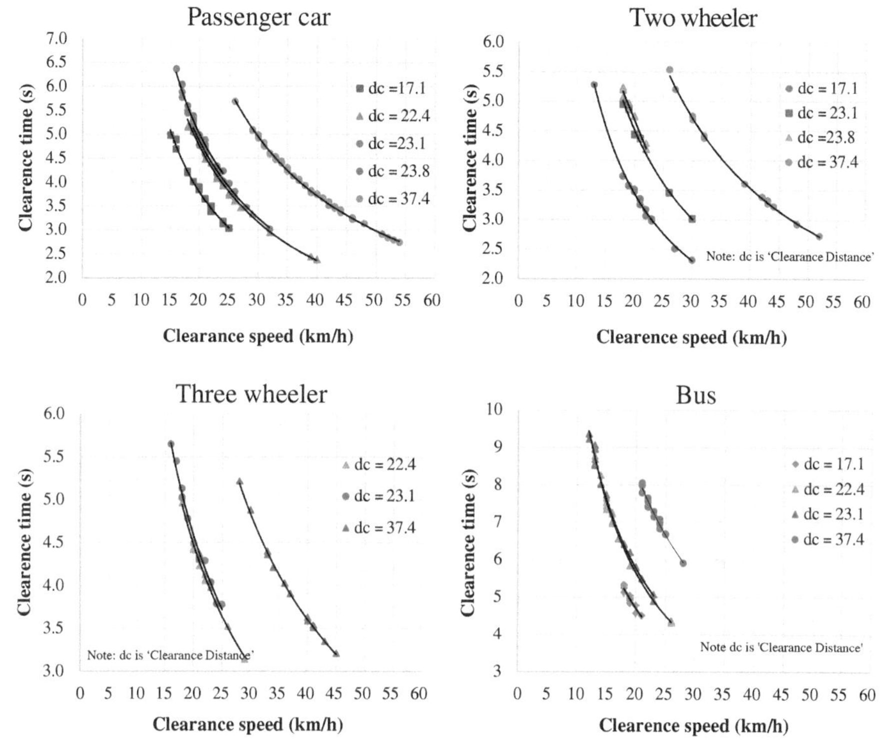

Figure 12.5 Clearance time vs clearance speed for different clearing distances.

12.5.7 Lost time

This indicates the time during which the intersection is not effectively utilized for any movement. As mentioned earlier in Section 12.5.4, when the green phase starts, the drivers take time to react and accelerate, resulting in some amount of time not being used for the effective movement of traffic, resulting in a start-up lost time. At the end of the green phase the traffic flow takes some time to stop and extend to a few seconds of the yellow phase, resulting in clearance time. The total lost time is the sum of the two aforementioned components.

12.6 Performance measurement of signalized intersection (Indian Highway Capacity Manual INDO-HCM, 2018)

Delay and queue length are the two most important performance measurement criteria for signalized intersections. Properly designed traffic signals reduce motorists' delay by providing the common road space periodically to vehicular traffic from

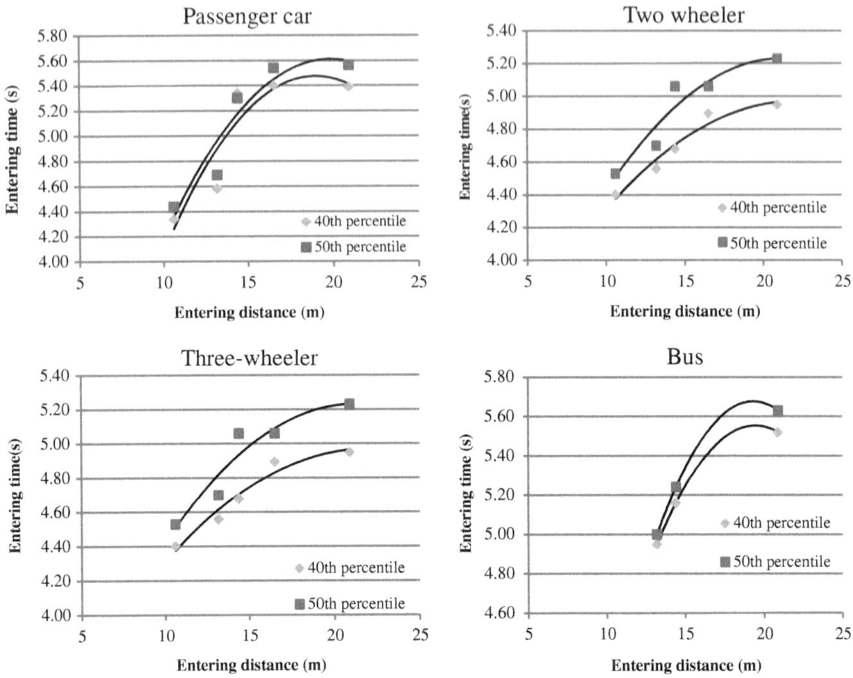

Figure 12.6 Relationship between entering time and entering distances.

different approaches and minimizing overall travel time of vehicles through intersections. Vehicle composition, geometry of the intersection, driver behavior, time of the day, type of vehicle, and availability of space are the factors which govern delay at a signalized intersection. The efficiency and quality of traffic operation at signal-controlled intersections are assessed in terms of total delay caused to an individual vehicle. It is an indicator of the inconvenience caused to the road users and a performance measure for the intersection.

12.6.1 Control delay model

Among different methods of data collection, a mid-size vehicle fitted with GPS device, which would aptly represent the variety of vehicle sizes and speeds predominant in heterogeneous traffic, is the most suitable one for Indian conditions (Bhuyan, 2009).

On comparing field delay values with prominent delay models prevalent in homogeneous traffic, even though inconsistencies are observed, the HCM 2010 model is found to be in close agreement with Indian conditions, and hence, many of the researches on delay in these conditions have adopted the theoretical form of the same. Eq. (12.14) gives the HCM 2010 delay model.

$$d = d_1 PF + d_2 + d_3 \tag{12.14}$$

where d is the average control delay (s/vehicle); d_1 is the average uniform delay per vehicle; d_2 is the average incremental delay per vehicle; d_3 is the average delay due to the pre-existing queue; PF is the progression adjustment factor.

However, most of the signalized intersections are found to be operating at under-saturated conditions and the presence of pre-existing queues is rare. This renders the term d_3 negligible and d_2 was retained in the same way as in the HCM 2010 model. Therefore, the modified delay model takes the form shown in Eq. (12.15) (Othayoth, 2018).

$$d = \alpha\, d_1 + d_2 \tag{12.15}$$

where d is the average control delay (s/PCU); α is the calibration parameter; d_1 is the average uniform delay (s/PCU); d_2 is the average incremental delay (s/PCU).

The value of α, unlike the progression factor in HCM 2010, takes a constant value. This is calibrated by optimizing to minimize the errors between the field-observed and estimated values of delay, and a value of 0.9 is observed. As sophisticated strategies are needed to obtain control delay from the field, it is obtained from stopped delay using a multiplicative factor, which is 1.3 in HCM 2010. Based on findings from studies (Mousa, 2002; Quiroga and Bullock, 1999), it is understood that driver behavior and site characteristics influenced delay characteristics significantly. This necessitates the estimation of a stopped delay-to-control delay conversion factor for Indian conditions. Regression analysis establishes that stopped delay at intersections in India is about 84% of control delay, and hence, a conversion factor of 1.19 can be adopted.

12.6.2 Intersection delay

If there are movement groups that are operated in different phases on an approach, the delay for an approach is computed by aggregating the individual delays of these movement groups as given in Eq. (12.16).

$$d_A = \frac{\sum d_m \times V_m}{\sum V_m} \tag{12.16}$$

where d_A = average control delay of approach A (s/PCU); d_m = average control delay for movement group m (s/PCU); V_m = volume of the movement group (m).

Intersection delay can be calculated as the weighted average of delay for each approach, given as seen in Eq. (12.17).

$$d_I = \frac{\sum d_A \times V_A}{\sum V_A} \tag{12.17}$$

where d_I = average control delay for the intersection (s/PCU); d_A = delay for the approach A; V_A = volume on approach A.

Table 12.10 Delay studies carried by different researchers in Indian condition

Methodology adopted	Remarks	Authors
Probabilistic approach	Expanded uniform series using Taylors series	Arasan and Jagadeesh (1995)
Modified Webster method	Developed additional adjustment factor based on observed delay to obtain delay using Modified Webster method	Raval and Gundaliya (2012)
Field measured linear model	Platoon ratio to overall delay has been considered to develop the delay model	Saha et al. (2017a,b)
Linear regression modeling of GPS-based vehicle trajectory data and field observed delay data	Proposed delay model for heterogeneous traffic by modifying the HCM 2010 delay model and proposed stop delay to control delay correction factor	Othayoth and Rao (2017)

Table 12.11 Level of service criteria for signalized intersections

Level of service (LoS)	Control delay (s/PCU)
A	<20
B	20–40
C	40–65
D	65–95
E	95–130
F	>130

Source: Indian Highway Capacity Manual (INDO-HCM), 2018. CSIR-Central Road Research Institute (CRRI), New Delhi.

Table 12.10 summarizes some of the significant contributions made by a few researchers in the direction of delay measurement at signalized intersections for mixed traffic conditions in India.

12.6.3 Level of service estimation

The control delay is considered as the service measure for the estimation of level of service at a signalized intersection. The levels of service and their corresponding control delay evolved are presented in Table 12.11.

Further, Othayoth and Rao (2017) carried out a user perception-based study on “Factors Influencing Level of Service for Motorized Vehicles at Signalized Intersection under Mixed Traffic Conditions.” These authors observed that waiting time is the most important factor while perceiving the level of service at a signalized intersection. Waiting is always considered as a negative experience, which makes the users frustrated.

12.7 Working example

The provided working example is the most commonly practiced and base adopted for design of traffic signal cycle under heterogeneous traffic conditions in India, which is documented and reported in the traffic signal design guidelines of Indian Road Congress (IRC:93-1985).

At a right-angled intersection of two roads, Road 1 has four lanes with a total width of 12.0 m and Road 2 has two lanes with a total width of 6.6 m. The volumes of traffic approaching the intersection during design hours are 900 and 743 PCU/h on the two approaches of Road 1 and 278 and 1802 PCU/h on the two approaches of Road 2. Design signal timings include:

Width of Road 1 = 12.0 m or total four lanes, with two lanes in each direction;
Width of Road 2 = 6.6 m or total two lanes, with one lane in each direction;
Approach volumes on Road 1 = 900 and 743 PCU/h;
Approach volumes on Road 2 = 278 and 180 PCU/h;
Pedestrian walking speed = 1.2 m/s;
Design traffic on Road 2 = the higher of the two-approach volumes per lane 900/2 = 450 PCU/h;
Design traffic on Road 2 = 278 PCU/h.

Step 1: Pedestrian crossing time
Pedestrian green time for Road 1 $= \frac{12.0}{1.2} + 7.0 = 17\ \text{s}$
Pedestrian green time for Road 1 $= \frac{6.6}{1.2} + 7.0 = 12.5\ \text{s}$
Step 2: Minimum green time for traffic
Minimum green time for vehicles on Road 2, G2 = 17.0 s
Minimum green time for Road 1, G1 $= 17 \times \frac{450}{278} = 27.5\ \text{s}$
Step 3: Revised green time for traffic signals
Adding 2.0 s each toward clearance amber and 2.0 s intergreen period for each phase, total cycle time required = (2 + 17 + 2) + (2 + 27.5 + 2) = 52.5 s

Signal cycle time may be conveniently set in multiples of 5 s and so the cycle time = 55 s.

The extra time of 55.0 − 52.5 = 2.5 s per cycle may be apportioned to the green times of Road 1 and Road 2, as 1.5 and 1.0 s, respectively.

Therefore, adopt G1 = 27.5 + 1.5 = 29.0 s
G2 = 17.0 + 1.0 = 18.0 s

Step 4: Check for clearing the vehicles arrived during the green phase
Vehicle arrivals per lane per cycle on Road 1 = 450/55 = 8.2 PCU/cycle

Minimum green time required per cycle to clear vehicles on Road 1 = 6 + (8.2 − 1.0)2 = 20.4 s (less than 29.0 s and therefore accepted)

Vehicle arrivals per lane per cycle on Road 2 = 278/55 = 5.1 PCU/cycle

Minimum green time for clearing vehicles on Road 2 = 6 + (5.1 − 1.0)2 = 14.2 s (less than 18.0 s)

As the green times already provided for the two roads by pedestrian crossing criteria in Step (2) earlier are higher than these values (29.0 and 18.0 s), the earlier design values are alright.

Step 5: Check for optimum signal cycle by Webster's equation

Lost time per cycle = (amber time + intergreen time + time lost for initial delay of first vehicle) for two phases = $(2+2+4)\times 2 = 16$ s.

$$\text{Saturation flow for Road-1 of width 6 m} = 525 \times 6 = 3150\ \text{PCU}/h$$

$$\text{Saturation flow for Road-2 of width 3.3 m} = 1850\ \text{PCU for 3.0 m wide road} + \frac{40 \times 3}{5} = 1874\ \text{PCU}/h$$

$$y_1 = \frac{900}{3150} = 0.286 \quad \text{and} \quad y_2 = \frac{278}{1874} = 0.148$$

$$Y = 0.286 + 0.148 = 0.434$$

Optimum signal cycle time, $C_0 = \frac{1.5L+5}{1-Y} = \frac{1.5 \times 16 + 5}{1-0.434} = 51.2$ s

Therefore, the cycle time of 55 s designed earlier is acceptable.

Phase diagram for Road 1 and Road 2

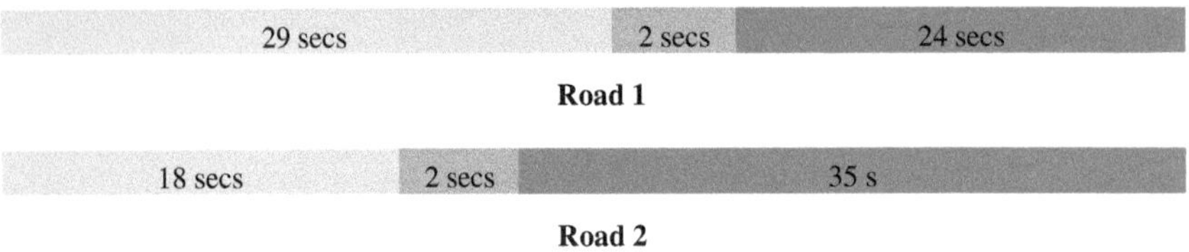

Acknowledgment

The authors would like to express their sincere gratitude to Central Road Research Institute (CRRI), New Delhi, for their immense support and motivation for bringing this chapter into its present form.

Bibliography

Ali, S.S.M., George, B., Vanajakshi, L., Venkatraman, J., 2012. A multiple inductive loop vehicle detection system for heterogeneous and lane-less traffic. IEEE Trans. Instrum. Meas. 61 (5), 1353–1360.

Anusha, C.S., Verma, A., Kavitha, G., 2013. Effects of two-wheelers on saturation flow at signalized intersections in developing countries. J. Transp. Eng. 139 (5), 448–457.

Arasan, V., Kashani, S., 2003. Modeling platoon dispersal pattern of heterogeneous road traffic. Transp. Res. Rec. J. Transp. Res. Board 1852, 175–182.

Arasan, T.V., Jagadeesh, K., 1995. Effect of heterogeneity of traffic on delay at signalized intersections. J. Transp. Eng. 121 (5), 397–404.

Arasan, V.T., Boltze, M., Subramanyam, S., 2006a. Determination of intergreen intervals in signal-time settings for heterogeneous traffic. Indian Highways 34 (8), 47.

Arasan, T.V., Vedagiri, P., 2006b. Estimation of saturation flow of heterogeneous traffic using computer simulation. Proceedings 20th European Conference on Modelling and Simulation, ECMC, Bonn, Germany.

Best Practices for Traffic Signal Operations in India, 2016. IBI Consultancy India Pvt. Ltd. Retrieved from <https://smartnet.niua.org/sites/default/files/resources/Best%20Practices%20for%20Traffic%20Signal%20Operations%20in%20India.pdf>.

Bhuyan, P.K., 2009. *Defining Level of Service Criteria for Urban Streets in Indian Context* (Ph.D. thesis). Indian Institute of Technology Bombay, Mumbai, India.

Biswas, S., Chakroborty, S., Ghosh, I., Chandra, S., 2018. Saturation flow model for signalized intersection under mixed traffic condition. Transp. Res. Rec. J. Transp. Res. Board https://doi.org/10.1177/0361198118777407.

Chakroborty, P., Das, A., 2014. Principles of Transportation Engineering. PHI Learning Private Limited.

Chand, S., Gupta, N., Velmurugan, S., 2016. Development of saturation flow model at signalized intersection for heterogenous condition. Transp. Res. Proced. 25 (2017), 1662–1671.

Gazis, D., Herman, R., Maradudin, A., 1960. The problem of the amber signal light in traffic flow. Oper. Res. 8 (1), 112–132.

German Road and Transportation Research Association (FGSV), 2003. Guidelinesfor Traffic Signals – English Version of Edition 1992. German Road and Transportation Research Association (FGSV), Cologne, Germany.

Indian Highway Capacity Manual (INDO-HCM), 2018. CSIR-Central Road Research Institute (CRRI), New Delhi.

Indian Road Congress, 1994. Guidelines for the Design of at Grade Intersections in Rural and Urban Areas. IRC SP, 41-1994.

Indian Road Congress, 1985. Guidelines on Design and Installation of Road Traffic Signals. IRC:93-1985.

Kadiyali, L.R., 2013. Traffic Engineering and Transport Planning. Khanna.

Kedia, A., 2014. Design of Intergreen Time for Signalized Intersections for Mixed Traffic. B.Tech thesis, IIT Kharagpur.

Khanna, S.K., Justo, C.E.G., Veeraragavan, A., 2014. Highway Engineering. Nem Chand&Bros.

Kulakarni, R., Chepuri, A., Arkatkar, S., Joshi, G., 2017. Saturation flow estimation at signalised intersections under heterogenous traffic conditions. In: 4th Conference of Transportation Research Group of India (CTRG 2017) IIT Bombay, 2017.

Mathew, T., 2014. Design Principles of Traffic Signal. NPTEL, Ministry of Human Resource Development (MHRD), Govt. of India.

Mousa, R.M., 2002. Analysis and modeling of measured delays at isolated signalized intersections. J. Transp. Eng. 128 (4), 347–354.

Othayoth, D., 2018. Modeling user perceived level of service and delay at signalized intersection under heterogeneous traffic condition. Ph.D. thesis. Indian Institute of Technology Bombay, Mumbai, India.

Othayoth, D., Rao, K.V.K., 2017. Factors influencing level of service for motorized vehicles at signalized intersection under mixed traffic condition. Transp. Dev. Econ. 3 (16). Available from: https://doi.org.10.1007/s40890-017-0045-x.

Patel, P., Dhamaniya, A., Katti, B.K., 2015. Effect of mixed traffic characteristics on saturation flow and passenger car units at signalised intersections. Eur. Transp./Trasporti Europei 2015 59, Paper n4.

Quiroga, C.A., Bullock, D., 1999. Measuring control delay at signalized intersections. J. Transp. Eng. 125 (4), 271–280.

Radhakrishnan, P., Mathew, T.V., 2011. Passenger car units and saturation flow models for highly heterogeneous traffic at urban signalised intersections. Transportmetrica 7 (2), 141–162.

Raval, N.G., Gundaliya, P.J., 2012. Modification of Webster's delay formula using modified saturation flow model for non-lane based heterogeneous traffic condition. Highway Res. J. 5 (1).

Ravikumar, P., Mathew, T.V., 2011. Vehicle-actuated signal controller for heterogeneous traffic having limited lane discipline. Inst. Transp. Eng. ITE J. 81 (5), 44.

Retzko, H.-G., Boltze, M., 1987. Timing of inter-green periods at signalized intersections: the German method. ITE Journal 57 (9), 23–26.

Saha, A., Chandra, S., Ghosh, I., 2017a. Saturation flow estimation at signalized intersections under mixed traffic conditions, WIT Transactions on the Built Environment, vol. 176. WIT Press, <www.witpress.com>, ISSN 1743-3509 (on-line) doi:10.2495/UT170331.

Saha, A., Chandra, S., Ghosh, I., 2017b. Delay at signalized intersections under mixed traffic condition. J. Transp. Eng. A: Syst. 143 (8), 04017041. Available from: https://doi.org/10.1061/JTEPBS.0000070.

South Korea

13

Jin-Tae Kim[1] and Youngho Kim[2]
[1]Korea National University of Transportation, Uiwang, Republic of Korea,
[2]Korea Transport Institute, Sejong, Republic of Korea

13.1 Background

This chapter presents general information and different concepts embedded in the traffic signal systems officially deployed in South Korea. The traffic signal control systems in South Korea do not yet have enough experience and interference to catch up with the international standard due to historical and geographical conditions. It has its own unique elements and techniques which are different from other countries.

13.1.1 National administrative rules

Two different national administrations, the National Police Agency and local governments, are involved with traffic signal control (The Korean Ministry of Government Legislation, 2017a,b). The National Police Agency is legally in charge of regulation, such as decision making on traffic signal installation and the changing of signal timing parameters. On the other hand, local governments oversee the construction, operation, and maintenance in accordance with the law as the owners of the systems in their respective jurisdictions.

Local governments have the right to develop the engineering design, budget plan, and execution of the system, but they do not have the right to make decisions to implement in the field. The National Police Agency has the right to implement in the field, but it has no right to handle budgets related to their decisions. The efficient collaboration between local governments and the National Police Agency can produce an effective combination of budget and regulation, and make the systems productive. It has been an important key factor over the past decades and the weakest part of the traffic signal systems in South Korea.

It has been found in practice that the National Police Agency and local governments work tightly regarding political issues but not regarding engineering techniques. For this reason, the progress of traffic signal control systems has

Global Practices on Road Traffic Signal Control. DOI: https://doi.org/10.1016/B978-0-12-815302-4.00013-3

been slow in South Korea. In addition, Korean highways cannot be linked to other countries due to its geographical location: the peninsula is surrounded by the sea and North Korea, so no connection is possible. Due to such circumstances, in the past, the Korean traffic signal control system has evolved based on its own interests and contains different design elements compared to the universal system.

The National Police Agency publishes and periodically updates the Manual of Traffic Signal Installation and Management (MTSIM) and Traffic Signal Controller Standard Specification (TSCSS), ensuring the uniformity of traffic signals nationwide (Korean National Police Agency, 2011, 2015). These two manuals are the national standards for traffic signal control systems. Strict standardization of the systems has prevented the introduction of new technologies and investment (Kim, 2004). Such provisions include the legal standardization of algorithms and the phase sequence (The Korean Ministry of Government Legislation, 2017b; Korean National Police Agency, 2015). It has induced the reduction of supply cost, which is at its lowest, and this has discouraged investment and thus diminished the size of the traffic signal industry.

13.1.2 History of traffic signal control

The history of traffic signal control systems in South Korea started with importing foreign systems after Korean independence in the 1940s or after the Korean War in the 1950s—no records are available. South Korea developed its first traffic signal controller by copying the visible functions of foreign signal controllers, without having prior knowledge of traffic engineering in the 1970s and the 1980s (Korea Road Traffic Authority, 2017). The current Korean local controllers and communication protocols declared as the national standards in the 2000s were from the Korean traffic responsive control (TRC) systems, the Cycle Offset Split Models of Seoul (COSMOS), which were prepared in the 1990s. Its industrial history is relatively short compared to other nations.

For that reason, the technical background behind the traffic signal control systems has been a little far from the fundamentals widely accepted in other countries from the perspective of traffic engineering. Several traffic engineers have attempted to resolve the technical shortcomings of the systems through their research (Kim, 2004; Chang et al., 2004; Lee et al., 2011; Bin and Ahn, 2007). These have not yet changed due to the strict policies of the National Police Agency and political involvement in the industry.

Almost all signalized intersections found in South Korea are currently in fixed-time operation—there are 43,600 local controllers in South Korea (Korean National Police Agency, 2018). Several TRC systems deployed in many cities in the 2000s have shifted their operational mode to time-of-day (TOD) operation, due to maintenance problems experienced in practice.

13.2 State-of-the-art traffic signal control

13.2.1 Signal heads and indications

The meanings of traffic signal lights in South Korea are slightly different from those addressed by the United Nations (UN) in the Convention on Road Signs and Signals (United Nations, 2006). The meanings of red, yellow, and yellow flashing signals are the same as those introduced in Chapter 2, Principles of road traffic signal control. However, the meanings of the following signal lights are different in South Korea from those defined in Chapter 2, Principles of road traffic signal control. It should be noted that the right-hand traffic rule applies in South Korea, and thus the exclusive left-turn phase is generally considered if necessary.

GREEN light is a signal for allowing vehicles to go through as well as turn right in South Korea. A separate green arrow light is used for left turn.

GREEN ARROW light is an exclusive signal for allowing vehicles to turn in the direction that the arrow indicates. The left-turn green arrow light is widely used as the exclusive left-turn signal.

RED FLASHING light in South Korea is a signal for allowing vehicles to proceed after a complete stop when there is no conflicting traffic, while it means to stop until the red flashing signal ends, as addressed by the UN Convention (United Nations, 2006).

13.2.1.1 Vehicular signals

The number of lenses assigned on a signal face in South Korea is somewhat different from those described by the UN Convention (United Nations, 2006). The UN Convention suggests using three lenses only for a single signal face—red, yellow, and green. By accepting it as one of the standards, the Korean Road Traffic Act (The Korean Ministry of Government Legislation, 2017a) additionally declared to use quadruple-lens signals: red, yellow, left turn green arrow, and green. Such quadruple-lens signal faces are widely used at signalized intersections in South Korea. The signal display sequence for a single vehicular movement should always be green, yellow, and red. No other light, such as countdown vehicular signals, should be used for additional purposes.

The quadruple-lens signal is the single device that handles two different movements "right-of-way" through and left turns. It is similar to the five-lens signal used in the United States. It sometimes creates ambiguity to drivers since both directional movements share the red and the yellow lenses. Fig. 13.1 presents the shape of the quadruple-lens signal face, where the remaining two lenses are the solid green and the yellow from the right.

Fig. 13.1 shows an example of the exclusive protected left-turn case. A quadruple-lens signal displays the red and green colors simultaneously—a conflicting combination of colors on a single face. The red means to stop for vehicles

Figure 13.1 The combined use of red and green for protected left turns on a quadruple-lens signal face.

Figure 13.2 The combined use of red and green for protected left turns on a three-lens signal face.

passing through, and the green-arrow allows a protected left turn. This combinatory traffic light has been nationally accepted as the de facto standard for exclusive protected left-turn lights. Such a combination is also used even on three-leg signalized intersections, where no through movement exists, as shown in Fig. 13.2. It utilizes the color combination, that is, displaying red and green at the same time on a single signal face as a color combination, rather than just green. The yellow lens (the middle lens) turns on when the left-turn phase ends.

It was found that drivers need additional time to figure out which color implies which direction (Bae and Kim, 2014). This combinatory use of traffic signals often does not deliver its meaning and increases drivers' reaction time. Such an increase in reaction time becomes significant when the combination of a yellow signal is involved. It was recently reported that such combinatory signals force drivers to use the temporal brain lobe (figuring out its combinatorial meaning) rather than the frontal lobe (recognizing the color itself) (Kim, 2016). The different partial use of the brain by drivers is the reason for the increased reaction time.

13.2.1.2 Pedestrian signals

The Korean Road Traffic Act (The Korean Ministry of Government Legislation, 2017b) addressed that the shape of Korean pedestrian traffic lights consists of two colors: red and green (see Fig. 13.3). Triple-lens pedestrian signals suggested in the UN Convention are not officially used in Korea. Supplemental countdown pedestrian signals are selectively used if necessary; one of two different pedestrian countdown signal types (the hourglass method and the numbering method) is selectively used.

The pedestrian signal systems display flashing green when the remaining pedestrian green is less than the designed pedestrian crossing time. There have been

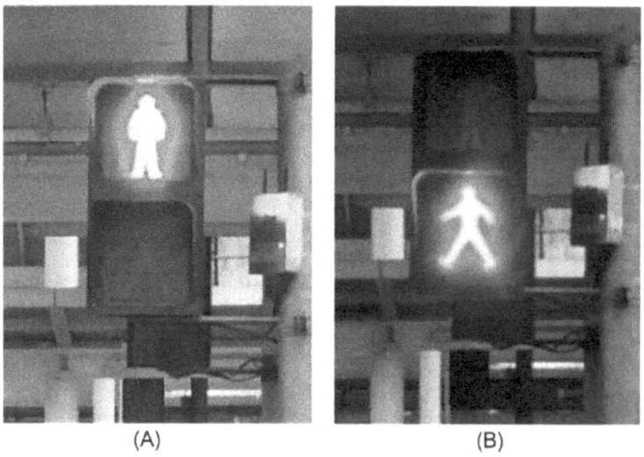

Figure 13.3 The Korean pedestrian signals. (A) Pedestrian red signal; (B) pedestrian green signal.

Figure 13.4 The Korean bicycle signals.

controversies concerning the time interval setting for flashing green, depending on the legal wording in the traffic laws. The length of flashing green for pedestrians is currently identical to the designed pedestrian crossing time.

13.2.1.3 Bicycle signals

There is no heavy bicycle traffic in South Korea, and thus the bicycle signal (see Fig. 13.4) is occasionally placed only where many bicyclists are observed.

Bicyclist treatment is as simple as pedestrian treatment. The Koran bicycle signal light turns on and off simultaneously with the pedestrian signal lights in Korea. Left-turn bicyclists should cross the streets twice—one for each pedestrian crossing.

13.2.2 Traffic signal warrants

The National Police Agency presented the traffic signal warrants, consisting of only four different categories: traffic volume, pedestrian volume, the number of accidents, and school zone (Korean National Police Agency, 2011). In practice, the installation of traffic signals is often determined by such criteria as official political agreement. Those warrants are as follows.

13.2.2.1 Traffic volume criteria

Traffic signals shall be installed at intersections where the hourly volume exceeds the values defined in Table 13.1 for more than 8 hours during a weekday. Those hours may or may not be necessarily consecutive. The traffic volume counts on major and minor streets should consider the same periods.

13.2.2.2 Pedestrian volume criteria

Traffic signals shall be installed at intersections where the number of pedestrians or the bidirectional vehicular volume during a weekday exceeds 600 vehicles/h or 150 persons/h, respectively.

13.2.2.3 School zone criteria

Traffic signals shall be placed at the nearest pedestrian crossing area from the school gate where (1) traffic volume exceeds 60 vehicles/h, and (2) no traffic signal exists within 300 m from the main gate.

13.2.2.4 Traffic accident criteria

Traffic signals shall be provided if there are more than five traffic accidents per year within 50 m of the designated location.

Table 13.1 Traffic volume standards

Number of lanes		Traffic volume (vehicles/h)	
Major approaches	**Minor approaches**	**Bidirectional volume**	**Heavy directional volume on minor approaches**
1	1	500	150
2 or more	1	600	150
2 or more	2 or more	600	200
1	2 or more	500	200

(A) (B)

Figure 13.5 Different placement of traffic signals. (A) Far-side traffic signals; (B) near-side traffic signals.

13.2.3 Traffic signal location

The location of traffic signal lights is of more importance at the far side of intersections in South Korea. Fig. 13.5A and B presents cases of far-side and near-side installations, respectively. During the last decade, installation of traffic signal lights at both the near and far sides of intersections has been officially promoted. A technical report indicated that the stop-line compliance rate among drivers increased where traffic signals are placed only at the near side of an intersection without the far-side signals (Kim, 2015). Currently, cases where there are far-side signals (Korean National Police Agency, 2005) are commonly found in practice but this is expected to change to both-side signals (Korean National Police Agency, 2011).

13.2.4 Directional movement treatment

Despite various operational options available from an engineering perspective, South Korea has consistently applied the turning-movement treatment rules.

13.2.4.1 Right turns

Right-turn-on-red (RTOR) treatment is dominantly used based on the decision of the National Police Agency. There is no "no-turn-on-red" sign legally defined (The Korean Ministry of Government Legislation, 2017b). When no right turn is needed during red, an exclusive vehicular right-turn signal (see Fig. 13.6) should be installed to display red to the direction.

13.2.4.2 Left turns

Protected left-turn treatment is used widely based on the decision of the National Police Agency. The protected left-turn phase has been implemented even at small intersections, rather than the permissive left-turn phase. Protected left-turn treatment is even applied at intersections where a simple two-phase operation is preferable from a traffic engineering perspective. The phase sequence is also guided to be simple and uniform. For example, it was a primary regulation to implement the lagging left-turn phase (The Korean Ministry of Government Legislation, 2017b). Permissive or protected left turns should be selected depending on the conditions

Figure 13.6 Exclusive right-turn signals used for the case of right-turn-on-red operation.

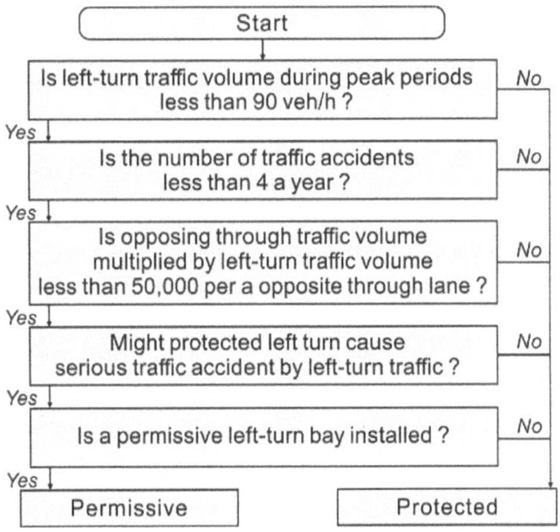

Figure 13.7 Selection process of permissive or protected left-turn treatment.

shown in Fig. 13.7. In practice, however, protected left-turn treatment is always preferred by the National Police Agency.

The most common left-turn signal phasing in South Korea is the protected-lagging phase. Various types of left-turn signal phasing can be used depending on the traffic volume of each movement, such as protected leading left, protected lagging left, simultaneous through and left, etc.

13.2.5 Traffic signal controllers

According to national statistics (Korean National Police Agency, 2018), 43,600 local controllers are currently deployed in Korea.

13.2.5.1 Types of local controllers

The types of Korean local controllers can be divided into three groups, as shown in Table 13.2. These include "primitive," "electronic," and "standard" in local terminology in Korea (Korea Road Traffic Authority, 2017). Among the Korean controllers currently in use, 45% are "primitive," 41% are "electronic," and the remaining 14% are "standard." For the convenience of readers, they are denoted as Types I, II and III, respectively, in this chapter.

Type I is the oldest controller developed in the 1970s. The Type I signal controller supports fixed-time signal operation only with a single signal timing plan. It does not support TOD operation, which utilizes multiple plans. Type I signal controllers work in a standalone manner, thus, it does not have the functionality to communicate with the other controllers. Therefore, it cannot be implemented for corridor progression, nor can it be connected with a traffic management center (TMC).

Type II is the second version of traffic signal controller developed in the 1980s. It has the functionality to deal with multiple signal timing plans and data sharing with others through online communication. Type II can support TOD operation and can connect with TMC or local controllers nearby. This controller can be implemented for signal group operation, such as signal timing progression.

For time synchronization among the component intersections in a group, various techniques have been implemented, for example, GPS clock synchronization and radio-based broadcasting clock synchronization. Type II signal controllers do not support actuated traffic operation.

Type III controllers are the recent version of the Korean traffic signal controller developed in the 1990s. The National Police Agency declared it in 2004 as a national standard. Whenever Type I or II controllers need to be replaced, it is mandatory to replace them with Type III controllers. This controller can support various traffic signal control options, for example, TOD operation, traffic actuated operation, and traffic situation responsive operation.

13.2.5.2 Phase control scheme

Domestic technical documents and manuals show that the Korean control system is based on the dual-ring phase representation scheme, but in fact it uses a single-ring phase representation scheme. This is due to using the "signal map method"

Table 13.2 Functionality comparison of different types of Korean traffic controllers

Index	Types	Design year	Functionality			
			Single plan	Multiple plan	Communication	Actuation
I	Primitive	1970s	Supported	Not supported	Not supported	Not supported
II	Electronic	1980s	Supported	Supported	Supported	Not supported
III	Standard	1990s +	Supported	Supported	Supported	Supported

embedded in the Korean local controller. The method converts combinatorial dual-ring-based phase operation into a straightforward single operation, which loses the advantage of using a dual-ring scheme. This limitation has hindered the normal progress of the Korean traffic signal control system.

13.2.5.3 Signal light control

The Korean traffic signal control scheme has been in operation on an approach basis, which is different from the dual-ring phase scheme. A set of internal devices, called a load switch unit (LSU) attached to the controller, controls the set of traffic lights as shown in Fig. 13.8. The set of lights includes pedestrian red (PR), pedestrian green (PG), red (R), yellow (Y), left-turn green arrow (A), and green (G). There is another type of LSU that was newly introduced in 2010 for the use of three-lens signals.

A single LSU controls six traffic signal lights. In general, four LSU cards are used to control the traffic signal set placed at a four-leg intersection. This Korean method to control phase activation is fundamentally different from the conventional method of dual-ring phases. Unlike foreign schemes that process traffic signal light control based on dual-ring-based software, South Korea is using it based on the LSU hardware.

13.2.5.4 Online communication

The online communication between local controllers and TMC has used a wired telecommunication network. It has been demonstrated that the cost of wire-based communication infrastructure is much higher, and it is hard to find faults in the network for maintenance. This communication infrastructure has recently changed to wireless communication by utilizing the third-generation (3G) long-term evolution (RTE) technique, which allows for long-distance communication without additional costs.

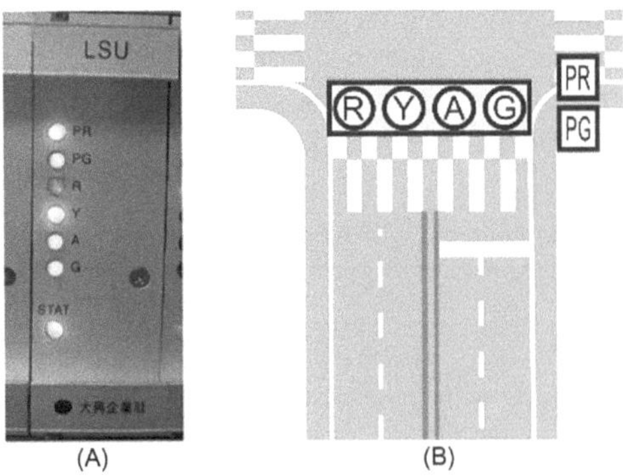

Figure 13.8 The LSU board used with the local controller. (A) LSU board; (B) LSU layout.

13.2.6 Software tools

Software tools used for traffic signal timing design in South Korea include the TRANSYT-7F, SYNCHRO, and the Korean Highway Capacity Software program (KHCS), providing the function of signal timing design based on a heuristic searching method and the Korean Highway Capacity Manual (KHCM). The tool most widely used in practice in South Korea is KHCS. Microscopic simulation tools used for the purpose of final evaluation include VISSIM, CORSIM, and PARAMICS.

13.2.7 Traffic signal operation in practice

There are three types of traffic signal control options available in South Korea: fixed-time control, traffic-actuated control, and TRC control. The TRC systems previously deployed in Korea shifted their operation to the fixed-time TOD mode due to the high maintenance costs of loop detectors. No local government in Korea currently uses TRC operation. Due to the difficulty of maintenance, local governments rarely use the traffic-actuated operation either. The problems related to this will be discussed in Section 13.2.7.1.

13.2.7.1 Fixed-time traffic signal control

The fixed-time control method is the traffic signal control method that is used nationwide. Several TRC systems and traffic-actuated systems deployed in various cities have shifted their operational mode to TOD operation, due to maintenance problems in practice. The fixed-time control method is then mostly (may be over 99%) used with the traffic controllers in South Korea, and 74.5% of those are coordinated (Korean National Police Agency, 2018).

Eighty-six percent of local controllers (Types I and II) in Korea can support fixed-time control only in a captive manner. A large portion of them can only use a single cycle during operation when Type I is involved. This situation applies to Type III controllers as well. Despite the advanced functionality, Type III has even been placed at intersections where fixed-time operation is supposed to be applied.

TOD plans should be managed periodically as the traffic patterns change. Korean local governments and the National Police Agency, however, tend to underestimate its importance in practice. They do not run the public traffic signal control division utilizing their own specialists but order services from private consulting firms. Several local governments, such as the local government of Seoul, have periodically placed the order for TOD plan management each year to let the consulting firms take care of it whenever necessary. On the other hand, the remainder of the local governments update the TOD plans only if public complaints exist.

When designing the fixed-time signal timing parameters, the delay minimization strategy is dominantly used in the industry. This is due to the working guide to check if the results support the design criteria of the level of service declared by the KHCM. Other design strategies have rarely been used.

For the evaluation of traffic signal timing, many local governments have requested microscopic simulation runs and fancy animation streaming data thereof. While VISSIM animation data are mostly required by local governments, the CORSIM simulation runs have been used selectively for proper evaluation. In practice, the evaluations have hardly been objective in administrative practice.

For TOD operation, there are multiple signal timing plans and scenario plans that field engineers can utilize for a single local controller. A total of 16 signal timing plan tables can be prepared and selectively used for single-day operation. Up to five scenario plans can be stored and selectively utilized for weekday operation. A total of 30 additional scenario plans can be stored and selectively utilized for holiday operations for 1 year. Special scenario plans can be stored additionally for exceptional implementation of right-of-way regulation.

Despite the number of plans and operation scenarios storable at a local controller, field engineers have used a fixed portion of those in practice. The number of TOD signal timing tables used for an ordinary day is usually three with Types II and III controllers: AM peak, PM peak, and nonpeak hours. This is due to limitations in funds and the shortage of concern regarding proper operation in practice.

The TOD signal timing maintenance and adjustment process begins only when there are either public complaints or official events (local government projects) taking place. No official signal timing maintenance guidelines, such as a guideline on maintenance cycle, are available in South Korea as of yet.

13.2.7.2 Other

Traffic-actuated control is used at signalized intersections in rural areas where fixed-time operation is unsuitable. Most of the intersections on the national rural highways are congested during weekends. Traffic that flows in rural areas is mostly moving either toward the urban or the rural directions. It usually thus penetrates the areas. There is a significant difference in the balance of traffic demand: between heavy through and light left turns, major and minor streets, and through and left-turn phases.

The traffic-actuated signal operation method has been independently studied in South Korea. The Korean traffic actuated operation method is different from that described in general traffic engineering books. While the conventional system deals with eight dual-ring phases, the Korean system focuses on different domestic single-ring-based phase schemes. Its design is based on supplemental studies conducted on previous types of local controllers.

Traffic-actuated operation in South Korea deals with the leading left-turn phases only while reserving the through phase for the rest time operation. The leading left-turn phase on either major or minor streets can be controlled as actuated phases by assigning the detector data. For that reason, traffic-actuated operation is only available with a leading left-turn phase in Korea—the lagging through phase uses the rest of the green. It cannot support lagging left turns since the Korean local controller does not support the dual-entry mode of operation associated with a barrier, in the United States (US) National Electronic Manufacturers Association (NEMA) perspective.

The traffic-actuated control method was introduced into practice through a national project conducted in the 1990s but failed to be maintained in the field. The real-time traffic-responsive control method was also introduced in the 1990s but failed to be maintained in the field as well. The failure of those was due to failure of loop detector maintenance.

The City of Seoul developed COSMOS, a real-time signal responsive control system. It can adjust cycle lengths and split of critical intersections that is usually the largest intersection among the coordinated progression group and conventional loop detectors are placed at. Other intersections in the group can store up to 16 pre-designed fixed-time plans applied selectively depending on the decision from the critical intersections.

The target operation volume-to-capacity ratio of the COSMOS algorithm is about 0.95, which can be effective when the operational traffic condition is growing toward saturation since it mitigates the sudden expansion of congestion (Kim et al., 2008). However, it may be ineffective during nonpeak hours as it tries to increase the operational condition up to such a condition. Frequent loop detector failure due to public works has given rise to maintenance problems regarding feasibility of investment among local governments as the cost is relatively high when compared with the proper hours for which the sophisticated traffic signal control operation works.

The Korean National Police Agency (KNPA) is currently developing the Smartest Model of Adaptive Road Traffic (SMART) Signal. It is a new traffic signal control system utilizing Intelligent Transport Systems techniques, including travel time data that surrogate other measurements from the sensory data. Such data include travel times of directional movements captured vehicle by vehicle, which are sampled based on the availability of capturing the wireless electronic signals from the media access control (MAC) address of personal belongings placed inside of the vehicle by which the equipment access point (AP) embedded in the local controller collects.

The national 5-year project was initiated in 2015 and is in its fourth phase. This traffic signal system is for the metropolitan highway network where traffic demand exists in general, and thus fixed-time signal operation is preferred. The system uses travel time information as the measure of performance, and reflects the advanced technical changes from the recent remarkable progress in the field of information technology, which was unavailable in the past. The SMART Signal solves the maintenance problems of the conventional loop detectors and increases the convenience of deployment.

13.3 Signal timing design

This section presents an introduction to the method of designing traffic signal control parameters, especially for fixed-time operation in South Korea. Fig. 13.9 illustrates a flowchart of the computational procedure for designing the traffic

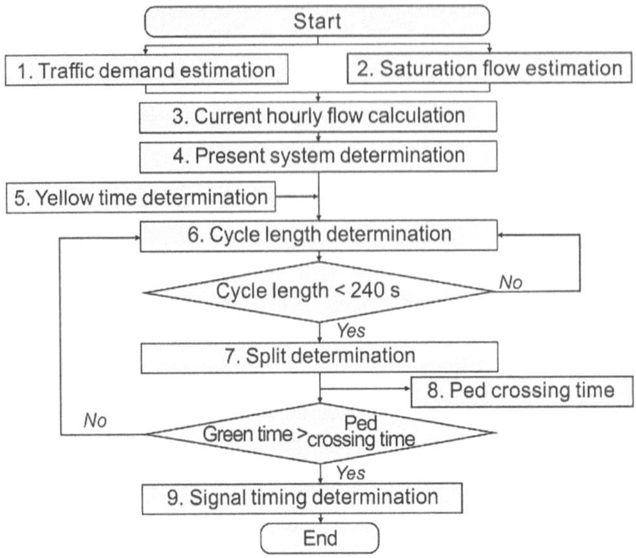

Figure 13.9 Signal timing design flow chart.

signal timing parameters in South Korea. For the signal timing design, the KHCM evaluation methodology is not mandatory but mostly referred to in practice without specific design guidelines relating to signal timing.

13.3.1 Minimum green time

Since no actuation is involved with the through phase in the Korean systems, the minimum green time of the through phase is set to the pedestrian crossing time. When there is no pedestrian crossing involved, the minimum green time of the through phase is mostly determined based on engineering-based judgment in practice by referring to the geometric and traffic conditions in the field. It is set to 10 seconds for the left-turn phase.

The pedestrian crossing time is computed based on the distance, the design walking speed of pedestrians, and the additional time. In general, 1 m/s of pedestrian walking speed should be considered when designed (Korean National Police Agency, 2011). For the intersections located in either school zones or elderly zones, the design pedestrian walking speed becomes 0.8 m/s. The length of flashing green for pedestrians is set to the time required to cross the distance with such a design pedestrian walking speed. For additional time, 4–7 seconds is recommended when designing (Korean National Police Agency, 2011). In practice, 7 seconds of additional time is generally used.

13.3.2 Intergreen time

Yellow time is computed based on the following equation:

$$Y = t_0 + \frac{v}{2a} + \frac{(w + L)}{v}$$

where Y = yellow time (seconds); t_0 = reaction time (default 1.0 second); v = vehicle approaching speed (m/s); a = deceleration rate (m/s^2); w = crossing distance (m); L = length of a vehicle (m).

In practice, 3 seconds of intergreen time is widely used at intersections: 3 seconds of yellow time and 0 seconds of all-red time. Four seconds of intergreen time is used at critical intersections where safety issues are highlighted in urban areas—3 seconds of yellow time and 1 second of all-red time. Five seconds of intergreen time is occasionally used at intersections where high cruising speed is expected.

13.3.3 Lost time

Three seconds of total lost time is considered for each phase in general in the signal timing design procedure, which consists of 2 seconds of start-up lost time and 1 second of clearance lost time.

13.3.4 Phase sequence

The number of phases, regarding the single-ring phase representation scheme for fixed-time operation, is determined by considering (1) the number of approaches, (2) the geometric condition of the intersection, and (3) traffic volume of each movement. The phase sequences have been designed by minimizing the sum of critical movement to introduce the minimum cycle length.

In practice, either the simultaneous protected-leading or protected-lagging left-turn phase sequence is dominantly used in South Korea (four-phase operation in terms of the single ring scheme). The protected lagging left-turn phase sequence is politically treated as the national standard based on the national traffic law (The Korean Ministry of Government Legislation, 2017b) in Korea. The National Police Agency has been firmly against this in order to further progress. Recently, the protected-permitted and permitted-protected left-turn (PPLT) treatments have been attempted in practice for test purposes. Occasionally, an exclusive all-pedestrian green signal phase is used with diagonal pedestrian-crossing lane markings

13.3.5 Saturation flow rate and flow ratio

The ideal saturation flow rate is calculated as 2200 pcphpl (pcu/h/lane) in Korea by employing the saturation headway of 1.63 seconds. The Korean HCM provides a standard procedure to calculate actual saturation flow rate considering adjustment factors, including the number and widths of the lanes, grades, weather, etc.

However, the laborious calculation of the saturation flow rate often does not bring many benefits to practical works. Therefore, the saturation flow rate is typically set to be 2000 pcphpl for through movements and 1800 pcphpl for turning movements at signalized intersections.

Using the saturation flow rate mentioned earlier, the flow ratio for each movement is calculated by traffic volume divided by the saturation flow rate. The flow ratio is used to specify the portion of cycle time to be allocated for the movement. The flow ratio is calculated separately for the through movement and the left-turn movement, if each movement has its own lane. However, if the left-turn movement and the through movement share a same lane, consolidated single values for the saturation flow rate and the flow ratio should be determined.

13.3.6 Cycle length

In general, the width of major urban arterial roads in Korea is broad, and so the pedestrian crossing time is long. The long pedestrian crossing times have lengthened the minimum green time of the associated phases at intersections. The minimum cycle length is the same as the sum of the minimum green time of each phase and is determined to be shorter than 240 seconds. Also, implementation of protected left-turn phases increases the total number of phases and the cycle length as well. Due to this, it has increased the cycle length and thus the initial queue lengths on approaches as well.

The optimal cycle length to minimize congestion at isolated intersections has been calculated based on Webster's formula. Webster's formula is applied to the calculation of optimal cycle lengths for nonsaturated intersections whose volume–capacity ratio is less than 1.0. Otherwise, the cycle length will be abnormally long. The cycle length is practically determined by the values in a range from the optimal cycle length multiplied by a constant 0.75–1.5. The values in this range usually do not have pronounced impacts on the delay. The cycle length of the midblock crossings is typically determined to be larger than double the pedestrian crossing time.

As indicated earlier, the cycle length utilized in Korea is relatively long and increases up to 150–240 seconds. It has been discussed in the Korean Traffic Engineering Academic Society that such a long cycle length would contribute to the aggressive driving tendency of drivers—drivers show a tendency to run red lights to avoid a long red light stopping delay.

13.3.7 Split

Once the cycle length has been determined, green time is allocated to each phase. Before the allocation of green time, minimum green time should be determined depending on the existence of pedestrian crossings. If a pedestrian crossing exists, the minimum green time should be determined considering the time required for pedestrians to cross an intersection safely, as long as there being no conflict with any movement occurring at the same time.

If the pedestrian crossing is not considered, the minimum green time is usually set to be 15 seconds for through movements and 5 seconds for turning movements. To allocate green time for each phase, the total displayed green time per cycle should be calculated by extracting total lost time from the cycle length.

Total displayed green time is allocated in proportion to flow ratios of critical movements for each phase. If any green time for each phase is shorter than the minimum green time, the green time is increased incrementally until the green time for each phase is longer than the minimum green time. The amount of increase is 5 seconds if the cycle length is shorter than 90 seconds, and 10 seconds if the cycle length is longer than 100 seconds.

13.4 Working example

The computation of traffic signal timing design in South Korea follows the critical movement analysis method, which is well known and thus available in ordinary traffic engineering books. The followings describes a short computation process mostly used for an isolated intersection in South Korea. Fig. 13.10 shows the geometric and traffic conditions at an example intersection.

Step 1: Gather information geometry, traffic, and control conditions

Basic information on the subject intersection is shown in Fig. 13.10. If field measurement is unavailable, the saturation flow rate is assumed to be 2000 and 1800 pcphpl for through and turning movements, respectively, with the ordinary geometric condition. Total lost time is assumed to be 3 seconds (2 seconds of start-up lost time and 1 seconds of clearance lost time).

Step 2: Determine the control strategy

For the isolated signalized intersection, the conventional delay minimization strategy is usually applied for the cycle length design. The volume-to-saturation flow ratio equalization strategy is commonly used for the split.

Step 3: Determine the phase sequence

Simultaneous lagging left-turn phase sequence should be considered firstly by strictly referring to the national traffic law (The Korean Ministry of Government Legislation, 2017b) in practice. The simultaneous through phase is followed by the simultaneous left-turn phase. For example: single ring phase 1—eastbound through and westbound through (WBTH); single ring phase 2—eastbound left turn and westbound left turn (WBLT); single ring phase 3—northbound through (NBTH) and southbound through; single ring phase 4—northbound left turn (NBLT) and southbound left turn.

Step 4: Determine the intergreen time

Three seconds of yellow time is generally considered in urban areas. It increases to 4 seconds in rural areas, where free-flow speed is relatively high. One second of all-red clearance interval is used when considered necessary.

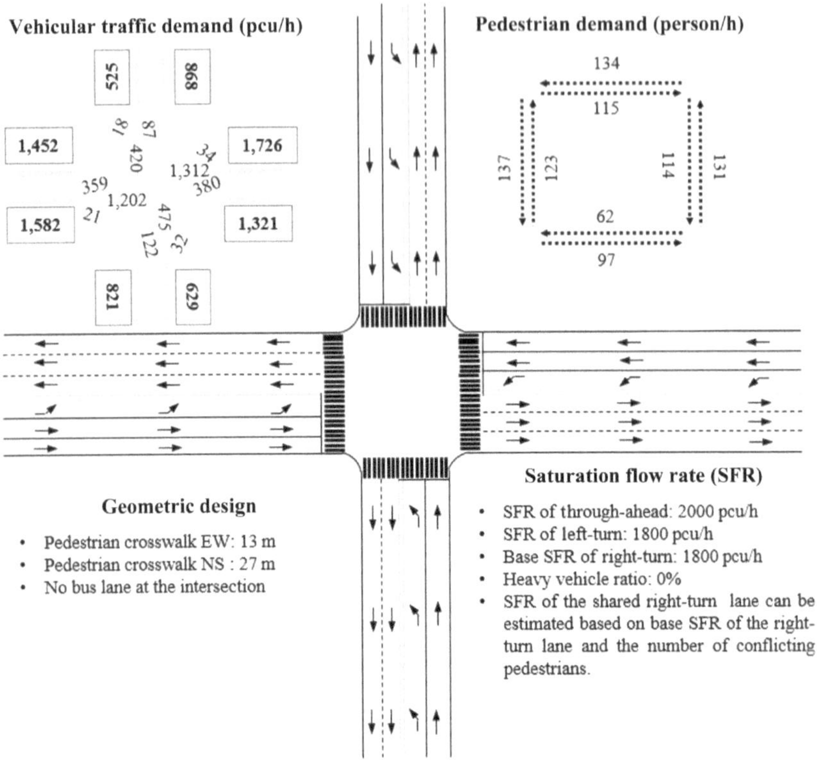

Figure 13.10 Example intersection.

Step 5: Determine the critical movements

Compare sets of volume-to-saturation flow ratio of the movements assigned to the phases. The critical movement per each phase is the one with the highest volume-to-saturation flow ratio value. The critical movements are found to be WBTH, WBLT, NBTH, and NBLT. The sum of the critical volume-to-saturation flow ratios of those critical movements is 0.8444 [= 0.3280 (38.84%) + 0.2111 (25.00%) + 0.2375 (28.13%) + 0.0678 (8.03%)].

Step 6: Determine the lost time

Total 3 seconds are considered as the total lost time per a single phase in practice; 2 seconds of start-up lost time and 1 second of ending lost time. Total 12 seconds of lost time from the four phases per a cycle should be considered for the design of a cycle length.

Step 7: Determine the cycle length

Compute a cycle length and split by applying Webster's optimal cycle length equation and the control strategy of volume-to-saturation flow ratio equalization. The cycle length for the intersection should be 150 seconds; $(1.5 \times 12 + 5) \div (1-0.8444) = 147.80 \approx 150$.

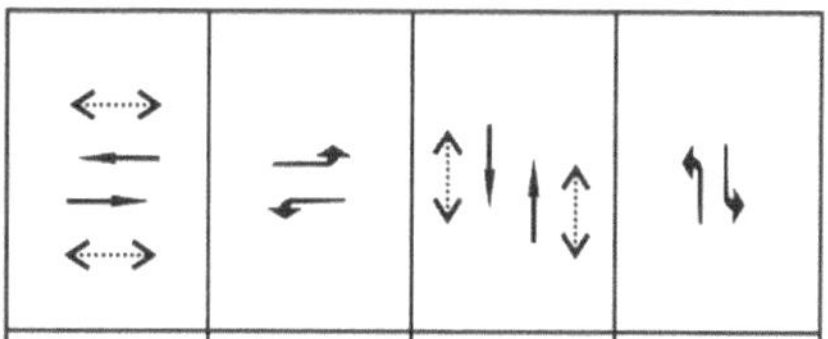

	Phase 1	Phase 2	Phase 3	Phase 4
Vehicle phase	58 s	38 s	42 s	12 s
green time	55 s	35 s	39 s	9 s
Intergreen time	3 s	3 s	3 s	3 s
Pedestrian phase	20 s		34 s	
Solid green	7 s		7 s	
Flashing green	13 s		27 s	

Figure 13.11 Signal timings designed for example intersection.

Step 8: Determine the pedestrian green time

Pedestrian green times are determined based on the length of the pedestrian crossing area by assuming 1.0 m/s of pedestrian walking speed and 7 seconds of clearance time in general. For the school zone and the elderly zone, 0.8 m/s of pedestrian walking speed is applied. The computed time is then treated as the minimum green of the concurrent through phase.

No exclusive bicycle traffic signal time computation procedure is necessary in Korea since those are set as equal to one of the pedestrian signal timings.

The pedestrian green time for the northbound and the southbound pedestrian crossings would be 34 seconds; 7 seconds (the maximum value of the initial time; 4–7 seconds) + 27 seconds (27 m/1.0 m/s). The one for the eastbound and westbound pedestrian crossings would be 20 seconds; 7 seconds (the maximum value of the initial time; 4–7 seconds) + 13 seconds (13 m/1.0 m/s).

Step 9: Determine the split

Compute the split based on the proportion of the critical volume-to-saturation flow ratios: 38.84%, 25.00%, 28.13%, and 8.03%. The phase lengths of the phases would then be 58, 38, 42, and 12 seconds, respectively (see Fig. 13.11).

The green times of the EW and NS through phases satisfy the minimum pedestrian crossing time computed earlier. Pedestrian green time generally starts with the EW- or NS-through phase, simultaneously.

Bibliography

Bae, H.S., Kim, J.T., 2014. Study on driver's perception reaction times to conventional and unconventional types of four-head traffic signals—comparison of non-declarative and declarative memories of drivers. In: The Conference of the Korean Society of Intelligent Transport Systems, Proceedings, pp. 229–234 (in Korean)

Bin, M.Y., Ahn, H.J., 2007. A study on improving traffic signal operation in Gyeonggi-do. Policy Report 2007-17, Gyeonggi Research Institute (in Korean).

Chang, K.Y., Lee, J.Y., Kim, J.T., Chang, M.S., 2004. A comparison study of the signal systems in the Korean road traffic law and the convention on road signs and signals. J. Korean Soc. Transp. Eng. 22 (1), 83–94. in Korean.

Jung, K.B., Kim, J.T., Lee, D.J., 2007. The study on the correlation between traffic signal location and stop line observance. Int. J. Highw. Eng. 9 (1), 29–38. in Korean.

Kim, J.T., 2004. Development of administrative work scheme based on understanding of the Korean Traffic Signal Operations, Korea Transport Institute. Policy Report 2004–2013 (in Korean).

Kim, J.T., 2015. Effects of traffic signal locations on stop line compliance of vehicle drivers in Korea. In: Proceedings of the Eastern Asia Society for Transportation Studies, vol. 10.

Kim, J.T., Lee, J., Chang, M., 2008. Neural-network-based cycle length design for real-time traffic control. Can. J. Civil Eng. 35 (4), 370–378.

Kim, J.Y., 2016. Study on safety and efficiency of different traffic signal types using drivers' recursive and non-recursive brain memories – focusing on comparison between three-head and four-head traffic signals, Master thesis, Department of Transportation Systems Engineering, Graduate School of Transportation, Korea National University of Transportation. (in Korean).

Korea Road Traffic Authority, 2017. Traffic safety control devices, <https://www.koroad.or.kr> (accessed in 2017) (in Korean).

Korean National Police Agency, 2005. Manual of Traffic Signal Installation and Management. Korea Road Traffic Authority (in Korean).

Korean National Police Agency, 2011. Manual of Traffic Signal Installation and Management. Korea Road Traffic Authority (in Korean).

Korean National Police Agency, 2015. Traffic Signal Controller Standard Specification, Version NPA-TSC-2010-R14. Korea Road Traffic Authority (in Korean).

Korean National Police Agency, 2018. Traffic Signal Control Statistics. Division of Traffic Operation (in Korean).

Lee, H.W., Joo, D.H., Hyun, C.S., Park, B.H., Kim, D.H., 2011. Study on headways at signalized intersections before and after installation of red arrow signal. J. Korean Soc. Transp. Eng. 29 (6), 57–65. in Korean.

The Korean Ministry of Government Legislation, 2017a. Road Traffic Act Enforcement Rule. National Law Information Center, <http://www.law.go.kr/>, (accessed in 2017) (in Korean).

The Korean Ministry of Government Legislation, 2017b. Road Traffic Act. National Law Information Center, <http://www.law.go.kr/>, (accessed in 2017) (in Korean).

United Nations, 2006. Convention on Road Signs and Signals. United Nations Economic Commission for Europe.

The State of Qatar and the United Arab Emirates

14

Wael Alhajyaseen[1] *and Ghassan Abu-Lebdeh*[2]
[1]Qatar Transportation and Traffic Safety Center, College of Engineering, Qatar University, Doha, State of Qatar, [2]Department of Civil Engineering, College of Engineering, American University of Sharjah, Sharjah, United Arab Emirates

14.1 Background

The State of Qatar and the United Arab Emirates (UAE) share many commonalities in the practice of traffic control, including traffic signals. This is in large part due to their close geographic proximity, common history, language, and culture, and recent rapid and simultaneous development and modernization. However, the origins of traffic signal practices in the two countries are different: Qatar's practices are based on UK standards and guidelines, while the origin of traffic signal control in UAE is largely, but not exclusively, formed using US documents and standards. Even within the UAE, the practices employed in different Emirates can be significantly dissimilar and unique, as will be noted in later parts of this chapter. Within the state of Qatar on the other hand, practices are quite similar without significant regional differences. The main municipalities with the largest population within the State of Qatar are the cities of Ad Dawhah (Doha), Al Rayyan, Al Wakrah, Umm Salal, and Al Khor. These areas combined comprise more than 90% of the country's population, with the major concentration being in and around the city of Doha. This is the main reason for the observed uniformity in design and operation of signal control. In rural areas, stop-controlled intersections and roundabouts are the most common control types. There is far more autonomy in internal and municipal affairs within the various emirates that comprise the UAE; hence, more diversity in signal control practice is visible. One important fact is that many of the standards both in the State of Qatar and the UAE were put in place by private consulting entities, mainly originated from the UK in the early days (1970s and 1980s), but recently an increasing number of US and Australian entities. The vast majority of signalized intersections are located in the larger and more influential emirates of Dubai, Abu Dhabi, Sharjah, and the Emirate of Ajman, which together account for 92% of the UAE population and 95.3% of the UAE area. There are few signals in the remaining emirates of Ras Al-Khaimah, Fujairah, and Umm al-Quwain, because these emirates are largely rural, with the exception of the capital cities. Nevertheless, in the urbanized capital cities traffic signal control at roundabouts is common and widely practiced. Due to the political structure of the union of the

Global Practices on Road Traffic Signal Control. DOI: https://doi.org/10.1016/B978-0-12-815302-4.00014-5

emirates, the various emirates maintain significant autonomy in running their internal affairs, including their engineering practices. The federal government's role in these areas is only advisory and acts more as a facilitator. The federal government may fund major capital interemirate projects, but rarely dictates or enforces standards.

Traffic signal design in Qatar has historically been based on the Qatar Highway Design Manual (QHDM, 1997a) and the Qatar Traffic Control Manual (QTCM, 1997b). Both of these manuals were developed largely based on the UK Highways Agency Design Manual for Roads and Bridges (DMRB) at the time. New versions of these manuals are available (2015). Guidance for the design and management of traffic signals in the State of Qatar has more recently been set out in the documents noted later.

In the case of the UAE, the manuals used in Dubai and Abu Dhabi have their roots in standard US traffic signal timing and analysis documents (TRB (Transportation Research Board), 2018, 2011). The various emirates maintain separate manuals for traffic signal design. Besides, the history of development of manuals and procedures is not as structured, nor did it follow the same timeline for the different emirates. However, the variation in documents does not necessarily mean the practices and applications are fundamentally different; the dissimilarity is only present in nonfundamental operational aspects, some of which will be highlighted in later parts of this chapter. In fact, the commonalties between the emirates, with Abu Dhabi, Dubai, Sharjah, and Ajman in specific, are evident and function to maintain sufficient uniformity and limit confusion among drivers. The State of Qatar sequence of signal control documents is:

- Interim Advice Note 017 (2013)—Traffic Signal Standard Details and Drawing Requirements (Ashghal Public Works Authority Expressway Programme);
- Qatar Highway Design Manual (QHDM, 2015b), Volume 1, Part 8: Design for Signalized Intersections;
- Qatar Traffic Control Manual (QTCM, 2015a), Volume 1, Part 5: Traffic Signals;
- Ashghal Public Works Authority (2016) Policy Statement: Traffic Signal Design Guidelines Supplement.

The QHDM and QTCM were released in July 2015. The QHDM describes the geometric design of intersections so that they can be signalized. The design of traffic signal installations, including guidance on the method of installation, signal control, and display sequences, is addressed in the QTCM. The Public Works Authority (Ashghal) Policy Statement of Traffic Signal Design was adopted in December 2016. The purpose of this document is to provide additional guidance to the users of the QHDM and QTCM, with further detailed design components to help facilitate the development of consistent traffic signal design plans.

Nonetheless, in reality many traffic signal designs incorporate local legacy practices, such as the inclusion of a flashing green phase before the yellow interval. In parts of the UAE, some legacy practices still persist; for example, pregreen yellow, which is used to prepare drivers for the onset of the green interval, is still used at many intersections in Sharjah.

For new traffic signals, the typical design and implementation process involves the following steps:

- Conceptual design undertaken by the Ministry of Transport and Communications (MOTC), Land Traffic Planning Department (LTPD), or by the Public Works Authority (Ashghal) in partnership with MOTC. The conceptual design typically involves using future (2031) traffic flow data from a strategic demand model and knowledge of the road hierarchy to determine the most suitable intersection type and phasing sequence for the peak hours.
- Preliminary and detailed design is then taken on by the Public Works Authority (Ashghal), Roads Design Department, or by a project design consultant. The detailed design typically defines the safety-critical features, such as equipment locations, signal phase sequences, and intergreen timings.
- Construction, contractors, and supervising consultants in service of the Road Controlling Authority will typically implement the signals, finalize the traffic signal configuration details, and enter the operational parameters per design into the signal controller. They will also set the initial signal timings, based on the actual traffic flows rather than modeled by future traffic demands.
- As-built drawings and phasing details are then provided to Ashghal Roads Maintenance Department (RMD) for maintenance and the Ashghal Traffic Control Centre (TCC) for operations.
- The RMD and TCC will then amend the phasing if necessary to suit on-site operation and traffic flow observations.

A number of major road infrastructure projects, which include traffic signals, are developed by Programme Management Consultancies (PMC) on behalf of the Road Controlling Authority. Therefore, the detailed design of traffic signals is typically undertaken by external design consultancies and it is reviewed and consented by the PMC. On occasion, the design of upgrades or conversions from intersections like roundabout to traffic signal, as part of operating or improving the road network, will be undertaken in-house by a combination of the Ashghal Roads Design Department and Roads Maintenance Department. A similar procedure as noted earlier is followed in various emirates in the UAE. However, operations of signal systems can be exclusively the domain of the respective police department, as in Abu Dhabi, or the domain of shared police, municipality, and private consultancies, as in Dubai.

There are currently 287 traffic signals throughout Qatar and this number continues to increase due to the rapidly developing infrastructure. Among these, 127 traffic signals are linked to the Traffic Control Centre, where they are managed and operated using SCATS. Most of the signals that are managed by SCATS are networked to other signals along various corridors. Base signal phasing plans are entered into SCATS based on the design configuration and then operated with a mix of vehicle activation and adaptive control. Green-wave configurations are established along corridors with multiple signals linked and coordinated with the most critical intersection, which is determined based on capacity.

The remaining 160 signals, that are not yet linked to SCATS and the TCC, are operated in isolation under fully vehicle-activated mode using the standard design phasing that is entered into the controller, based on actuated signal control.

However, some intersections are temporarily operated using a fixed-time sequence, including when there are no detectors, during communication faults, or during the final stages of construction just before the detectors are installed but after the signals are brought into service. In the UAE, the city of Dubai has the most signals among all emirates. The traffic signal system in Dubai (over 800 signals in total) is controlled using the SCOOT system, although many of the advanced functionalities within the system are not used. A SCOOT control system is also used in Abu Dhabi, which has over 125 signals. In the four most urbanized emirates, Dubai, Abu Dhabi, Sharjah, and Ajman, traffic signals are continuously being added, due to rapid infrastructural developments or as a result of converting roundabouts into signalized intersections.

Installation in the field consists of signal poles located overhead in the far side of intersection approaches. Additional signal heads are installed on the side to supplement the role of the overhead ones. Deviation from this standard practice in the State of Qatar and the UAE may be seen at old intersections in older parts of the cities or at intersections that once were a roundabout. Furthermore, the traffic signal heads are mounted vertically following common international practice, where the red circle indication is positioned at the top, then below is the yellow circle indication, and the green circle indication at the bottom. Arrow indications are typically used for protected turn phases, serving exclusively designated turn lanes.

14.2 Control strategy selection

The QHDM (2015b) of the State of Qatar specifies that traffic signals should be considered in association with the route classification requirements, set out in Part 2 of the QHDM, regarding planning. In the UAE, installation of traffic signals follows the procedure based on the signal warrants in the US Manual of Uniform Traffic Control Devices (MUTCD). In many cases, traffic signals were installed to replace congested roundabouts on locations where severe congestion prompted the change of control. In most of those cases no formal signal warrant evaluation was conducted, as it was assumed that a traffic signal was warranted based on volumes alone. Since traffic signal studies and installation are usually undertaken by private consultancies in the UAE, signal norms and practices typical from the United Kingdom, Australia, and the United States are sometimes simultaneously apparent in the UAE.

The QHDM, as are the practices in Dubai, Abu Dhabi, and Sharjah, specifies that traffic signals should be assessed by a needs-based study that should include the following factors:

- Feasibility and desirability of alternative intersection solutions: priority, roundabout, and grade-separated;
- Road classification and expected traffic speeds;
- Crash records for an existing intersection;
- Whether the intersection is within an urban traffic control area and, if so, its proximity to other intersections or roundabouts.

Section 1.1 of the QTCM, Part 5, outlines the following criteria that should be met before a traffic signal project is implemented:

- Criterion 1—Intersection Vehicular Traffic Flow Volumes;
- Criterion 2—Pedestrian Flow Volumes;
- Criterion 3—School or Civic Amenity Access;
- Criterion 4—Coordination and Management of Traffic Flow;
- Criterion 5—Known Crash Locations;
- Speed Criteria.

It is recommended that a traffic study be carried out to collect sufficient data to indicate typical vehicular movements, including turning counts. Following the satisfaction of one of the aforementioned criteria, traffic modeling would be used to examine the extent to which traffic signals would address the problem. The QHDM and QTCM specify that traffic signals are not recommended on roads with speed limits exceeding 80 km/h. It is noted, however, that there remains a legacy of some signalized intersections installed on roads with 100 km/h speed limits, prior to the release of the QHDM (2015b). Most of the higher-speed highways in the UAE were initially constructed with roundabouts (as opposed to traffic signals) to provide access to and from intersecting roads. Nevertheless, most, if not all of those roundabouts, especially in urbanized areas of the emirates, have been converted to grade-separated intersections in parallel with widening and modernizing those high-speed roads. This latter is particularly true in the emirates of Abu Dhabi, Dubai, and Sharjah.

However, in practice, because of the high traffic flows, many intersections that are considered for potential traffic signal installation readily meet the QHDM and QTCM warrants. For that reason, planning decisions around the adoption of traffic signals are often made based on engineering judgment and whether or not the additional cost and terrestrial requirements for a grade-separated interchange can be justified. Generally, the design of traffic signals includes the collection of traffic survey data, such as turning counts and traffic modeling, using relevant capacity analysis software. Usually, SYNCHRO Studio is used to draft the general signal plan.

In the State of Qatar, as outlined earlier, signal phasing plans are entered into SCATS, after which the signals are operated via SCATS, utilizing relevant phasing plans at various times of the day with a mix of vehicle activation and adaptive control. Green-wave configurations are established along corridors with multiple signals, linked to and coordinated with the most critical intersection. The critical intersection is determined based on the volume to capacity ratio. Signals operating in isolation, which are not yet linked to SCATS or the TCC, are configured using standard design phasing and timings within the standalone controller based on vehicle activation. All traffic signals in Qatar are fitted with loops for vehicle activation.

Most signals in Dubai and Abu Dhabi are centrally controlled through the SCOOT system and nearly all signals are equipped with detection systems that are based on image processing instead of inductive loops. In Sharjah, there is central control, but timing adjustments are made based on visual inspection of queues via cameras. The control strategies in the three emirates are different owing largely to specific technical expertise of staff and the topography and age of the road network. In the case of Abu Dhabi, which uses a gird system, control with green waves is implemented and noticeable by road users, with the exception of rush hours, when most signals operate at or above capacity. In the case of Dubai, although the SCOOT system is in operation, the lack of a grid system combined with the old layout of the network, thus inconsistent and nonuniform block dimension, renders use of green waves impractical.

Approximately half of the 800 major signals in Dubai are integrated into and centrally controlled by the SCOOT control system, regardless of centrality or closeness to other signals. Whether signals are operated in isolation or as part of a larger group is decided based on the intensity of traffic at the subject signal (volume to saturation flow ratio) and also based on proximity to other signals. Signals that are more than 1.6 km apart (approximately 1 mile) are treated and timed as isolated. Even when controlled as groups, signals in Dubai are set to self-adjacent based on localized traffic conditions such as headways. Similar procedures are followed in Ajman. In the case of Abu Dubai, coordination of signals along corridors is the default. The grid system and uniformity of block lengths resulting in uniform signal separation makes coordination schemes a default choice. In the case of Sharjah, signals are usually treated as isolated, in part due to long separation between signals and persistent heavy flow during peak hours.

14.3 Traffic signal installation and operation procedures

There is not one uniform and nationwide procedure of installing and operating traffic signals in the UAE. As noted earlier, different emirates exercise total control of signals within their jurisdictions. Actual design and analysis of signals is usually undertaken by private consultancies, which are mostly British, American, and Australian companies. In-house signal design and analysis is rare in the four more urbanized emirates of Abu Dhabi, Dubai, Sharjah, and Ajman. Although not always stated explicitly, signal timing follows recommendations in US standard signal control and analysis manuals (TRB (Transportation Research Board), 2011, 2018). In Abu Dhabi, Dubai, and Sharjah initial cycle lengths and green intervals are decided based on flow ratios (adjusted-volume-to-saturation-flow-rate) of different phases. Adjustments and/or extensions to interval lengths (cycle length) are made based on localized conditions, such as closeness to special traffic generators or nearby intersections. Pedestrians are usually served in their own protected modes and pedestrian crossing is all multistage. All-pedestrian phases are not applied in the State of Qatar or the UAE. Permitted traffic turns that share green with pedestrian traffic are not used, nor are vehicular left turns permitted. Long cycle lengths are accepted and regularly used, especially at centrally located intersections.

The following is a summary of key signal-timing practices that are common in the State of Qatar and the UAE, which are distinctly more conservative compared to practices in the United States:

1. No permitted left turns (all left turns are protected);
2. No turn-on-red. Free right turns, also known as filtered right turn in the UAE, are often used at roundabouts that are converted to signalized intersection with sufficient space available;
3. No right turns against pedestrian traffic;
4. Bicycles and pedestrians are served simultaneously in protected phases jointly with non-conflicting vehicle movements.

14.3.1 Necessary documents or input information for signal timing

The Public Works Authority (Ashghal) Policy Statement on Traffic Signal Design (2016) in the State of Qatar, similar to the less-than-uniform practice in the four more urbanized emirates of the UAE (Abu Dhabi, Dubai, Sharjah, and Ajman), requires that the following documents be developed and provided as part of each new traffic signal design:

- Traffic signal layout;
- Phasing diagram;
- Ducting details;
- Cabling details;
- Gantry and pole details;
- Typical standard details (covering electrical and mechanical components);
- Traffic detour plans and alternate routes (when applicable).

14.3.2 Calculation process of signal timing plans

The critical movement analysis method is used to design signal timing in most cases, similarly to both the State of Qatar and the UAE. Using this method, demand volumes are adjusted for temporal peaking characteristics and special lane utilization variations. The critical-volume-to-saturation-flow ratios for the phases are subsequently estimated before an initial cycle length is computed.

14.4 Signal phasing

The typical phase sequence in the State of Qatar comprises a green traffic signal followed by a flashing green traffic signal, which is displayed for 3 seconds to warn approaching drivers of the imminent change before another 3 seconds yellow interval. The yellow interval is then followed by an all-red interval that is 2 seconds minimum, but the actual time is calculated based on the distance between each set of conflicting vehicles, as documented in Ashghal Policy Statement: Traffic Signal

Design Guidelines Supplement (2016). In general, the practice on the yellow and the all-red intervals in the UAE is similar to the values noted earlier for the State of Qatar, yet in Abu Dhabi the yellow is 3 seconds, while the all-red is 2 seconds.

The flashing green traffic signal time in addition to the yellow interval has the same meaning as the static green signal, in terms of right-of-way in which drivers are permitted to proceed across the stop line. It is important to highlight that the flashing green signal is not considered part of the yellow plus all-red period (using US terminology), which is also referred to as the intergreen interval.

Fig. 14.1A shows the standard traffic signal sequence from green to red (QTCM, 2015a) in the State of Qatar. The QTCM states also that at some existing installations where the roadway is 80 km/h or above, the sequence should include a flashing green period prior to the yellow as shown in Fig. 14.1B. In practice, traffic signals in the State of Qatar adopt a flashing green period for 3 seconds before changing to yellow. Flashing green prior to yellow is only common in a few of the emirates, respectively, Abu Dhabi, Dubai, and Ajman, but even within those emirates the practice is not universal. On the other hand, 3 seconds is the common and universal value used for all-red. Although, the QTCM (2015a) recommends not applying the flashing green to new signals and will phase it out over time. However, there has been no specific implementation program developed for this yet, but new and refurbished sites do not use a flashing green period anymore. In the UAE, the implementation of flashing green prior to the onset of yellow is a common practice; for instance, in Abu Dhabi it is used as a measure to reduce red light running. When applied, the flashing green is a 3-second interval followed by another 3-second yellow interval. The standard red to green sequence in the State of Qatar is direct without any intermediate indication. On the other hand, in parts of the UAE, especially in Sharjah, a short yellow interval of 2 seconds preceding the start of green is currently in use, particularly at signalized intersections that are adjacent to roundabouts and at signalized roundabouts. The use of signals within roundabouts is common in the State of Qatar and in a few of the emirates, notably in Sharjah, in order to regulate in-flows during peak times to prevent traffic of heavier approaches from dominating circulation within the roundabout.

In the State of Qatar and the UAE, permitted left turns at traffic signals are not allowed. Multiple two and three left-turn lanes are common at large signalized

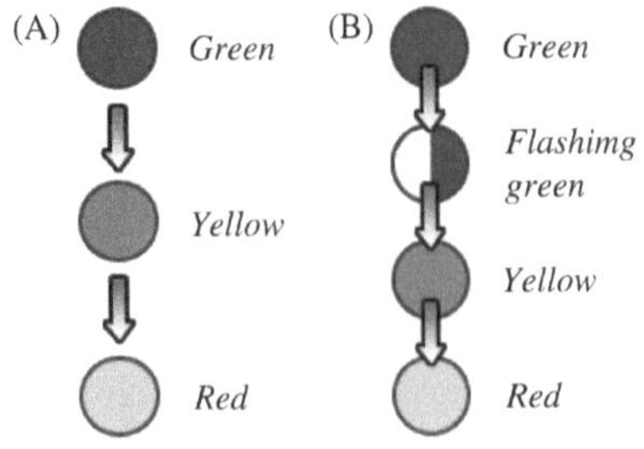

Figure 14.1 Signal indication sequence at the end of the green interval. (A) Standard sequence (QTCM, 2015a). (B) Sequence in practice.

intersections throughout the State of Qatar. Dubai and to a lesser extent Abu Dhabi follow a similar practice, often with two exclusive left-turn lanes and also three exclusive left-turn lanes are not unusual. Arrow indications are typically used when signaling is directed exclusively to turn lanes, as shown in Fig. 14.2. However, in practice, the signal indication for the through and the right-turn traffic does not use arrows, instead a solid green ball is used to indicate use for and priority to through traffic or through plus right traffic. Furthermore, the majority of signalized intersections in the State of Qatar and the UAE have separate free right-turn lanes with splitter islands in between, particularly the new intersections or those that have previously been major roundabouts. In these cases right turns are not considered in the signal phasing plan.

Regarding phasing structure; approach-based phasing, which is also known as split-phasing (Fig. 14.3A) and lead–lead and lag–lag phasing plans (respectively, Fig. 14.3B and C) are the predominant phasing schemes in the State of Qatar and the UAE. Phasing for protected left turns includes leading and lagging phasing, split phasing, and diamond phasing (Fig. 14.3D). In leading left turns (Fig. 14.3B) a protected left turn precedes the through movement of traffic, while in lagging left turns (Fig. 14.3C) the protected left turn follows up the through movement of traffic. The leading and lagging left turn phasing noted in Fig. 14.3 are common in the emirates of Abu Dhabi, Dubai, Sharjah, and Ajman. It is important to note that the majority of the signalized intersections in the State of Qatar and the UAE have channelized free right turns, as visible in Fig. 14.4. This is the reason why the presented phasing plans throughout this chapter do not include right-turn movements.

Aiming to achieve better progression at coordinated signalized intersections in the State of Qatar, it is common to have a leading left turn in one direction and a lagging left turn in the other direction, which is commonly referred to as the "lead-lag" phasing. On approaches when there is no sufficient space for adequate separation between opposing vehicles that carry out simultaneous left-turn maneuvers, the geometry dictates the adoption of split phasing. In contrast, lead-lag phasing for opposing approaches is not practiced in the UAE. The most common phasing treatments used in the State of Qatar are split phasing and lead-lag phasing and there is a limited number of signalized intersections with diamond phasing.

(A) (B)

Figure 14.2 Traffic movement arrows in signal head indications. (A) Schematic drawing. (B) Photo of a typical approach of signalized intersections.

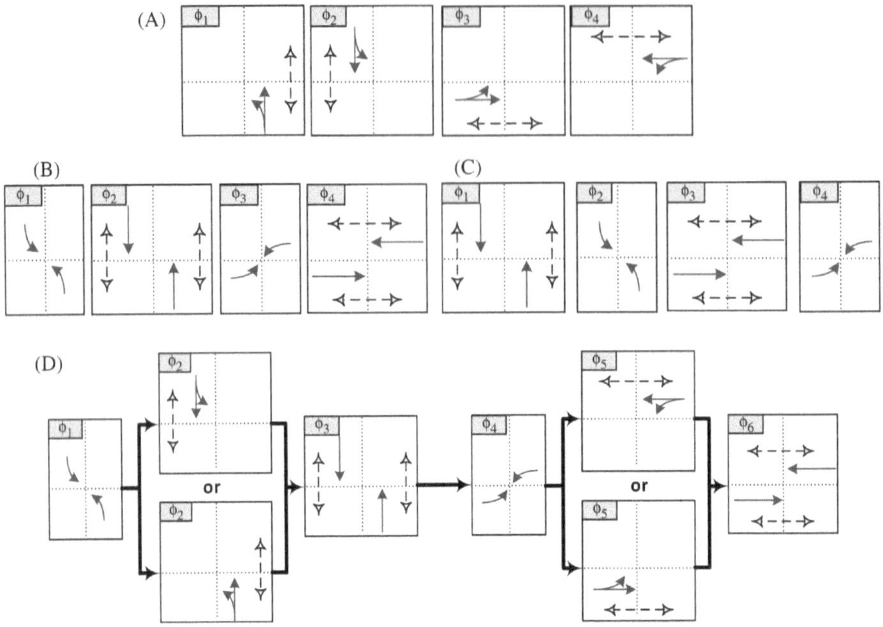

Figure 14.3 Common phasing plans. (A) Split phasing. (B) Leading left turn phasing. (C) Lagging left turn phasing. (D) Basic single diamond phasing.

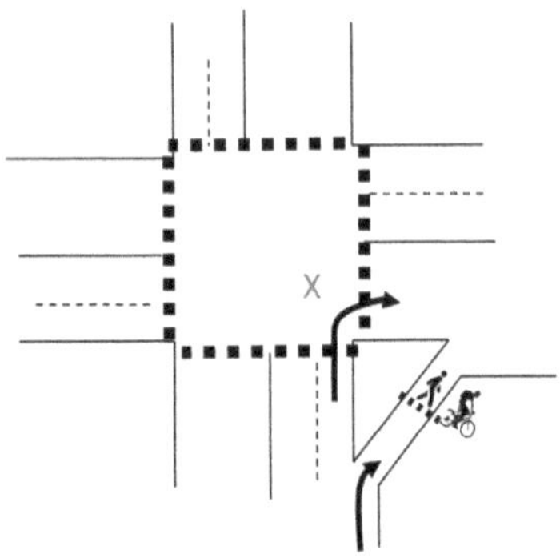

Figure 14.4 Filtered right turn with pedestrian crossing.

Right turns are typically segregated by splitter islands, unsignalized and configured either with or without merging acceleration lanes. There are few signalized right turns that adopt two-indication signals (yellow and red). The original rationale for two-indication signals on right turns is to not give the false impression of right-of-way if the right turn is give-way (yield) controlled at the exit point and there is no merging acceleration lane. This configuration was adopted from similar installations in Australia.

14.4.1 Treatments of bicycles and pedestrians

In general, pedestrian demand is low at most locations within the State of Qatar, except for some intersections that are close to shopping centers or recreational areas. However, with the heavy investment in public transport, especially the modern metro system with a vast smart bus feeder system, it is expected that pedestrian activities will significantly increase. Pedestrian activities are particularly heavy in urbanized central areas of Abu Dhabi, Dubai, and Ajman. Safe accommodation of pedestrians and cyclists is fundamental and currently integrated into signal design in the UAE.

Pedestrian or cycle crossings are often very long at large traffic signal intersections in both the UAE and Qatar. It is common that signalized intersections are equipped with push buttons to activate pedestrian phases (Muley et al., 2017). If the push button is not pressed, no green indication will be given to pedestrians, even if they have no conflict with the current vehicle phase. Sometimes there is provision for button activation half way across the crossing at the median island; however, the central islands are often very narrow for pedestrians. Most often in Qatar, the signalized intersections are configured to enable pedestrians to cross the entire intersection, but two-stage crossing is also accommodated, mainly for late-starter pedestrians, which can lead to quite large cycle times. In the UAE exclusive (protected) pedestrian phases are the norm, except for crossing the path of a filtered right traffic movement, as indicated by Fig. 14.4. In this case, pedestrians and cyclists have the right-of-way and vehicular traffic must yield. Multistage crossing is more common at large intersections in the UAE.

There are currently a small number of mid-block signalized pedestrian crossings in the State of Qatar. These are typically configured as a two-stage crossing. Mid-block pedestrian signalized crossings in the UAE are not common, but they are applied when walking time or distance to intersection signals is too long, because 5 minutes walking is too harsh in the hot and humid UAE weather. When used, typically to cross major arterials with sufficiently wide medians, they are designed to be two-stage crossings. Where space is available, pedestrian bridges are installed over heavy-vehicular-flow arterials in places where mid-block crossing is necessary but would be disruptive to vehicular traffic. Several of those pedestrian bridges are currently nearing completion in the cities of Doha and Sharjah.

14.4.2 Treatment of transit priority

For public transport systems requiring dedicated right-of-way, such as tram and light rail, a distinct traffic control system is provided according to QTCM (2015a) (Part 5, Section 7.1). Nevertheless, such systems are not yet implemented in the State of Qatar, since there are no tram or light rail systems so far. In the UAE, only the city of Dubai has tram services, which share the right-of-way with automobiles (Arabian Business, 2014). Operationally, the tram has priority over other traffic; however, it functions only in "protected" phases and does not share green time with other traffic. Additional rigid penalties are in place as a means to deter against jumping red lights. Besides integrating tram priority at signalized intersections, complementary new rules and signs were introduced when the tram service started in 2014 in order to ensure the safety and efficiency of operations of both the tram and other vehicular traffic (DEB Dubai Expt Blog, 2014).

Regular transit signal priority (TSP) capabilities to prioritize buses and emergency vehicles is in place but not operational in both Dubai and Abu Dhabi (Al Khateeb and Abdulfatah, 2017; The National, 2016). The TSP system in Abu Dhabi is fairly new and is currently used only for prioritizing civil defense vehicles. Nonetheless, the TSP technology in Dubai is not operational due to fairly heavy traffic flow on major corridors, congestion, and the infrequent presence of transit buses. Emergency vehicle preemption (EVP) systems are implemented in the city of Doha to give immediate priority to emergency vehicles while terminating or reducing the priorities for all conflicting movements.

14.5 Parameter determination

In general, in the state of Qatar, the QTCM (2015a) specifies the minimum signal timing requirements as summarized in Table 14.1. The yellow and all-red time is made up of the vehicle change interval and the vehicle clearance interval. The vehicle change interval is the yellow period that follows the green signal and precedes the red signal. QTCM (2015a) also recommends calculating the yellow period based on the speed of approaching traffic and the distance required to clear the intersection. Although similar in purpose, change (yellow) intervals in the UAE are calculated so as to eliminate the dilemma zone, based on the speed, deceleration, perception–reaction time, and width of the intersection. A 3 seconds yellow time period is the default value.

For all new signal installations, the yellow interval is 3 seconds. At a few existing intersections, the yellow period may have been set to a time longer than 3 seconds, depending on the speed limit of the approach. In practice, yellow is left as a standard 3 seconds and any additional clearance time required is applied as all-red. This is thought to be safer and provides more consistency.

Following the yellow interval, a short all-red period is displayed to allow vehicles already in the intersection to clear before the next phase starts. The all-red interval is based on the time needed for vehicles to travel the necessary distance within the intersection to clear conflict points. QTCM (2015a) states that a

Table 14.1 Minimum traffic signal timings

Period	Minimum timing (s)
Traffic green minimum	7
Green arrow minimum (left turning)	4
Yellow minimum	3

minimum all-red time of 2 seconds should be used, similar to the values used in the UAE. However, the Public Works Authority Roads Maintenance Department and Traffic Control Centre of the State of Qatar have indicated that all-red times of 1 and 2 seconds are used that may be extended to 3 or 4 seconds for very large intersections.

Table 14.2 outlines the yellow and all-red times based on different conflict clearance distances as proposed by QTCM (2015a) and practiced in the field. These are proposed based on an operational speed of 80 km/h since this is the most common speed limit in all arterials and collectors. It is observed that at some intersections all-red intervals are very long, which is mainly attributed to the extremely wide intersections that require long clearance times. Despite QTCM (2015a) stating that designs that lead to yellow and all-red times greater than 9 seconds should be avoided, indicated by the red highlighted area in Table 14.2, in reality yellow and all-red times are typically 7 or 8 seconds and may even be 9 or 10 seconds for larger intersections.

It is important to mention that a flashing green indication, mostly lasting 3 seconds, is commonly used throughout the State of Qatar and the UAE before the yellow period. But QTCM (2015a) states that the flashing green indication should not be used at any new or newly maintained signalized intersection and that there shall be a steady yellow lasting a minimum of 3 seconds. Hence, it is expected that the flashing green indication will phase out in the State of Qatar but not in the UAE.

Regarding pedestrian signal setting, QTCM (2015a) specifies pedestrian signal timing provisions for standalone signals and intelligent pedestrian crossings. The pedestrian signal indication is provided in two parts; green time (pedestrian green: PG) and clearance time (pedestrian flashing green: PFG). PG for standalone crossings varies between 4 seconds and 12 seconds and is typically provided as 6 seconds and 7 seconds. PG is determined based on the pedestrian flow, crossing distance, availability of small-sized median, and the presence of a high proportion of elderly or disabled pedestrians. However, QTCM (2015a) does not provide a clear procedure on how to estimate PG considering the previous factors. In fact, PG is usually estimated according to the judgment of the traffic engineer. Typically,

Table 14.2 Proposed clearance times (yellow and all-red) for different conflict clearance distances

Clearance distance (m)	Yellow and all-red time (s)
9	5
10–18	6
19–27	7
28–37	8
38–46	9
47–55	10
56–64	11
65–73	12

PFG or clearance time is calculated based on the crosswalk length and crossing speed. The crossing speed is assumed as 1.2 m/s. However, it is observed that many signalized intersections in the State of Qatar have shorter PFG intervals than recommended by QTCM (2015a). Furthermore, it should be noted that most signalized intersections in Doha city are equipped with push buttons to activate the pedestrian signal. These push buttons do not alter the intersection phasing plan to provide pedestrians the right-of-way, but they only give a sign to the controller to turn on the pedestrian green indication when the relevant phase starts.

In the UAE, the steady WALK pedestrian green, which is equivalent to PG, is based on an initial startup time between 3 seconds and 4 seconds plus a duration based on the width of the sidewalk and the number of pedestrians to be served in a cycle. The flashing red DO NOT WALK signal indication, which is equivalent to PFG, is strictly based on the length of the crosswalk and the average pedestrian speed, usually taken between 1.2 and 1.4 m/s. Because of the diverse ethnicities and languages, WALK and DO NOT WALK are displayed using an animated digital image of pedestrians as opposed to words. The WALK period is marked by an animated image of a walking pedestrian in normal speed, while DO NOT WALK is marked by an animated image of a hurrying pedestrian.

14.5.1 Cycle length

Phasing plans are typically designed for a cycle length between 150 seconds and 240 seconds. QTCM (2015a) does not recommend specific procedures to estimate cycle length, but suggests deriving it from suitable modeling software. In practice, SCATS will define the cycle length and adapts this based on traffic volumes. Typically, most signalized intersections in Qatar operate with a cycle length over 150 seconds. In Abu Dhabi and Dubai, SCOOT optimizes signal timing and changes green times based on detected traffic. SCOOT calculates the time of each cycle based on the number of detected vehicles through sensors, which are mostly cameras, and has minimum and maximum timings that are selected based on the traffic demand level. For example, a maximum length of 255 seconds is used for red and a minimum of 7 seconds is applied for green. Actual durations of different intervals are not set in stone, but the volume of traffic dictates the green or red time allowed. For isolated intersections, Synchro Studio is used to estimate the optimum signal timings. In the State of Qatar, some of the isolated traffic signalized intersections may have cycle lengths greater than 240 seconds, depending on how they are configured within the controller in response to onsite traffic flow observations. Operators at the traffic control center may also manually shorten or extend some phases at certain times of the day, in response to onsite observations or specific requests to provide "continuous flow" during special events.

14.5.2 Saturation flow rate

QTCM (2015a) does not specify or recommend values for the saturation flow rates for different traffic movements. It recommends the estimation of the saturation flow rate for each site, using either the methodology published by the United Kingdom's Transport Research Laboratory Report RR67 (Kimber et al., 1986) or the method used in the United States of America as outlined in the Highway Capacity Manual (2011). In general, the recommendations from US documents are followed in the State of Qatar and the UAE. The ideal/base saturation rate used is 1900 veh/h/lane. The actual saturation flow rate will depend on the adjustments that would reflect the prevailing conditions. For through traffic, usually it does not go significantly below 1900 veh/h/lane, while for left-turn traffic it is about 1500 veh/h/lane and for shared left-turn and through lanes it is between 1600 veh/h/lane and 1800 veh/h/lane.

14.5.3 Lost time

There are no studies or reports stating the values of the start-up and clearance-lost times used in practice in the State of Qatar. However, QTCM (2015a) recommends the estimation of the lost time for each site as part of the signal design process. It proposed the use of Eq. (14.1) for the estimation of start-up lost time.

$$L = t_2 - \frac{7200}{S} \tag{14.1}$$

where L is the lost time (s), t_2 is the time taken for the first two vehicles to cross the stop line, and S is the saturation flow rate (veh/h). In the State of Qatar, an average of 6 s/phase is usually used in the settings of signal control. This long lost time is attributed to the assumption made by practitioners that part of the vehicle flashing green interval is lost time. However, in the UAE the lost times are assumed on average to be 2–5 s/phase, with the lower bound of 2 s/phase mostly being used and being treated as a default value.

14.6 Special consideration

Signal displays at intersections typically include primary displays and far-side secondary signals on mast-arms. On smaller intersections it is often difficult for drivers who stopped at the intersection stop line to see the signal heads on the secondary mast arms. Therefore, multiple display arrangements are practiced in the UAE with all new intersections having multiple signal displays, both overhead and on the side.

Use of U-turns is common and almost a default practice. Consequently, multiple left plus U-turn lanes are implemented at large intersections in specific, of which many intersections were previously roundabouts. For example, referring to Fig. 14.2, in which the most left lane will function as a defacto U-turn lane while the second-left lane will function as defacto left-turn lane. Exclusive signalized U-turns at intersections or mid-blocks with unique signal indication are applied at several locations in Doha city. Fig. 14.5 shows the layout of exclusive U-turn lanes and their signal indication head.

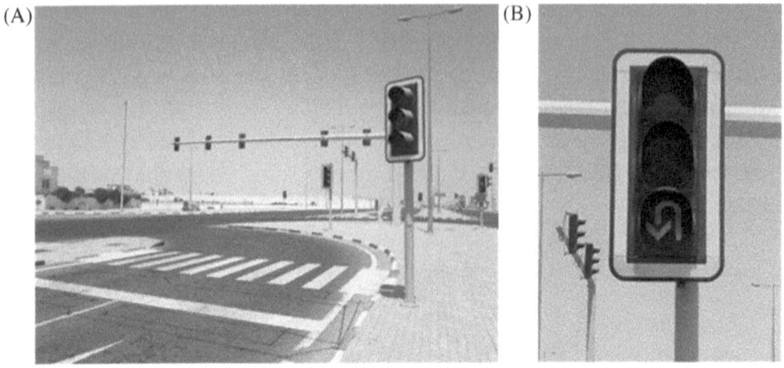

Figure 14.5 Exclusive U-turn traffic signals in the State of Qatar. (A) U-turn layout. (B) U-turn traffic signal head.

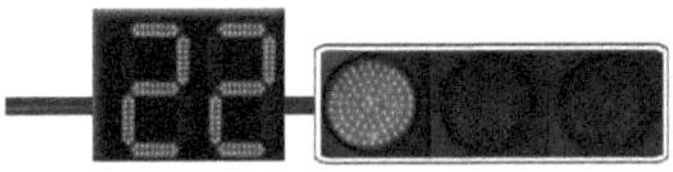

Figure 14.6 Countdown traffic signal in Ajman.
Source: From Credit: Gulf News, 2013. Ajman: countdown traffic timer cuts accidents at intersections (accessed 27.02.18).

Figure 14.7 Pedestrian countdown signal in Abu Dhabi.
Source: From https://www.flickr.com/photos/zia_buet01/24242640439.

Countdown signals for both vehicular traffic and pedestrians are used in the UAE, but not in all emirates. Countdown signals for vehicular traffic are common in Ajman (Fig. 14.6). Pedestrian countdown signals (Fig. 14.7) are used in Abu Dhabi and Ajman, but are not realized in the other emirates. Pedestrian countdown signals are also used at a few intersections in the city of Doha in the State of Qatar.

14.7 Examples

The two examples presented later are of observed signal timings at two intersections in Doha city. They are quite similar to typical signalized intersections in the

UAE. As noted before, isolated signalized intersections in the State of Qatar which are not connected to the SCATS system are operated in isolation under fully vehicle-activated mode, using the standard design phasing that is entered into the controller. However, some intersections are temporarily operated using fixed time sequence, often when there are detector or communication faults or during the final stages of construction just before the detectors are installed but after the signals are brought into service. Synchro Studio is utilized to estimate cycle length and green splits, whereas Webster formula is not used to estimate cycle length.

14.7.1 Al Rufaa intersection, Doha, Qatar

Fig. 14.8 shows the layout of the Al Rufaa intersection and the observed peak hour vehicle and pedestrian demands. The signal control of the intersection is not coordinated with any neighboring intersection. The intersection is operated in full

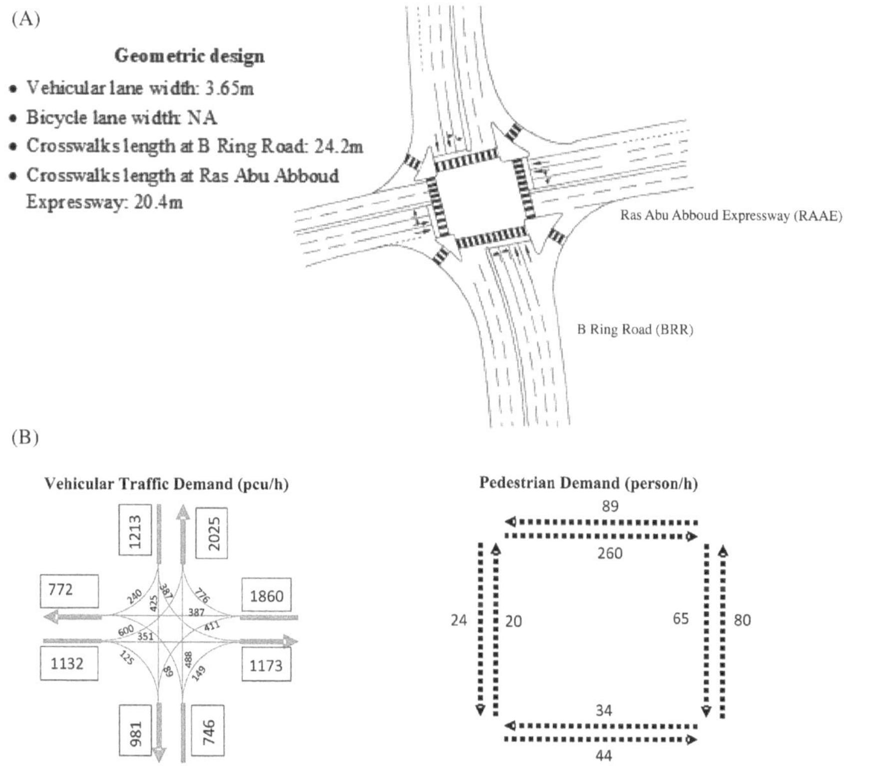

Figure 14.8 (A) Al Rufaa intersection layout; (B) vehicle and pedestrian peak hour demands (typical week day).

Table 14.3 Phasing plan and signal timing of Al Rufaa intersection

AL Rufaa Intersection		Signal phasing length (seconds)																				Cycle length (seconds)
		φ1					φ2					φ3					φ4					
		1	2	3	4	5	6	7	8	9	10	11	12	13	14	15	16	17	18	19	20	
	Vehicle																					
	Pedestrian																					
	Vehicle																					
	Pedestrian																					
	Vehicle																					
	Pedestrian																					192
	Vehicle																					
	Pedestrian																					
Timings	Vehicle movements	40		3	3	2	40		3	3	2	40		3	3	2	40		3	3	2	
	Pedestrians	8	12	-	-	-	8	12	-	-	-	8	12	-	-	-	8	12	-	-	-	
Phasing Plan		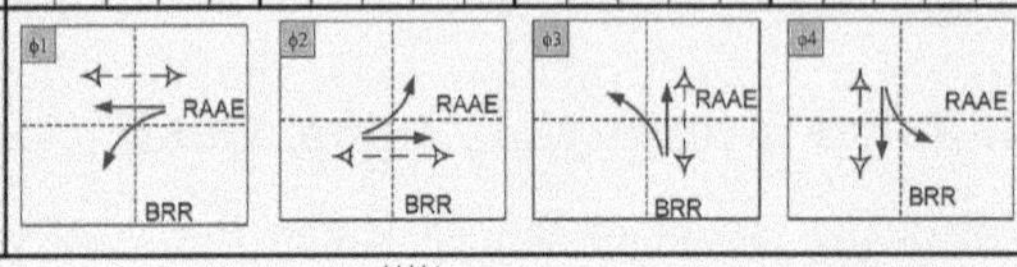																				

—— Green ········ Flashing Green /\/\ Yellow |||| Pedestrian flashing green ▬ Red

actuation mode. Inductive detector loops are installed at all entry lanes of each approach. The cycle length is not fixed and changes are based on the activation of green interval extension calls from the loop detectors. In the UAE, detection is mainly done using cameras and image-processing protocols and the use of inductive loops is rare, especially at newly constructed intersections. Table 14.3 shows the implemented phasing plan and sample of the signal timing information of the Al Rufaa intersection. The phasing scheme is split-phasing. During peak hours, cycle length is observed to be between 190 seconds and 200 seconds. Pedestrian demand is low and there are no cyclists. PG interval at all approaches is set as 8 seconds, according to the recommended pedestrian green time in HTSM (2015). However, the pedestrian flashing green PFG interval is set as 12 seconds, equally at all approaches, being significantly shorter than the recommended period following the QHDM (2015b) procedure, which is 21 seconds for phases 1 and 2 [(24.2 m)/(1.2 m/s)], and 17 seconds for phases 3 and 4 [(20.4 m)/(1.2 m/s)]. It is improtant to note that, based on the assigned green times for vehicular movements on all phases, there is enough time to provide pedestrians with the required clearance time (PFG).

14.7.2 Al Meena intersection, Doha, Qatar

Fig. 14.9 shows the layout of the Al Meena intersection and the observed peak hour vehicle and pedestrian demands. Similar to the Al Rufaa intersection, Al Meena intersection is isolated and the signal control is fully actuated, as inductive loop detectors are installed in all approaches. The cycle length is not fixed but the phasing plan is fixed throughout the day. Table 14.4 shows the implemented phasing plan and a sample of the signal timing information. During peak hours, the cycle length is observed to be between 220 seconds and 240 seconds. Pedestrian and cyclist demands are low. PG interval at all approaches is set at

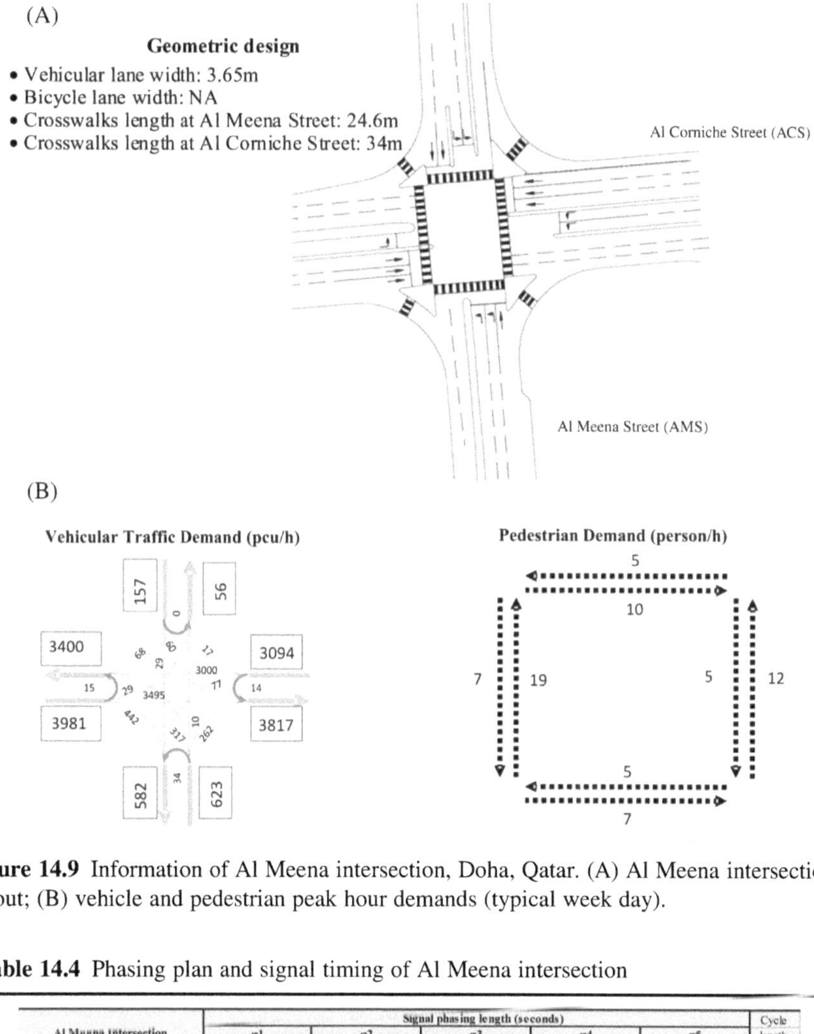

Figure 14.9 Information of Al Meena intersection, Doha, Qatar. (A) Al Meena intersection layout; (B) vehicle and pedestrian peak hour demands (typical week day).

Table 14.4 Phasing plan and signal timing of Al Meena intersection

Al Meena Intersection		Signal phasing length (seconds)																									Cycle length (seconds)
		φ1					φ2					φ3					φ4					φ5					
		1	2	3	4	5	6	7	8	9	10	11	12	13	14	15	16	17	18	19	20	21	22	23	24	25	
Vehicle																											239
Pedestrian																											
Vehicle																											
Pedestrian																											
Vehicle																											
Pedestrian																											
Vehicle																											
Pedestrian																											
Vehicle																											
Pedestrian																											
Timings	Vehicle movements	17		3	3	2	99		3	3	2	17		3	3	2	33		3	3	2	33		3	3	2	
	Pedestrians	-	-	-	-	-	10	9	-	-	-	-	-	-	-	-	10	17	-	-	-	10	17	-	-	-	
Phasing Plan																											

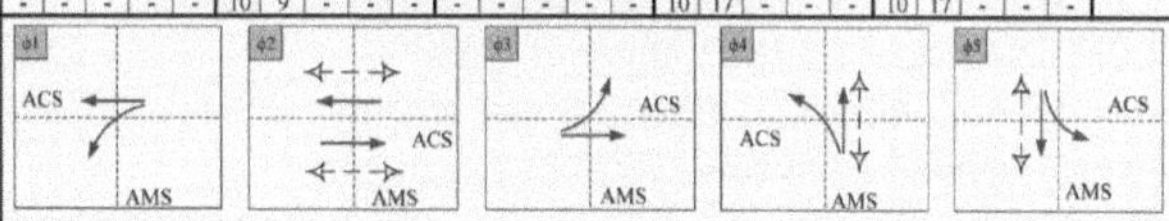

—— Green ------ Flashing Green /\/\ Yellow ||||| Pedestrian flashing green ▬ Red

10 seconds, according to the recommended pedestrian green time in HTSM (2015). However, the PFG interval is set at 9 seconds for phase 2 and 17 seconds for phases 4 and 5, which is significantly shorter than the recommended period following the QHDM (2015b) procedure, being 21 seconds for phase 2 [(24.6 m)/(1.2 m/s)], and 29 seconds for phases 4 and 5 [(34 m)/(1.2 m/s)]. It is improtant to note that based on the assigned green times for vehicular movements on all phases, there is enough time to provide pedestrians with the required clearance time (PFG).

Disclaimer

The views expressed in this chapter are solely those of the authors. The governmental agencies mentioned in the text assume no liability for the accuracy of the information or practices and standards presented in the chapter.

Bibliography

Al Khateeb, H., Abdulfatah, A., 2017. The impact of transit signal priority: case study in Dubai, UAE, vol. 120, 2017. In: International Conference on Advances in Sustainable Construction Materials & Civil Engineering Systems (ASCMCES-17).

Arabian Business, 2014. Dubai Police, RTA reveal details of tram-related fines, penalties. <http://www.arabianbusiness.com/dubai-police-rta-reveal-details-of-tram-related-fines-penalties-569997.html> (accessed 25.02.18.).

DEB (Dubai Expt Blog), 2014. RTA issuing new traffic rules as Dubai Tram is set to roll in. <http://www.dubaiexpatblog.com/rta-issuing-new-traffic-rules-dubai-tram-set-roll.html> (accessed 23.02.18.).

Gulf News, 2013. Ajman: countdown traffic timer cuts accidents at intersections. <https://gulfnews.com/uae/transport/ajman-countdown-traffic-timer-cuts-accidents-at-intersections-1.1208641> (accessed 27.02.18.).

Kimber, R.M., McDonald, M., Hounsell, N.B., 1986. The prediction of Saturation Flows for Road Junctions Controlled by Traffic Signals. Transport and Road Research Laboratory (TRRL) (67), <https://trid.trb.org/view.aspx?id = 237965>.

Ministry of Municipality and Environment (MME), 1997a. Qatar Highway Design Manual (QHDM), first ed. MME, State of Qatar.

Ministry of Municipality and Environment (MME), 1997b. Qatar Traffic Control Manual (QTCM), first ed. MME, State of Qatar.

Ministry of Municipality and Environment (MME), 2015a. Qatar Traffic Control Manual (QTCM), second ed. MME, State of Qatar.

Ministry of Municipality and Environment (MME), 2015b. Qatar Highway Design Manual (QHDM), second ed. MME, State of Qatar.

Muley, D., Alhajyaseen, W., Kharbeche, M., 2017. An overview of pedestrian signal setting and implementation in the State of Qatar, Proced. Comput. Sci., vol. 109. Elsevier, pp. 545–552.

Policy Statement, 2016. Traffic Signal Design Guidelines Supplement. Public Works Authority, State of Qatar.

TRB (Transportation Research Board), 2011. Highway Capacity Manual 2010. The US National Academies of Sciences, Engineering, Medicine, Washington, DC, USA.

TRB (Transportation Research Board), 2018. Signal Timing Manual, second ed. <http://www.trb.org/Publications/Blurbs/173121.aspx>.

The National, 2010. Timers on all Corniche crossings. <https://www.thenational.ae/uae/timers-on-all-corniche-crossings-1.500192> (accessed 20.02.18).

The National, 2016. Abu Dhabi to test emergency vehicle signal priority system. <www.thenational.ae/uae/transport/abu-dhabi-to-test-emergency-vehicle-signal-priority-system-1.164034> (accessed 21.02.18.).

Initial comparative analysis of international practice in road traffic signal control

15

Keshuang Tang[1], *Manfred Boltze*[2], *Zong Tian*[3] *and Hideki Nakamura*[4]
[1]College of Transportation Engineering, Tongji University, Shanghai, P.R. China, [2]Institute for Transport Planning and Traffic Engineering, Technische Universität Darmstadt, Darmstadt, Germany, [3]Department of Civil and Environmental Engineering, University of Nevada, Reno, Reno, NV, United States, [4]Department of Civil and Environmental Engineering, Nagoya University, Nagoya, Japan

15.1 Introduction

The country reports included in this book offer a profound basis for sharing knowledge of traffic signal control practices in different countries. Providing a comprehensive comparison of the various signal control aspects throughout the globe has proved to be a challenging task, and any recommendations and conclusions would require further research after the publication of this book. However, within this chapter, it was possible to review the previous chapters and well-adopted guidelines (FHWA, 2008, 2012; TRB, 2010; FGSV, 2001, 2015; UK DOT, 2006; AUSTROADS, 2003; JSTE, 2018) for some selected aspects. Such an initial comparison among the selected countries is intended to provide a global picture of the general practices on fixed-time and isolated signal control for international readers. The compared items include the number of traffic signals, signal control modes, types of signal controllers and systems, signal displays, signal warrants, stage structure and sequence, some critical signal timing parameters, signal timing procedures, and performance evaluation. Common and distinct points are summarized and their advantages and disadvantages as well as underlying reasons are also discussed from a global perspective.

Note that the comparisons are highlights mostly based on the presented contents in previous chapters, and they may not cover all situations in the concerned regions or countries comprehensively. Interested readers are encouraged to read previous chapters and related guidelines for more information.

15.2 Number of traffic signals

The quantities and densities of traffic signals (i.e., signalized intersections or signalized pedestrian crossings) in the covered countries are given in Table 15.1, together

Global Practices on Road Traffic Signal Control. DOI: https://doi.org/10.1016/B978-0-12-815302-4.00015-7

Table 15.1 Number and density of traffic signals

Countries	Number	Density (traffic signals/1000 inhabitants)	First traffic signal implemented
The United States	Overall: 330,000	Overall: 1.009	Cleveland, 1914
Canada	Overall: 25,000	Overall: 0.70	–
Germany	Berlin: 2100, Karlsruhe: 310, Darmstadt: 160	Overall: approximately 1.0, Berlin: 0.618, Karlsruhe: 1.000, Darmstadt: 1.000	Hamburg/Berlin, 1920s
Austria	Vienna: 1300, Graz: 295	Overall: approximately 1.0, Vienna: 0.7222, Graz: 1.017	Vienna, 1920s
The United Kingdom	Overall: >8000, Greater London: 3000	Overall: >0.120, Greater London: 0.342	London, 1868
France	Overall: >30,000	Overall: 0.5–1.2	Paris, 1912
Switzerland	Usually 100–150 traffic signals per city	Overall: 0.5–1.0	–
Turkey	28 cities: 7089	28 cities: 0.129	Istanbul, 1929
Australia	Overall: >10,000	Overall: >0.40	Melbourne, 1928
Japan	Overall: 171,000	Overall: 1.35	Tokyo, 1930
China	Overall: >100,000	Overall: >0.072	Shanghai, 1920s
India	Bangalore: 400	Bangalore: 0.04	–
South Korea	Overall: 43,600	Overall: 0.847	1940s or 1950s
Qatar	Overall: 287	Overall: 0.112	1970s–1980s
UAE	Dubai: >800, Abu Dhabi: >125	Dubai: >0.258, Abu Dhabi: 0.053	1970s–1980s

with the place and time that the first traffic signal was implemented. It can be seen that the United States has the highest number of traffic signals, at 330,000, while Qatar has the lowest number, with 287. In terms of density, Japan has the highest with 1.35 signalized intersections per 1000 inhabitants, and the United States, Germany, Austria, France, Switzerland, and South Korea are almost at the same level, with nearly 1 traffic signal per 1000 inhabitants. In contrast, densities in the United Kingdom and Australia are approximately in the range of 0.1–0.5, and those in Turkey, China, India, Qatar, and UAE are much lower. While the overall density in Canada lower than that in the United States, a recent survey of four major metropolitan areas in Canada (Toronto, Vancouver, Waterloo, and Calgary) indicates a similar level to the United States, which is about 1.06. In addition, it is interesting to note that the world's first traffic signal powered by gas was implemented in London in 1868, and the first electronic traffic signals were implemented in Paris, France, in 1912, and later in Cleveland, United States, in 1914.

15.3 Signal control modes

The proportions of fixed-time control, actuated, and adaptive control in the covered countries are listed in Table 15.2. It was found that actuated or adaptive control is more common in most of the economically developed countries, including the United States, Canada, Germany, Austria, the United Kingdom, France, Switzerland, Australia, New Zealand, and Japan. In particular, adaptive control is in the majority in Qatar and UAE, as most signalized intersections are operated by SCATS and SCOOT in these two nations. It is also worth noting that countries with widespread use of signal countdown devices are more dependent on time-of-day (TOD) fixed-time control.

Fully actuated or actuated-coordinated control under TOD signal plans is common in the United States and Canada, while centralized adaptive control using complex signal control algorithms, such as SCATS, SCOOT, STREAMS, is typical in the United Kingdom, Australia, New Zealand, Qatar, and UAE. Rule-based actuated control is more popular in Germany, Austria, and Switzerland, and is often used to prioritize public transport. The majority of traffic signals in Japan are either under fixed-time control (48%) or adaptive (43%) control, and only a small portion is under fully actuated and isolated control (9%). Twenty-six percent of traffic signals are operated with semiactuated control mode in the 28 surveyed cities in

Table 15.2 Signal control modes

Countries	Fixed-time	Actuated or adaptive
The United States and Canada	Small portion	Majority (fully actuated or actuated-coordinated under TOD plans)
Germany and Austria	Small portion	Majority (mostly rule-based actuated control)
The United Kingdom	Small portion	Majority (mostly isolated actuated or centralized adaptive using SCOOT)
France	Small portion	Majority (similar to Germany)
Switzerland	Small portion	Majority (similar to Germany)
Turkey	62%	38% (semiactuated: 26%; fully actuated: 7%; centralized adaptive: 4%; and flash: 1%)
Australia and New Zealand	Small portion	Majority (mostly isolated actuated or centralized adaptive using SCATS and STREAMS)
Japan	48%	52% (centralized adaptive: 43%; isolated actuated: 9%)
China	95%	5% (isolated actuated or centralized adaptive using SCATS, etc.)
India	Majority	Small portion
South Korea	99%	1% (isolated actuated)
Qatar and UAE	Small portion	Majority (mostly isolated actuated or centralized adaptive using SCATS and SCOOT)

Turkey, together with 7% of fully actuated control, 4% of adaptive control, and 1% of flash control. In countries like China, India, and South Korea, fixed-time control remains predominant, though there is a very small portion of isolated actuated control or centralized adaptive control.

Note that some statistics presented in the table are not from comprehensive and precise surveys, but based on rough estimations by the authors of previous chapters.

15.4 Signal controllers and control systems

Globally speaking, the development of modern signal controllers dates back to the 1930s when they replaced manually controlled ones. As the first generation of signal controllers, fixed-time signal controllers had a short history in the United Kingdom, the United States, Germany, France, etc., and they were quickly replaced by actuated or adaptive signal controllers owing to rapid development of detector and computer technologies. Domestic signal controllers in Japan, China, and South Korea appeared later in the 1960s and 1970s. So far, all the covered countries except Qatar and UAE predominately use domestic signal controllers. Signal controllers have continuously evolved from isolated fixed-time, isolated actuated, centralized coordinated, to centralized adaptive, along with the advancements of computer, communication, and sensor technologies.

Major signal controller manufacturers in the covered countries are listed in Table 15.3. This table shows that, despite the many existing international and domestic competitors, Siemens is still the most successful global provider for signal controllers. It holds quite a few foreign markets in the United States, Austria, Switzerland, Turkey, China, Qatar, and UAE. However, its market shares have been significantly decreasing recently in a few countries including the United States and China, where domestic companies have quickly emerged.

Notably, since the publication of new signal controller standards ATC (Advanced Traffic Controller) in the United States in 2006 and OCIT (Open Communication Interface for Road Traffic Control Systems) in the EU in 2008, connectivity between signal controllers, detectors, as well as vehicles and infrastructures has attracted a great deal of attention from the industry. These standards provide an open hardware and software platform, which can support wide intelligent transportation systems (ITS) applications.

Signal controller manufacturers usually need to provide customers with signal control software to facilitate the utilization of their controllers. Signal control systems emerged from such a context. In addition, a number of third-party companies and research institutes have also made tremendous efforts in developing independent control systems. Major systems currently being implemented in the covered countries are listed in Table 15.3. Those systems can be classified into three broad categories, that is, rule-based, model-based, and mixture. For instance, most systems such as BALANCE, TACTICS, and VS-PLUS implemented in Germany, Austria, and Switzerland are basically rule-based, though some of them also have

Table 15.3 Major signal controller manufacturers and systems

Countries	Signal controller manufacturers	Signal control systems
The United States and Canada	Econolite, McCain, Naztec/ Trafficware, Peek, Siemens USA, Intelight, etc.	Centracs, QuicNet, ATMS, TACTICS, MaxView, TransSuite, etc.
Germany and Austria	Siemens, Swarco, AVT Stoye, etc.	BALANCE, MOTION, VS-Plus, etc.
The United Kingdom	Swarco, EDS, SRL, etc.	SCOOT, LINSIG, OSCADY PRO, etc.
France	Fareco, Aximum, Lacroix Traffic, Polyvelec, etc.	CRONS, PRODYN, GLOCALE, CLAIRE, GERFAUT II, GERTRUDE, etc.
Switzerland	VRGA, Bergauer AG, Siemens, Swarco, etc.	VS-PLUS, FESA, TS200, etc.
Turkey	Isbak, Tek Teknotel, Siemens, Rayennur, Tankes, ISSD, etc.	ATAK, CHAOS, CYCLOPS, etc.
Australia and New Zealand	Tyco Traffic and Transportation, Quick Turn Circuits, Aldridge Traffic Controllers, etc.	SCATS, STREAMS, SIDRA INTERSECTION
Japan	Nippon Signal, Sumitomo Electric Industries, Panasonic, Kyosan Electric Mfg., etc.	MODERATO, etc.
China	Qingdao HISENSE, Nanjing LES, Shanghai BAOKANG, Lianyungang JARI, Anhui KELI, SIEMENS China, etc.	HICON, NATS, MiTCO, JR-MATH, KELI-UTC, SCOOT, SCATS, etc.
India	CDAC, CMS, BELL, Keltron, Envoyse Electronics, Swarco, Simense, etc.	WiTraC, UCON, etc.
South Korea	Korea Electric Traffic, Suhdol Electric and Communication, etc.	TRC, COSMOS, SMART, etc.
Qatar and UAE	Siemens, etc.	SCATS, SCOOT, etc.

model-driven functions (rarely used). Other systems such as Centracs, ATMS, SCOOT, SCATS, CRONOS, PRODYN, STREAMS, and MODERATO implemented in the United States, the United Kingdom, France, Australia, and Japan are rather model-based or a mixture of rule-based and model-based.

Though a variety of centralized adaptive control systems have been developed and implemented in a number of countries, their practical effectiveness and cost–benefit have been argued over for a long time in industry and academia. These systems are generally considered to be capable of handling highly fluctuated traffic flow at the network level, largely relying on good-quality traffic detectors and experienced engineers. However, due to the failure of traffic detectors and lack of experienced knowledgeable people, as well as complex dynamics of traffic flow, the capabilities and functionalities of these systems cannot be fully utilized in many cases, and thus their practical performance has not yet been very convincing generally. On the other hand, many academics and practitioners believe that well-designed fixed-time signal control, together with well-defined corridor coordination, is a good alternative in terms of cost and benefit, given limited resources for installing advanced signal controllers and systems as well as hiring experienced engineers for signal timing design and maintenance. In addition, the isolated actuated signal control with simple control logic is also considered as an efficient means for rural intersections. In some countries, such as the United States and Germany, actuated-coordination control has also proved to be quite effective.

15.5 Signal displays

15.5.1 Lens color and indication sequence

Types and meanings of static red, yellow, and green signals for motor vehicles indicated by circular green balls or arrow symbols are similar in the covered countries, which are globally unified based on the UN standard. It is a common principle that circular green balls are used for permitted movements and the arrow symbols can only be used for protected movements. Nevertheless, improper use of arrow symbols still exists in practice in some countries. The main differences between the covered counties lie in signal indication sequence, especially during the transition intervals of stages. Table 15.4 presents typical signal indication sequences for motor vehicles and pedestrians prevailing in the covered countries.

With respect to motor vehicle signals, a red-yellow signal before the onset of green is used to warn drivers of start-up in most European countries such as Germany, the United Kingdom, Austria, and Switzerland. The time duration of the red-yellow signal is often 1–2 seconds. This seems to be efficient in reducing start-up lost time and is also more compatible with actuated or adaptive signal control, compared with the red countdown signals. A flashing green signal before the onset of yellow is usually used to indicate the end of a green phase in Austria, China, Qatar, and UAE, but it is generally prohibited or rarely adopted in most of the other countries. The time duration of the flashing green is usually in the range of 3–6 seconds, and is a constant 4 seconds in Austria. Previous studies have indicated that such a flashing green might be helpful for approaching drivers at intersections with an insufficient length of yellow time or a high proportion of large vehicles (Köll et al., 2004; Tang et al., 2016). However, it also increases the option zone (i.e., type 2 dilemma zone) at the same time. Thus it is generally not recommended, except at those special types of intersections mentioned earlier.

Meanwhile, green and red countdown timers are quite popular in countries like Turkey, China, Qatar, and UAE. In general, signal countdown timers for vehicles are not widely accepted internationally, as they may induce aggressive passing of drivers, and more importantly it is hard to be implemented together with actuated or adaptive control. Moreover, a few special signal indications exist in some

Table 15.4 Typical signal indication sequences

Countries	Motor vehicles	Pedestrians
The United States and Canada	Red→green→yellow→red	Red→green→flashing red→red
Germany	Red→red and yellow (1 s)→ green→yellow→red	Red→green→red
Austria	Red→red and yellow (1 s)→ green→flashing green (4 s)→yellow→red	Red→green→flashing green (4 s)→red
The United Kingdom	Red→red and yellow (2 s)→ green→yellow→red	Red→green→flashing red→red
France	Red→green→yellow→red	Red→green→red
Switzerland	Red→red and yellow (1 s)→ green→yellow→red	Red→green→red, Red→green→flashing green→red, Red→green→yellow→red
Turkey	Red→green→yellow→red	Red→green→red
Australia and New Zealand	Red→green→yellow→red	Red→green→flashing red→red, red→green→yellow countdown→red
Japan	Red→green→yellow→red	Red→green→flashing green→red
China	Red→green→yellow→red, red→green→flashing green→yellow→red	Red→green→flashing green→red
India	Red→green→yellow→red	Red→green→red
South Korea	Red→green→yellow→red	Red→green→flashing green→red
Qatar and UAE	Red→green→flashing green→yellow→red	Red→green→flashing green→red

countries. For instance, a flashing yellow signal or a flashing red signal is additionally utilized in the case of permitted/protected left turns in the United States. As for public transport vehicles (buses and trams) using the road, some countries also use specific signals, for example, Germany.

With respect to pedestrian signals, the basic symbols are also similar in the covered countries, which are a walking pedestrian (for green) and a standing pedestrian (for red). However, the combination of a hand symbol (orange) and a walking pedestrian symbol (white) is adopted in North America. Unlike the signal indication sequence for motor vehicles, a flashing green signal before the onset of red for the clearance of pedestrians is widely accepted in many countries, for example, Austria, Switzerland, Japan, China, South Korea, Qatar, and UAE. Meanwhile, a red flashing signal is common in the United States, Australia, and New Zealand, and in Switzerland, a yellow signal is also used before the onset of red in addition to flashing green. However, it was reported in several countries that some pedestrians still run into the crosswalk during the clearance interval. This is partly because pedestrians do not correctly understand the meaning of the flashing green. Thus it is important to make efforts in educating and advertising the meaning of the clearance interval for pedestrians, no matter which type of clearance signal is implemented.

Green and red countdown timers for pedestrians are widely implemented all over the world. However, the displays of countdown timers are slightly different from each other. For instance, the number of red or green dots (not exactly the remaining number of seconds) displayed on the signal face indicates the remaining red or green time in Japan. The number of seconds for remaining red or green time is directly displayed on the signal face in Australia, New Zealand, Turkey, China, Qatar, and UAE. In addition, separate LED screens for countdown timers can often be seen in those countries as well. A yellow countdown timer for pedestrians is especially used in the United States and Australia. In summary, the signal countdown timer for pedestrians is popular in many countries. A full countdown timer is not very compatible with the actuated or adaptive control. However, since the clearance interval of pedestrians is much longer than that of motor vehicles and it must be provided in any modes of signal control, the signal countdown timer can start from the beginning of the clearance interval, for example, a partial countdown timer.

Dedicated signals for cyclists indicated by a bicycle symbol exist in several countries, such as Germany, Austria, Switzerland, South Korea, and China. In European countries, bicycles are basically treated as vehicles, and thus its signal indication sequence is the same as that for motor vehicles. However, bicycles are often grouped together with pedestrians in China and South Korea, and thus its signal indication sequence is more like that of pedestrians.

15.5.2 Arrangement and placement

The arrangement of signal displays varies a little from country to country. Signal displays for motor vehicles are basically vertical alignment in European and Oceanian countries as well as Qatar and UAE, while a mixture of vertical and horizontal alignment prevails in North America, Japan, South Korea, China, and India. A vertical alignment for pedestrian signals is most common in the covered countries.

The placement of vehicular signal heads can be grouped into the far-side pattern (i.e., signal heads placed at the opposite exit) and the near-side pattern (i.e., signal heads placed near to the stop-line of approach). Setting patterns of signal displays in the covered countries are listed in Table 15.5. It can be seen that Germany, Austria, the United Kingdom, Switzerland, Turkey, Australia, New Zealand, Qatar, and UAE prefer the near-side pattern. In addition, one cluster of signal heads is often separately installed for one specific lane or movement, that is, lane-based (or movement-based), as the vision zone of drivers is comparatively narrow in the case of a near-side pattern. Meanwhile, supplemental far-side signal faces can also be set for turning traffic streams at corners, especially at large intersections, in order to improve the visual field of turning drivers after they enter the intersection.

Table 15.5 Typical placement of signal displays

Countries	Near-side	Far-side	Approach-based	Lane-based
The United States and Canada		√	√	
Germany and Austria	√			√
The United Kingdom	√			√
France	√			√
Switzerland	√			√
Turkey	√			√
Australia and New Zealand	√			√
Japan		√	√	
China		√	√	
India		√	√	
South Korea		√	√	
Qatar and UAE	√			√

On the other hand, vehicular signal heads are commonly placed at the far side in the United States, Canada, Japan, China, India, and South Korea. One cluster of signal heads is installed for one specific approach, that is, approach-based. Meanwhile, it is also possible to add supplemental near-side signal faces, if the distance between the stop-line and its corresponding signal heads is too large (e.g., greater than 40 m) or vehicle speed at the approach is too high (e.g., greater than 50 km/h). For instance, a near-side signal face must be provided if the distance between the stop-line and its signal heads is greater than 180 ft. (i.e., 54.9 m) in the United States.

In general, the far-side pattern is often compatible with the approach-based signal settings, while the near-side pattern is more feasible for lane-based signal settings. In some countries, both far-side and near-side signal displays can be concurrently installed at large intersections or high-speed intersections with limited visual fields for drivers. So far, no strong evidence supports that those patterns or styles will make significant differences to intersection safety and efficiency. Basically, any of the patterns or styles is applicable, as long as the drivers' visibility to traffic signals can be ensured. However, it is recommended that signal setting patterns should be unified across a nation as much as possible, as drivers may become confused and make the wrong decisions, given different patterns or styles at different sites.

15.6 Signal warrants

Major criteria to install a traffic signal are fairly similar in the various countries, while threshold values of specific requirements vary. Major criteria include vehicular traffic volume, pedestrian volume, number of accidents in past years, and location of the intersection (e.g., school zone, hospital zone, etc.). Meanwhile, the functionality of the intersection in the road network, special needs for public transport priority or signal coordination, construction and maintenance costs, available physical space for waiting road users, and environmental pollution are also taken into consideration in some countries. If any one of the above criteria is satisfied, an installation of a traffic signal is justified in most countries.

Regarding traffic volumes, motor vehicles and pedestrians are comprehensively considered in North America, including the 8-hour, 4-hour and peak-hour motor vehicle volumes of the main road and the crossing road, and the 4-hour and peak-hour mixed traffic volumes consisting of pedestrian and two-way vehicular traffic volume of the main road. Other countries follow a similar practice to North America; however, a few countries such as Australia, Japan, China, and India only use the 8-hour traffic volume or the peak-hour traffic volume. In addition, the traffic volume thresholds are slightly different. For example, India and China apply comparably higher threshold peak-hour traffic volumes than the United States.

In terms of number of accidents, the minimum requirement for most countries such as the United States, France, Turkey, South Korea, India, and Qatar, is five accidents within the past year. In addition to the frequency of traffic accidents, the severity of traffic accidents is also considered in some countries, for example, Germany, Austria, Australia, New Zealand, Japan, and China. For instance, Japan has proposed a threshold number of two accidents with casualties in the past year. Instead of number of accidents within the past year, an average number of accidents within the past 3 years is also applied in some countries such as Australia and China. For instance, given more than five accidents or one fatal accident averagely within the past 3 years, a traffic signal should be installed in China. If more than 3 two-way or traffic casualties occurred per year over the past 3 years, a traffic signal should be installed in Australia.

However, it is also important to note that signal control is not the best way for traffic management at intersections under some conditions. Successful practices in Germany, France, and the United Kingdom have indicated that a compact roundabout is a good alternative in terms of safety and operational efficiency for intersections with low pedestrian/bicycle volumes and a low or medium level of vehicular traffic demand.

15.7 Stage structure

As explained in Chapter 2, Principles of road traffic signal control, there are some inconsistencies in terminology and definition of phase and stage across the world. The terms phase and stage in this section follow the definitions given in Chapter 2, Principles of road traffic signal control, which might not be the same as those in other chapters. Remarkable differences arise in stage structure among the covered countries, due to distinct signal control philosophy, signal timing procedures, and available functionalities of signal controllers.

In North America, stages are usually formed based on the National Electrical Manufacturers Association (NEMA) ring-barrier structure, in which every left-turn or through movement is considered as a single phase. A typical four-leg intersection usually needs a maximum number of eight phases, and as a result a maximum number of six stages including two overlapping ones could be generated for a typical four-leg intersection. In practice, the number of stages is often four to six for typical four-leg intersections and three to four for three-leg intersections. It could further increase for five-leg or unconventional intersections.

In Germany, Austria, and Switzerland, stages are formed based on a flexible combination of compatible movements controlled by exclusive signal groups, with the constraints of intergreen times. The best stage plans, that is, groups of compatible movements, that aim to balance safety (reducing traffic conflicts by a larger number of stages) and efficiency (maximizing capacity by a small number of stages), are selected based on trial calculations. The number of stages could exceed six for a typical four-leg intersection in some cases, as there is no barrier constraint used in North America. In the case of public transport priority, the number of stages could be even larger. Though such a method requires more computational effort, it is capable of tailoring stage plans for each specific intersection, according to intersection geometry, traffic flow characteristics, public transport priority, etc.

In Australia and New Zealand, a unique method of critical movement search diagram is developed to identify critical movements, based on capacity analysis. Stages are then formed based on the critical path connected by the critical movements. The objective of this method is to maximize the number of overlaps even though it will increase the number of stages, as maximizing the number of overlaps will reduce the total time required to satisfy the capacity requirements of all critical movements at the intersection.

In the United Kingdom, Japan, China, Turkey, and South Korea, the conventional symmetric stage structure with two to four stages is commonly seen, which is generally determined based on the Webster method (Webster, 1958). Such a stage structure can facilitate quick signal timing calculations and is easy to understand and follow for road users. However, it is not very efficient in handling unbalanced traffic demand and complicated traffic conflicts. Special stage plans such as overlapping stages are often adopted as countermeasures for those situations.

In general, it is hard to conclude which structure is the best as each has its own advantages and disadvantages, and these depend on application conditions. As recent signal controllers in most countries can realize most types of stage plans, regardless of stage structure, the decision regarding stage structures should be made largely based on signal control philosophy.

15.8 Stage sequence and turning treatments

Due to the distinct stage structures discussed earlier, various stage sequences and turning treatments can also be noticed in the covered countries, as highlighted in

Table 15.6 Typical treatments of turning movements

Countries	Left-turning treatment	Right-turning treatment
The United States and Canada	• Leading, lagging • Protected, permitted, protected/permitted	• Right-turn-on-red permitted • Protected right turn
Germany and Austria	• Leading, lagging (common) • Protected, permitted, protected-permitted	• Right-turn-on-red prohibited (only permitted with a static green arrow) • Protected, permitted (common)
The United Kingdom[a]	• Left-turn-on-red prohibited • Protected left turn	• Leading, lagging • Protected, permitted
France	• Leading, lagging • Protected, permitted	• Right-turn-on-red permitted • Protected right turn
Switzerland	• Leading, lagging • Protected, permitted	• Right-turn-on-red prohibited • Protected right turn
Turkey	• Lagging • Protected, permitted	• Right-turn-on-red permitted • Protected right turn
Australia and New Zealand	• Leading, lagging • Protected, permitted	• Right-turn-on-red permitted • Protected right turn
Japan[a]	• Left-turn-on-red prohibited • Protected left turn	• Lagging • Protected, permitted, permitted-protected
China	• Lagging • Protected, permitted	• Right-turn-on-red permitted • Prohibited/permitted • Protected
India[a]	Left-turn-on-red permitted	• Lagging • Protected, permitted
South Korea	• Lagging • Protected, permitted	• Right-turn-on-red permitted • Protected right turn
Qatar and UAE	• Leading, lagging • Protected, permitted	• Right-turn-on-red permitted • Protected right turn

[a]Nations with left-hand traffic rule.

Table 15.6. To avoid misunderstandings, it needs to be mentioned first that the left-hand traffic rule is adopted in the United Kingdom, Japan, and India, unlike the other covered countries.

In North America, the protected, permitted, and protected/permitted left turns are sophisticatedly determined, dependent on intersection geometry, traffic flow conditions, and safety requirements. In addition, there is no general preference for either leading left turn or lagging left turn, and the decision of stage sequence is made case by case, based on the evaluation of safety and efficiency for all the alternatives. Right-turn-on-red (RTOR) is generally permitted, while the protected right turn is also implemented in some situations.

Germany, Austria, Switzerland, Australia, and New Zealand are singled out for the treatment of turning movements being integrated into the signal timing procedure and not constrained by a predefined stage structure. Both protected-and-permitted left turns are allowable whenever feasible and it is determined based on safety and capacity considerations. Similar to North America, there is no special preference for the leading or lagging left turn. In addition, RTOR is basically prohibited in Germany, Austria, and Switzerland, and both protected-and-permitted right turns are implemented dependent upon local conditions; however, under carefully specified conditions, RTOR may be allowed by a special additional sign, that is, a static green arrow.

In the United Kingdom, Japan, Turkey, China, South Korea, and India, where the conventional symmetric stage structure is dominant, the lagging left turn (right turn in the case of left-hand traffic rule) is a default alternative except in the United Kingdom, regardless of protected or permitted. The leading left turn (right turn in the case of left-hand traffic rule) is only considered for special conditions such as coordination, short queue storage space, and actuated control. In addition, the permitted-and-protected right turn is also applied in Japan. Furthermore, RTOR (LTOR in the case of left-hand traffic rule) is permitted except in Japan and the United Kingdom, while the protected right turn (left turn in the case of left-hand traffic rule) is also applied mostly at those large intersections with high demand of conflicting bicycles or pedestrians. It is unique that the prohibited-and-permitted right turn has been occasionally adopted in some Chinese cities such as Shanghai. More specifically, the start of right-turn green is usually postponed by 10–15 seconds, in order to mitigate the conflicts between the right-turning traffic and the densely waiting bicycles and pedestrians at the beginning of vehicular stages.

Overall, both leading and lagging left turns (right turn in the case of left-hand traffic rule) have been adopted all over the world. Many studies have investigated the advantages and disadvantages of leading and lagging left turns in combination with protected, permitted, and protected/permitted left turns, from the perspectives of safety and efficiency (FHWA, 2008). Though quite a few countries tend to use the lagging left turn for various reasons, it is recommended that the selection of leading or lagging should not be limited in general, as dynamic use of left turns can enable more flexible stage plans. However, special attention should be given to the signal display sequence if a protected/permitted (protected-and-permitted or permitted-and-protected) left turn is used, to avoid yellow trap and potential traffic conflicts at the change of signals. Successful practices can be seen in the United States and Japan in this regard.

It is suggested that RTOR (LTOR in the case of left-hand traffic rule) can be allowed at intersections with low conflicting pedestrian/bicycle volumes and should be prohibited at intersections with high conflicting pedestrian/bicycle volumes. Given high pedestrian/bicycle volumes, two promising alternatives to right-turn control could be considered: one is to postpone the start of right-turn green by a few seconds to allow the discharge of high-volume pedestrians/bicycles at the beginning of the stage (i.e., leading pedestrian interval), the other is to use the protected right turn. These two types of right-turn control have been successfully implemented in a few countries such as Germany and China.

15.9 Saturation flow rate

It is recommended in the covered countries that field measurement is preferred to obtain the saturation flow rate whenever possible, as the saturation flow rate can significantly vary from location to location. When field measurements are not feasible, practical saturation flow rates are usually estimated by adjusting the base saturation flow rates with factors reflecting the local conditions. Default base saturation flow rates for different movements in the covered countries are listed in Table 15.7. This shows that an 1800 pc/(h ln) for the protected right-turning and left-turning movements in exclusive lanes and a 1900 or 2000 pc/(h ln) for through

Table 15.7 Ideal saturation flow rates

Countries	Left-turn movements	Through movements	Right-turn movements
The United States and Canada	Adjusted based on saturation flow rate of through movement	Cities with population $\geq 250{,}000$: 1900 pc/(h ln); otherwise, 1750 pc/(h ln)	Adjusted based on saturation flow rate of through movement
Germany and Austria	1800 pc/(h ln)	2000 pc/(h ln)	1800 pc/(h ln)
The United Kingdom	$s \approx 2080 - 140\delta_n - 42\Theta_G G + 100(w_l - 3.25)$where s = estimated saturation flow rate, pc/(h ln); $\delta_n = 1$ for a nearside lane, 0 otherwise; G = gradient in percent; $\Theta_G = 1$ for positive gradients, 1 otherwise; and w_l = lane width, m		
France	1800 pc/(h ln)	1800 pc/(h ln)	1800 pc/(h ln)
Switzerland	1800 pc/(h ln)	2000 pc/(h ln)	1800 pc/(h ln)
Turkey	1800 pc/(h ln)	1800 pc/(h ln)	1800 pc/(h ln)
Australia and New Zealand	1810 pc/(h ln)	1850 pc/(h ln)	1810 pc/(h ln)
Japan	1800 pc/(h ln)	2000 pc/(h ln)	1800 pc/(h ln)
China	1550 pc/(h ln)	1650 pc/(h ln)	1450 pc/(h ln)
India	$\text{USF}_0 = \begin{cases} 630, & \text{if } w < 7.0 \text{ m} \\ 1140 - 60w, & \text{if } 7.0 \text{ m} \leq w \leq 10.5 \text{ m} \\ 500, & w > 10.5 \text{ m} \end{cases}$		
	$\text{USF} = w \times \text{USF}_0 \times f_{bb} \times f_{br} \times f_{is}$ where USF_0 = unit base saturation flow rate, pc/h/m; USF = prevailing saturation flow rate, pc/h/m; w = effective width of approach (m); f_{bb} = adjustment factor for bus blockage; f_{br} = adjustment factor for blockage of through vehicles; f_{is} = adjustment factor for the initial surge of vehicles		
South Korea	1800 pc/(h ln)	2000 pc/(h ln)	1800 pc/(h ln)
Qatar and UAE	–	1900 pc/(h ln)	–

movements in exclusive lanes are often used. However, they are significantly lower in China, and slightly lower in Turkey, Qatar, and UAE.

Adjustment methods for saturation flow rate can be generally divided into two types. One is to directly correct the base saturation flow rates by multiplying reduction parameters in a range of 0–1. Affecting factors in consideration are fairly similar in the covered countries, including lane function, lane width, lane utilization, traffic composition, bus blockage, turning radius, gradient, conflicting pedestrian/bicycle volumes, etc. All the covered countries except the United Kingdom fall into this type. The other type is to use a linear regression method, while taking adjustment factors as the independent variables and saturation flow rate as the dependent variable. This type of method is particularly adopted in the United Kingdom, and an example for through movements without opposed turning traffic is given in the table.

It needs to be mentioned that the determination of prevailing saturation flow rate in India is fairly special compared to the other countries, due to the presence of heterogeneous traffic flows. Unit base saturation flow rates are given based on effective width of approach, as explained in the table. The prevailing saturation flow rates are then calculated based on unit base saturation flow rates and adjustment factors.

Despite existing well-developed highway capacity manuals in North America and Europe, domestic and local methods for the measurement and estimation of saturation flow rate are strongly encouraged, in particular for those countries where traffic flow compositions and driving behavior are remarkably different from North America and Europe. For instance, motorcycles, auto rickshaws, and two-wheelers constitute the major traffic flow in some Asian countries such as India, Vietnam, and Malaysia. Under such a context of traffic composition, those methods for saturation flow rate measurement and estimation recommended in North America and Europe may not work well, as those methods basically assume a dominant proportion of cars in the traffic flow. A good practice can be seen in India where new local methods have been developed and recommended in the national guideline.

15.10 Cycle time

Table 15.8 presents determination methods for the optimum cycle length and the minimum cycle length, as well as common ranges of cycle length in practice, in the covered countries. It is clear to see that most of the countries calculate the optimum cycle length and the minimum cycle length more or less based on Webster's method. Though the cycle length is obtained through signal control software in Qatar and UAE in practice, the underlying ideas are similar to Webster's theory.

Direct use of Webster's theory can be found in the United Kingdom, France, Turkey, China, India, and South Korea. Webster's formulas for the optimum and minimum cycle lengths are recommended in the technical guidelines in those countries, and signal timing procedures also follow the principle of equilibrium of

Table 15.8 Cycle length

Countries	Optimum cycle length (s)	Minimum cycle length (s)	Common range (s)
The United States and Canada	Determined by trial calculations considering a proper v/c ratio for critical vehicular traffic and sufficient green times for pedestrians	$C_{min} = \frac{L}{1-Y}$	[60, 140], >180 is rare
Germany and Austria	$c_0 = \frac{1.5L+5}{1-Y}$	30, $C_{min} = \frac{\sum_i t_{ig,i} + \sum_j t_{g,min,j}}{1-Y}$	[30, 90], in exceptions: the upper limit is 120
The United Kingdom	$c_0 = \frac{1.5L+5}{1-Y}$	–	[60, 120]
France	$c_0 = \frac{1.5L+5}{1-Y}$	–	[40, 90], in exceptions: the upper limit is 120
Switzerland	Similar to Germany	–	[60, 90], in exceptions: the upper limit is 120
Turkey	$c_0 = \frac{1.5L+5}{1-Y}$	40, $C_{min} = \frac{L}{1-Y}$	[40, 140], in exceptions: the upper limit is 250
Australia and New Zealand	$c_o = \frac{(1.4+k)L+6}{1-Y}$	$c_{min} = \frac{L}{1-U}$	$[c_{min}, \min(c_{max}, c_o)]$, c_{max} is recommended as 100–120 for a two-phase intersection, and 150–180 for sites with multiple phases and high traffic demand
Japan	Determined by trial calculations considering a proper v/c ratio for critical vehicular traffic and sufficient green times for pedestrians	$C_{min} = \frac{L}{1-\frac{Y}{0.9}}$	In principle $[C_{min}, 120]$, some exceptional cases: >180
China	$C_O = \frac{1.5L+5}{1-Y}$	30 s, $C_{min} = \frac{L}{1-Y}$	[90, 180] for mega cities, [90, 150] for medium and large cities, [60, 120] for small cities
India	$C_O = \frac{1.5L+5}{1-Y}$	$C_{min} = \frac{L}{1-Y}$	$[C_{min}, \min(c_{max}, c_o)]$, c_{max} is recommended as 120
South Korea	$C_O = \frac{1.5L+5}{1-Y}$	$C_{min} = \frac{L}{1-Y}$	[150, 240]
Qatar and UAE	Obtained from software (e.g., SCATS, SCOOT)		[150, 240]

C_O = the optimum cycle length (s); C_{min} = the minimum cycle length (s); L = the total lost time per cycle (s); Y = sum of maximum flow ratios; k = stop penalty parameter; U = sum of green volume/capacity (v/c) ratios for critical movements. Practically, the cycle length closest to the calculated optimum cycle length that can satisfy the requirements of minimum green times for pedestrians and vehicles is typically chosen as the final cycle length.

In the United States and Canada, since only a small portion of fixed-time isolated intersections exists and is mostly seen in downtown areas, cycle length for such intersections is mostly determined based on minimum pedestrian timing requirements. Where cycle length is driven by vehicular demands, an empirical-based critical movement analysis, which is consistent with Webster's principle of equilibrium of v/c ratios, is used. The final cycle length is often reached by trial calculations, considering a proper v/c ratio for vehicular traffic, for example, 0.95, and sufficient green times for pedestrians.

In Germany, Austria, and Switzerland, intergreen times play an important role in the determination of cycle length. The optimum cycle length is calculated from the sum of intergreen and the sum of decisive flow ratios per stage, while following the basic principle of the Webster method. The final cycle length is then determined by the critical path (i.e., maximum of the sum of green and intergreen time for vehicles and pedestrians, respectively). Webster's formula for the optimum cycle length is only optional for isolated rural intersections with low pedestrian and bike demand.

In Japan, Australia, and New Zealand, some adaptations of Webster's formulas for the optimum and minimum cycle lengths have been proposed in the guidelines. It is unique that a stop penalty parameter is incorporated into the calculation of the optimum cycle length in Australia and New Zealand, to represent a multiplicity of control objectives such as delay and fuel consumption. An initial cycle length is usually assumed to identify the critical movements based on a critical movement search diagram, and stage plans are then determined based on the critical path (i.e., a linkage of critical movements). The cycle length closest to the calculated optimum cycle length is eventually chosen. On the other hand, a v/c ratio of 0.9 is integrated into the calculation of the minimum cycle length for reserving capacity in Japan. Similar to North America, the final cycle length is often reached by trial calculations, considering a proper v/c ratio for vehicular traffic and sufficient green times for pedestrians.

As for common ranges of cycle length in practice, it is shown in the table that cycle lengths are much shorter in Germany, Austria, and Switzerland, that is, less than 120 seconds in most cases. This could be largely attributed to flexible signal control and special considerations for the needs of bicycles and pedestrians. Cycle lengths in the United States, Canada, the United Kingdom, Australia, and New Zealand are close, that is, roughly 60–150 seconds in most cases. However, cycle lengths are far greater in Japan, Turkey, China, India, South Korea, Qatar, and UAE, for example, approximately 90–240 seconds in many cases. This can be explained by the differences in intersection geometry, stage structure, and high pedestrian/bicycle demand to some degree. It should be mentioned that long cycle times are often applied for reasons of capacity for the motorized traffic. However, in several countries, the needs of nonmotorized traffic, specifically pedestrians, seem not to be sufficiently considered. In addition, pedestrian waiting times resulting from long cycle times may lead to red light running and severe safety problems. Long cycle times may also lead to unsaturated vehicle flow during long green times.

Practices in Germany, the United Kingdom, and China support that the waiting times of pedestrians should not exceed 90 seconds, and it is a key constraint in the determination of the cycle length. Therefore it is recommended that the cycle length should be set as short as possible, if the minimum green times can be satisfied. Given a constraint of 90 seconds as the maximum waiting time for pedestrians, the maximum cycle length should generally be less than 120–150 seconds. In addition, flexible stage plans such as overlapping and two-step crossings of pedestrians, can help reducing cycle length. Moreover, it should be noted that the applicability of Webster's formulas is usually limited to isolated intersections with random vehicle arrivals and low pedestrian/bicycle volumes. Thus it only acts as a reference when determining the practical cycle length in many countries.

15.11 Signal change and clearance intervals/intergreen times

According to Chapter 2, Principles of road traffic signal control, intergreen time is defined as the time between the end of green for one traffic stream and the beginning of the green for another stream, which is not compatible with the first stream. This concept is conventionally adopted in Germany, Austria, the United Kingdom, Switzerland, Australia, and New Zealand. By this definition, the intergreen time considers the detailed movements of the last vehicle clearing the intersection and the first vehicle entering the intersection. However, the concept of signal change and clearance intervals defined in the United States is more often used in the other countries. It normally includes a yellow time (i.e., change interval) and an all-red time (i.e., clearance interval). In recent years, the concept of intergreen time has also been introduced to China and India, to deal with complicated traffic conflicts in the context of heterogeneous traffic.

The objectives of signal change and clearance intervals are universal, and are eliminating the dilemma zone for approaching drivers and ensuring a safe clearance for road users through the intersection at the change of stages. Table 15.9 summarizes the determination methods of signal change and clearance intervals/intergreen times for motor vehicles (public transport not included) in the covered countries.

It can be seen that contributory factors for the design of yellow time include the perception–reaction time, the maximum acceptable deceleration rate, and approaching speed. On the other hand, the all-red time is a function of the clearing distance and the clearing speed. General calculation methods for yellow and all-red are fairly similar to each other. Major differences arise for the values of the parameters such as the perception–reaction time and the maximum acceptable deceleration rate, as well as the definition of clearing distance.

Table 15.9 Signal change and clearance time intervals for motorized vehicles

Countries	Signal change interval, t_y (s)	Clearance interval, t_{ar} (s)	Intergreen time, t_{ig} (s)
The United States and Canada	$t_y = \tau + \frac{V}{2(d + 9.8G)}$ (usually, $\tau = 1.0$ s, $d = 3.05$ m/s^2)	$t_{ar} = \frac{W + L}{V}$	–
Germany and Austria	3 s, if speed limit ≤ 50 km/h 4 s, if speed limit = 60 km/h; and 5 s, if speed limit ≥ 70 km/h	–	$t_{ig} = t_{cr} + t_{cl} - t_e$ $t_{cl} = \frac{L + l_{cl}}{v_{cl}}$, $t_e = \frac{l_e}{v_e}$
The United Kingdom	$t_y = \tau + \frac{V}{2d}$ (usually, standardized as 3 s)	–	$t_{ig} \in [5, 12]$ with a constant 3 s of t_y and a 2 s of yellow-and-red time, dependent on $(l_{cl} - l_e)$
France	$t_y = \tau + \frac{V}{2d}$ (usually, 3–5 s dependent upon speed limits)	$t_{ar} = \frac{W + L}{V}$ (usually, standardized as 2 s)	–
Switzerland	Same as Germany		
Turkey	Minimum: 2 s, 4 s for major intersections, 3 s for other intersections	Usually 2 s, ranging between 2 s and 8 s	–
Australia and New Zealand	$t_y = \tau + \frac{V}{2(d + 9.8G)}$ subject to $t_y \in [3, 6]$	$t_{ar} = \frac{W + L}{V}$ subject to $t_{ar} \in [1, 3]$	–
Japan	$t_y = \tau + \frac{V}{2d}$ (usually, $\tau = 0.7$ s; $d = 3.0$ m/s^2)	$t_{ar} = t_{cl} - t_e = \frac{W + L}{V} - t_e$ (usually, t_e is set as 0)	–
China	$t_y = \tau + \frac{V}{2d}$ (usually, 3 s for motor vehicles and 2 s for nonmotor vehicles, disregarding speed limit)	$t_{ar} = \frac{W + L}{V}, \in [0, 2]$ (usually, 1 s for small intersections, 2 s for large intersections, and sometimes 0 s)	An adapted German method is recommended for complicated intersections
India	$t_y = \tau + \frac{V}{2d}$ (usually, $\tau = 1.0$ s; $d = 2.53$, 1.78, 2.32, and 2.23 m/s^2 for passenger cars, auto rickshaws, two-wheelers, and buses, respectively)	–	Similar to Germany

(Continued)

Table 15.9 (Continued)

Countries	Signal change interval, t_y (s)	Clearance interval, t_{ar} (s)	Intergreen time, t_{ig} (s)
South Korea	$t_y = \tau + \frac{V}{2d}$ (usually standardized as 3 s)	$t_{ar} = \frac{W+L}{V}$ (usually, 0 s for ordinary intersections, 1 s for critical intersections, and 2 s for high-speed intersections)	–
Qatar and UAE	$t_y = \tau + \frac{V}{2d}$ (usually standardized as 3 s)	$t_{ar} = \frac{W+L}{V}$ (usually, 1–2 s for ordinary intersections, and 3–4 s for very large intersections)	–

t_y = yellow time (s); t_{ar} = all-red time (s); τ = perception–reaction time (s); V = approach speed (m/s); d = deceleration rate (m/s^2); W = intersection width (m); L = vehicle length (m); t_{ig} = intergreen time (s); t_{cr} = crossing time (s); t_{cl} = clearing time (s); t_e = entering time (s); l_{cl}/l_e = clearance/entering distance (s); V_{cl}/V_e = clearance/entering speed (m/s). In some countries, there are specific regulations for individual types of vehicles, for example, in Germany for public transport vehicles.

In addition, the yellow time and the all-red time in the United Kingdom, Turkey, China, South Korea, Qatar, and UAE, are often standardized as constant values for simplification in practice.

In Germany, Austria, and Switzerland, intergreen time is considered as a whole in the calculations, while the yellow time can also be separately determined based on the speed limit. The intergreen time can be computed for each combination of traffic streams if they have a joint conflict area. Thus it is considered as an efficient means of handling complicated traffic conflicts. In addition, it is a special case in the United Kingdom that several sets of intergreen times with a constant 3 seconds of yellow time and a constant 2 seconds of red-yellow time, dependent upon the difference of clearing distance and entering distance, are recommended.

In general, although the fundamental theory behind the design of signal change and clearance intervals/intergreen times is common, a variety of methods have been proposed under various traffic circumstances all over the world. Minor differences exist in the determination of yellow time. The all-red times generated by the American and Japanese methods are comparably long provided identical conditions, which is largely caused by the distinct stage structures. Flexible phasing and relatively low pedestrian demand at signalized intersections in the United Kingdom, Germany, Austria, Australia, and New Zealand enable the pedestrian green to end earlier than the vehicle green. Therefore only the conflict points of vehicular traffic need to be taken into consideration when determining clearance distance. However, the conventional stage structure widely used in the other countries usually switches off pedestrian green together with the vehicular traffic. Thus the clearance distance has to include intersection width.

15.12 Signal timing procedure

Despite all the above differences, signal timing procedures in the covered countries can be globally divided into four types in general according with their core steps and the philosophy it.

Type 1 is a critical-movement-based method with the NEMA ring-barrier constraint, which is mainly applied in the United States and Canada. Critical phase/movement pairs are identified according to the sum of phase volumes, and those critical phases must be in one side of the barrier in the same ring. Phase splits are allocated proportionally, depending on the volume ratios of the critical phases and the noncritical phases, respectively. It is noted that the sum of the duration of critical phase splits must equal that of the noncritical phase splits in each side of the barrier.

Type 2 is a group-based/movement-based method, which is used in Germany, Austria, and Switzerland. It is characterized by calculating intergreen times for all the conflicting flows and determining the stage sequence to minimize the total intergreen times. The purpose of the combination of signal groups in Germany and Austria is to minimize the critical sum of traffic stream volumes. Conversely, in other types of methods the stage structure and sequence are usually decided before calculating intergreen times.

Type 3 is a critical-movement-based method without the NEMA ring-barrier constraint, which is adopted in Australia and New Zealand. Two major tools, that is, the critical movement search table and the critical movement search diagram, are used to identify critical movements. The required passing time of all the movements is calculated and summed up for each path. The movements on the path with the largest required passing time to complete a cycle are the critical movements. Those critical movements should be rechecked if the cycle time is adjusted.

Type 4 is a stage-based method, which prevails in the United Kingdom, France, Turkey, Japan, China, India, South Korea, Qatar, and UAE. Stages are usually predefined according to intersection geometry and traffic volumes. Webster's method is mostly utilized with equal v/c ratios for all stages. Furthermore, signal timing plans are adjusted based on the requirements of the minimum green times for pedestrians and vehicles.

In general, type 1 can provide conveniences when dealing with standard intersections. However, it may have limitations when operating signalized intersections with complex traffic movements, multimodal operations, and unconventional geometric characteristics. Type 2 is the most flexible and could be more efficient in handling complicated intersections, public transport priority, and heterogeneous traffic flow, however it requires more computational effort. The capability and complexity of type 3 are between those of types 1 and 2. Type 4 is the easiest calculate; however, it is often inefficient in dealing with complicated traffic compositions and traffic demand patterns. Stage plans generated by types 1 and 4 are basically easier for road users to comply with, but green times may be redundant for noncritical traffic movements. Types 2 and 3 are capable of more efficiently allocating green times among the incompatible traffic movements, but a larger number of overlapping stages basically lead to a higher frequency of stage switching. It might affect traffic safety at the change of stages, if without disciplined road users and proper engineering countermeasures.

Hence, it is difficult to conclude which method is superior, as each has advantages and disadvantages, as discussed earlier. The signal timing procedure should be appropriately selected based on multiple factors, for example, signal control objectives, functionalities of signal controllers, special needs of public transport priority or signal coordination, traffic flow characteristics, and intersection geometry.

15.13 Performance evaluation

Performance indicators for the evaluation of signalized intersections include control delay, capacity, degree of saturation, queue length, number of stops, travel time, number of traffic accidents or conflicts, fuel consumption, and so on. Despite the multiplicity and diversity for performance evaluation, the primary criterion for quantifying the level-of-service (LOS) of signalized intersections is control delay from an international perspective. Table 15.10 presents the measures for LOS as well as the LOS criteria in the covered countries.

Table 15.10 LOS criteria for motorized vehicles at signalized intersections

Countries	Measure	LOS criteria					
		A	B	C	D	E	F
The United States and Canada	Average control delay per vehicle (s)	[0,10]	[11,20]	[21,35]	[36,55]	[56,80]	> 80
Germany and Austria	Average control delay per vehicle (s)	[0,20]	[21,35]	[36,50]	[51,70]	> 70	If demand exceeds capacity
The United Kingdom	Average control delay per vehicle (s)	[0,10]	[11,20]	[21,35]	[36,55]	[56,80]	> 80
Switzerland	Average control delay per vehicle (s)	[0,20]	[21,35]	[36,50]	[51,70]	[71,100]	> 100
Turkey	Average control delay per vehicle (s)	[0,10]	[11,20]	[21,35]	[36,55]	[56,80]	> 80
Australia and New Zealand	Average control delay per vehicle (s)	[0,14]	[15,28]	[29,42]	[43,56]	[57,70]	> 70
China	Average control delay per vehicle (s)	[0,10]	[11,20]	[21,35]	[36,55]	[56,80]	> 80
India	Average control delay per vehicle (s)	[0,20]	[20,40]	[40,65]	[65,95]	[95,130]	≥ 130
South Korea	Average control delay per vehicle (s)	[0,15]	[15,30]	[30,50]	[50,70]	[70,100]	> 100
Qatar and UAE	Average control delay per vehicle (s)	[0,10]	[11,20]	[21,35]	[36,55]	[56,80]	> 80

It was found that most countries select average control delay per vehicle as the LOS criteria for motorized traffic (public transport not included), but the ranges of threshold values are different. The United States, Canada, Turkey, and China adopt the same standard recommended in the HCM 2010. Notably, the threshold values are significantly higher (almost doubled) in India, due to a different perception of quality of service of road users. In addition, there are no LOS criteria for signalized intersections applied in practice in France and Japan. Different measures and LOS criteria are used for pedestrians and public transport in some countries, such as the United States, Germany, Austria, Switzerland, and China.

In addition, quality management for traffic signals has been proposed in Germany, Australia, and Switzerland. Quality management involves all stages of the life cycle of traffic signals, for example, planning, implementation, and operation. It aims at achieving continuously high quality regarding traffic safety, environmental impacts, and traffic flow. Some details of quality management for road traffic signals are given in Section 2.4 in Chapter 2, Principles of road traffic signal control.

Performance evaluation is fundamental for continuous monitoring and improvement of signal timing plans at signalized intersections. Comprehensive and accurate evaluation can help diagnosing signal control problems and providing insights for improvement. Thus it is not only essential for the stage of signal timing design, but also important for the stage of operation and maintenance. From a viewpoint of cost–benefit, efficient monitoring and improvement of existing signal timing plans is sometimes more effective than using advanced control systems and devices. It is thus recommended that government sectors responsible for road traffic signal control should arrange a stable budget for performance evaluation and continuous improvement of signal control schemes.

On the other hand, performance indicators should be properly selected based on signal control objectives, as too many indicators do not necessarily make sense for efficient improvement. As discussed in Chapter 2, Principles of road traffic signal control, the importance of delay, capacity, degree of saturation, queue length, number of stops, probability of queue spillback, fuel consumption, and traffic noise, is variable under various traffic circumstances. In addition, though average control delay is widely used for determining LOS, the multicriteria and multimodal impact assessment has gained increasing concerns, which simultaneously accounts for the safety, efficiency, environment, and energy of multiple traffic modes. For instance, control delay per person might be a good integrated indicator in the context of multimodal operations, as it can account for the number of people accommodated by each type of vehicle.

15.14 Conclusions

The results of the comparative analysis documented in this chapter clearly indicate that it is valuable to conduct further research based on international practice in road traffic signal control. In many cases, differences may be easily justified by different conditions, for example, in traffic composition, driving behavior, or simply history in traffic engineering. Nevertheless, there are also several items which are treated in different ways although conditions are almost the same. For example, in some countries, very long cycle times are allowed and used in practice, while others are setting significantly lower limits for this value. There are also clear indications that the consideration of various road user groups and optimization criteria varies significantly from country to country.

Specifically, for these aspects where current practices are different, it seems very valuable to have a further exchange of knowledge and experiences among experts from different countries. Experts are invited to utilize the relevant Special Interest Group (SIG C2) of the World Conference on Transport Research Society (WCTRS) as a platform for such exchange. The editors of this book hope that major benefits for road users around the globe may arise from discussions in that group. Involved experts might have a chance to influence and improve national standards for road traffic signal control in their country. And it is an encouraging vision to develop an international standard for road traffic signal control which helps all countries to keep their traffic signals up-to-date and to ensure safe, environmentally friendly, and efficient traffic operations.

Acknowledgments

As joint editors of this book, we would like to express sincere appreciations to the authors of each country chapter for their devoted efforts and contributions to this book. The editors are also very grateful to the Elsevier staff including Mr. Brain Romer, Ms. Lindsay Lawrence, Ms. Sheela Josy, and many others, for working closely with us to successfully publish this book. Special gratitude is expressed to Ms. Jiarong Yao, Ms. Yaning Wei, and Mr. Can Chen for their contributions in assembling key information from the country chapters.

Bibliography

Australasian Road Transport and Traffic Agencies (AUSTROADS), 2003. Guide to Traffic Engineering Practice Series: Traffic Signals. AUSTROADS, Australia.

Federal Highway Administration, US Department of Transportation (FHWA), 2008. Traffic Signal Timing Manual. FHWA.

Federal Highway Administration, 2012. Manual on Uniform Traffic Control Devices for Streets and Highways. US Department of Transportation.

FGSV, 2001. Handbuch fuer die Bemessung von Strassenverkehrsanlagen (Highway Capacity Manual). FGSV, Cologne (in German).

FGSV, 2015. Richtlinien fuer Lichtsignalanlagen – RiLSA (German Standard for Traffic Signals), edition 2015 FGSV-Verlag, Cologne.

JSTE-Japan Society of Traffic Engineers, 2018. Basic Manual on Planning, Design and Traffic Signal Control of At-Grade Intersections. JSTE, Tokyo (in Japanese).

Köll, H., Bader, M., Axhausen, K.W., 2004. Driver behaviour during flashing green before amber: a comparative study. Acc. Anal. Preven. 36 (2), 273–280.

Tang, K., Xu, Y., Wang, F., Oguchi, T., 2016. Exploring stop-go decision zones at rural high-speed intersections with flashing green signal and insufficient yellow time in China. Acc. Anal. Preven. 95 (B), 470–478.

Transportation Research Board, 2010. Highway Capacity Manual 2010. Transportation Research Board, Washington, D.C.

UK Department of Transport, 2006. Traffic Advisory Leaflet: General Principles of Traffic Control by Light Signals. UK Department of Transport, London.

Webster, F.V., 1958. Traffic signal settings, Road Research Technical Paper, vol. 39. HMSO, London.

Index

Note: Page numbers followed by "*f*" and "*t*" refer to figures and tables, respectively.

Kolibri 360 GmbH
Ferdinand-Jühlke-Straße 7
99095 Erfurt, Deutschland
produktsicherheit@kolibri360.de